Studies in Systems, Decision and Control

Volume 262

Series Editor

Janusz Kacprzyk, Systems Research Institute, Polish Academy of Sciences, Warsaw, Poland

The series “Studies in Systems, Decision and Control” (SSDC) covers both new developments and advances, as well as the state of the art, in the various areas of broadly perceived systems, decision making and control–quickly, up to date and with a high quality. The intent is to cover the theory, applications, and perspectives on the state of the art and future developments relevant to systems, decision making, control, complex processes and related areas, as embedded in the fields of engineering, computer science, physics, economics, social and life sciences, as well as the paradigms and methodologies behind them. The series contains monographs, textbooks, lecture notes and edited volumes in systems, decision making and control spanning the areas of Cyber-Physical Systems, Autonomous Systems, Sensor Networks, Control Systems, Energy Systems, Automotive Systems, Biological Systems, Vehicular Networking and Connected Vehicles, Aerospace Systems, Automation, Manufacturing, Smart Grids, Nonlinear Systems, Power Systems, Robotics, Social Systems, Economic Systems and other. Of particular value to both the contributors and the readership are the short publication timeframe and the world-wide distribution and exposure which enable both a wide and rapid dissemination of research output.

** Indexing: The books of this series are submitted to ISI, SCOPUS, DBLP, Ulrichs, MathSciNet, Current Mathematical Publications, Mathematical Reviews, Zentralblatt Math: MetaPress and Springerlink.

More information about this series at http://www.springer.com/series/13304

Mahmoud Pesaran Hajiabbas ·
Behnam Mohammadi-Ivatloo
Editors

Optimization of Power System Problems

Methods, Algorithms and MATLAB Codes

Springer

Editors
Mahmoud Pesaran Hajiabbas
Postdoctoral Fellow of Faculty of Electrical and Computer Engineering
University of Tabriz
Tabriz, Iran

Behnam Mohammadi-Ivatloo
Associate Professor of Faculty of Electrical and Computer Engineering
University of Tabriz
Tabriz, Iran

ISSN 2198-4182 ISSN 2198-4190 (electronic)
Studies in Systems, Decision and Control
ISBN 978-3-030-34049-0 ISBN 978-3-030-34050-6 (eBook)
https://doi.org/10.1007/978-3-030-34050-6

This Springer imprint is published by the registered company Springer Nature Switzerland AG
The registered company address is: Gewerbestrasse 11, 6330 Cham, Switzerland

Introduction

Motivation

Providing a reliable and secure power and energy system is one of the main challenges of the new era. The efficient operation of power systems contributes to decrease in fuel consumption and gas emission, conservation of natural resources, ensuring sustainability with better planning, and providing cleaner energy. The evolving modern optimization methods lead to more effective solutions and are promising for the continuously changing power system management, planning, and operation. One of the most favored tools of researchers and electric system developers for power system optimization is MATLAB software. Therefore, there has been an increased call for sharing the properly developed codes for power system optimization.

A Brief Overview of the Book Covered Topics

The book is suitable for dedicated and general audiences that include power system professionals, as well as researchers and developers from electrical power engineering and power system planning communities. It is expected that readers to be graduates of energy and power engineering degree programs having a basic mathematical background.

The Book Organization

The book is organized under two main sub-topics, comprising of power system optimal planning and configuration and power system optimal operation. Brief description of chapters' content is presented in the following paragraphs.

Chapter "Modelling for Composite Load Model Including Participation of Static and Dynamic Load" focuses on modeling for composite load model including participation of static and dynamic load. It is well recognized that voltage problems in power system are much affected through the connected loads. Different types of load can be modeled on their characteristic basis for computation of power system problems effectively. For different power system studies, especially in the area of power system optimization problems that includes voltage control with reactive power compensation, transfer function $\Delta Q/\Delta V$ of composite load is required. This chapter gives a detailed mathematical modeling to compute the reactive power response with small voltage perturbation for composite load. Composite load is defined as a combination of static and dynamic load model. To develop this composite load model, the exponential load is used as a static load model and induction motors are used as a dynamic load model in this chapter. To analyze the dynamics of induction motor load, fifth-, third-, and first-order models of induction motor are formulated and compared using differential equation solver in MATLAB coding. Since the decentralized areas have many small consumers which may consist large numbers of induction motors of small rating, it is not realistic to model either a single large rating unit or all small rating induction motors together that are placed in the system. In place of using single large rating induction motor, a group of motors are being considered, then aggregate model of induction motor is developed using law of energy conservation, and this aggregate model is used as a dynamic load model. Transfer function of composite load is derived in this chapter by successive derivation for exponential model of static load and for fifth- and third-order induction motor dynamic load models using state-space model.

Chapter "A Novel Forward-Backward Sweep Based Optimal DG Placement Approach in Radial Distribution Systems" presents a novel forward–backward sweep-based optimal DG placement approach in radial distribution systems. This chapter proposes a novel backward–forward sweep (BFS)-based methodology for optimal allocation of DG micro-plants in radial distribution systems aiming to minimize total real power losses. Voltage-permitted range limit and feeder capacity criterion are considered as optimization constraints. Simulation of BFS-based DG placement method is conducted on 33-bus distribution network to demonstrate its robustness and effectiveness in comparison with other procedures.

Chapter "Optimal Capacitor Placement in Distribution Systems Using a Backward-Forward Sweep Based Load Flow Method" investigates optimal capacitor placement in distribution systems using backward–forward sweep-based load flow method. This chapter aims to present a backward–forward sweep (BFS)-based algorithm for optimal allocation of shunt capacitors in distribution networks. Minimum value of real power losses is selected as objective function. Moreover, feeder current capacity and bus voltage magnitude limits are considered as optimization constraints. In addition, it is assumed that sizes of capacitors are known parameters. First capacitor is considered to be located at first bus of test system. Then, BFS load flow is run and objective function is saved as first row and first column component of a loss matrix. Secondly, first capacitor is assumed to be installed at bus 2 and BFS load flow is run to obtain objective function as second row and first column component of loss matrix. When all buses are assessed for installation of capacitor 1 and losses are calculated in each scenario, similar analyses are carried out and objectives are saved as second column of loss matrix. Same strategy is applied on other capacitors. Finally, a loss matrix is formed with a number of rows and columns equal to a number of buses and shunt capacitors, respectively. Best places for installation of capacitors are determined based on components of loss matrix. Simulation of BFS-based capacitor placement problem is conducted on 33-bus distribution network to demonstrate its robustness and effectiveness in comparison with other procedures.

Chapter "Optimal Capacitor Placement and Sizing in Distribution Networks" discusses optimal capacitor placement and sizing in distribution networks. Utilizing capacitor banks in order for local compensation of load reactive power is common in distribution networks. Using capacitors has positive effects on networks such as power and energy loss reduction, voltage deviation, and network harmonic reduction as well as improvement in network power factor. Capacitor placement is applied on the network in the form of single or multi-objective problems. Decreasing the total network loss is often the main reason for using capacitors in distribution networks. Capacitor placement approach involves the identification of location for capacitor placement and the size of the capacitor to be installed at the identified location. An optimization algorithm decides the location of the nodes where the capacitors should be placed. As we know, the capacitors are categorized into two main types of fixed and switchable capacitors. Selecting an appropriate type of capacitor is related to the topology of network, load value, and economic situation. They are also different from coding point of view. In this section, the model of coding is presented at first, and then, the approach of applying is described based on optimization algorithm. The capacitors are often used for peak loads, but they may be present in the network in off-peak due to the switching issues. The network voltage may be increased in off-peak with the presence of capacitors. Therefore, it is very important to consider both peak and off-peak in the capacitor sizing and placement problem. The proposed model is applied on IEEE 10 and 33-bus standard test cases in order to demonstrate the efficiency of the proposed model.

Chapter "Binary Group Search Optimization for Distribution Network Reconfiguration" studies binary group search optimization for distribution network reconfiguration. Total loss minimization is considered as the objective which is solved subject to system radial operation and power flow constraints. Here, the basics of GSO algorithm is presented first and then necessary modification for developing BGSO is discussed. The main part of this chapter deals with a source code, which expresses step-by-step implementation of BGSO method to optimal network reconfiguration problem. Needless to emphasize that the BGSO and associated source code presented in this chapter is a general engine that can be easily adjusted to any optimization problem with binary variables. In addition, the source code associated with the developed forward–backward sweep-based load flow study is also provided. The simulation studies are performed on different distribution networks to examine the scheme at various conditions and problem complexities. Comprehensive simulation studies conducted in this chapter verify effectiveness of the BGSO and developed source code for solving optimal distribution network reconfiguration problem.

Chapters "Combined Heat and Power Economic Dispatch Using Particle Swarm Optimization," "Combined Heat and Power Stochastic Dynamic Economic Dispatch Using Particle Swarm Optimization Considering Load and Wind Power Uncertainties," and "Economic Dispatch of Multiple-Chiller Plants Using Wild Goats Algorithm" exercises the combined heat and power economic dispatch using particle swarm optimization, the combined heat and power stochastic dynamic economic dispatch using particle swarm optimization considering load and wind power uncertainties, and the economic dispatch of multiple-chiller plants using wild goats algorithm, respectively.

Chapter "Optimization of Tilt Angle for Intercepting Maximum Solar Radiation for Power Generation" investigates the optimization of tilt angle for intercepting maximum solar radiation for power generation. The novelty is determination of optimum tilt angles (β_opt) for photovoltaic system at 11 different sites for Gujarat in India. The β_opt is searched for maximum incident solar radiation (SR). For calculation, SR values given by National Aeronautics and Space Administration (NASA) are utilized. It was found that the optimum tilt angle varies between 1° and 57° throughout the year in Gujarat, India. The monthly optimum tilt angle is maximum in December for different sites in Gujarat, India. This study is useful for industry and researcher to install PV system in India to generate maximum power.

Chapter "Probabilistic Power Flow Analysis of Distribution Systems Using Monte Carlo Simulations" analyzes the probabilistic power flow analysis of distribution systems using Monte Carlo simulations. This chapter aims to present a Monte Carlo simulation-based probabilistic power flow method for finding all critical buses against variations of active and reactive loads. In this approach, backward–forward sweep-based load flow is used to find optimal operating point of benchmark distribution grid in each scenario. The number of scenarios with bus voltage magnitude violation probability is used to cluster nodes into two critical and non-critical categories. Robustness and effectiveness of Monte Carlo-based probabilistic power flow algorithm are revealed by simulations on 33-bus radial distribution system.

Chapter "Long-Term Load Forecasting Approach Using Dynamic Feed-Forward Back-Propagation Artificial Neural Network" implements the long-term load forecasting approach using dynamic feed-forward back-propagation artificial neural network. This chapter presents a novel approach based on dynamic feed-forward back-propagation artificial neural network (FBP-ANN) for long-term forecasting of total electricity demand. A feed-forward back-propagation time series neural network consists of an input layer, hidden layers, and an output layer and is trained in three steps: a) Forward the input load data, b) compute and propagate the error backward, and c) update the weights. First, all examples of the training set are entered into the input nodes. The activation values of the input nodes are weighted and accumulated at each node in the hidden layer and transformed by an activation function into the node's activation value. It becomes an input into the nodes in the next layer, until eventually the output activation values are found. The training algorithm is used to find the weights that minimize mean squared error. The main characteristics of FBP-TSNN are the self-learning and self-organizing. The proposed algorithm is implemented on Iran's power network to prove its accuracy and effectiveness and compare with real historical data.

Chapter "Multi-objective Economic and Emission Dispatch Using MOICA: A Competitive Study" applies MOICA on multi-objective economic and emission dispatch using. The application of multi-objective imperialist competitive algorithm is investigated for solving economic and emission dispatch problem. It is aimed to minimize two conflicting objectives, economic and environmental, while satisfying the problem constraints. In addition, nonlinear characteristics of generators such as prohibited zone and ramp up/down limits are considered. To check applicability of the MOICA, it is applied to 12 h of IEEE 30-bus test system. Then, results of MOICA are compared with those derived by non-dominated sorting genetic algorithm and multi-objective particle swarm optimizer. The finding indicates that MOICA exhibits better performance.

Chapter "Voltage Control by Optimized Participation of Reactive Power Compensation Using Fixed Capacitor and STATCOM" integrates fixed capacitor and STATCOM to control voltage by optimized participation of reactive power compensation. Finally, chapter "Backward-Forward Sweep Based Power Flow Algorithm in Distribution Systems" employs backward–forward sweep-based power flow algorithm in distribution systems. To solve this problem, backward–forward sweep (BFS) load flow algorithm is presented by scholars. This chapter aims to present MATLAB codes of BFS power flow method in a benchmark distribution grid. Feeder capacity and voltage magnitude limit are considered in finding a good operating point for test grid. Input data such as bus and line information matrices are presented in MATLAB codes. Simulations are conducted on IEEE 33-bus radial distribution system. Feeder current, bus voltage magnitude, active and reactive power flowing in or out of buses, and total real power loss system are found as outputs of BFS load flow approach.

Contents

Modelling for Composite Load Model Including Participation of Static and Dynamic Load

Nitin Kumar Saxena and Ashwani Kumar

Abstract It is well recognized that voltage problems in power system is much affected through the connected loads. Different types of load can be modeled on their characteristics basis for computation of power system problems effectively. For different power system studies especially in the area of power system optimization problems that includes voltage control with reactive power compensation, transfer function $\Delta Q/\Delta V$ of composite load is required. This chapter gives a detailed mathematical modelling to compute the reactive power response with small voltage perturbation for composite load. Composite load is defined as a combination of static and dynamic load model. To develop this composite load model, the exponential load is used as a static load model and induction motors are used as a dynamic load model in this chapter. To analyze the dynamics of induction motor load, fifth, third and first order model of induction motor are formulated and compared using differential equations solver in MATLAB coding. Since the decentralized areas have many small consumers which may consist large numbers of induction motors of small rating, it is not realistic to model either a single large rating unit or all small rating induction motors together that are placed in the system. In place of using single large rating induction motor a group of motors are being considered and then aggregate model of induction motor is developed using law of energy conservation and this aggregate model is used as a dynamic load model. Transfer function of composite load is derived in this chapter by successive derivation for exponential model of static load and for fifth and third order induction motor dynamic load model using state space model.

Keywords Static load · Dynamic load · Composite load · Aggregate load · *ZIP* load model · Exponential load model · Induction motor load

N. K. Saxena (✉)
Electrical and Electronics Engineering, KIET Group of Institutions, Ghaziabad, India
e-mail: nitinsaxena.iitd@gmail.com

A. Kumar
Electrical Engineering Department, National Institute of Technology, Kurukshetra, India
e-mail: ashwa_ks@yahoo.co.in

M. Pesaran Hajiabbas and B. Mohammadi-Ivatloo (eds.),
Optimization of Power System Problems, Studies in Systems, Decision and Control 262,
https://doi.org/10.1007/978-3-030-34050-6_1

1 Introduction

Power system problems involve how the systems behave at generation, transmission and distribution side for varieties of changes in load, input and faults. The power system based problems can be classified into a broad range, depending on the interest of individual researchers and power engineers and hence every research has their own domain of interests depending on the available state, control and disturbance vectors. But the main task of every study is to establish secure, reliable, continuous, efficient, stable and economic power flow in power system. Many researchers and power engineers have carried out their researches in the area of optimization for power system problems too. Further, these studies can broadly be classified into the power systems problems associated with voltage stability, ancillary service, power quality, load forecasting, electricity pricing and many more depending on the researchers and power engineers' domain of interest.

Load characteristics have also significant impacts on the power system problems. When the load demand fluctuates, the voltage level also changes. Referring to 'Principal of Decoupling' in power system which states that real power has more affinity towards power angle or frequency while reactive power has more affinity toward voltage, impact of voltage can be better correlate with load reactive power change and can almost be neglected with load real power change. Hence, to control the power system voltage, an adequate reactive power must be available at load end. This adequate reactive power control in power system at load is termed as reactive power compensation. Without any compensation this voltage variation may go beyond the voltage permissible range and therefore such power would not be acceptable for the end users [1]. Reference [2] explains how the power system voltage and choice of compensation techniques significantly depends on selection of the load model and its parameters. In order to effectively analyze the power system problems, the loads need to be modeled accurately along with the other power system elements like transformers, transmission lines, generators etc. Power system planners and operators attempt to accurately model loads in order to analyze their systems. However, it is very difficult to exactly describe the loads in a mathematical model because loads consist several components and have very different dynamic characteristics [3]. The information/knowledge about load model parameters, that properly depict load behaviour during electric power system disturbances, enables proper power system planning, reliable prediction of prospective operating scenarios and provides for adequate control actions to be chosen in order to prevent undesired system behaviour and ultimately system instability. Accurate load modelling is important to correctly predict the response of the system to disturbances. With a poor load model we would need to operate the system with a higher safety margin [4]. Load models, which quantify real and reactive power responses to voltage and frequency disturbances, are generally divided in two groups—static load models (SLM) and dynamic load models (DLM). These SLM and DLM are classified according to the effect of the voltage on the load. If the load variation depends only on the instantaneous voltage input and is unrelated to the preceding voltage inputs, the static load model is used.

However, if the load characteristics are affected by all of the voltage inputs over time, the dynamic load model needs to be used [3].

Since the static and dynamic load their own existence in power system so a combine load model must be developed for power system studies. The model consisting static and dynamic both the load chrematistics together are called composite load. Since the distribution areas have many consumers which may consist large numbers of induction motors, it is not realistic to model every induction motor that is in the system. However, it is impractical to accurately represent each individual load due to the intense computation process involved. Hence, aggregate models or single unit models with minimum order of induction motor are needed to represent a group of motors. Appropriate dynamic load model aggregation reduces the computation time and provides a faster and efficient model derivation and parameters identification. It is found that the small-scale aggregation model gives acceptably accurate results than the large-scale aggregation model and is good for power system stability analysis [5]. Hence, in place of using single large rating induction motor a group of motors are being considered and then aggregate model of induction motor is developed using law of energy conservation [6].

Summarizing the all facts discussed above the outline of this chapter is to elaborate the detailed mathematics for composite load which includes static, dynamic and aggregation of load model. Since, the Principle of Decoupling in power system explains that the load reactive power is more influence with voltage compares to load real power; therefore, modeling of composite load is focused for direct coupling between reactive power and voltage only in this chapter. Hence, a transfer function of voltage changes with reactive power change is derived using state space equations for induction motor and then same is derived for the composite load.

2 Classification of Load Model

Distribution system has the most uncertain behaviour due to the different existing load characteristics. The load can be classified by different category as in Table 1.

The characteristics of each type of load is different depending on the participation of lighting load, small motor loads, large motor loads, transformer loads and other miscellaneous loads. A composite load can be developed by knowing the percentage of different types of load participation at substation. Table 1 explains the different classification of possible load in distribution system; still load characteristics are too diversified to define load behaviour. Power engineers also explain the load on the basis of diversity in which load curve and load duration curve are developed for a specified time on the basis of available statistical data. Load pattern of consumers defines by following terms irrespective of the Category of load.

Table 1 Classification of load on the basis of different category

Sr. No.	Category of load	Remarks for category
1	On the basis of circuit element available in load	Resistive load Capacitive load Inductive load
2	On the basis of voltage current relation in load: linear or non-linear load	Linear load Non-linear load
3	On the basis of load consumer category	Residential electrical loads Commercial electrical loads Industrial electrical loads Municipal/governmental electrical loads
4	On the basis of load group	Individual loads Area load
5	On the basis of load operation time	Continuous electrical loads Non-continuous electrical loads Intermittent duty electrical loads Periodic duty electrical loads Short time duty electrical loads Varying duty electrical loads
6	On the basis of electrical load number of phase	Single phase electrical loads Three phase electrical loads
7	On the basis of unit of rating of the load	Electrical loads in KVA Electrical loads in KW Electrical loads in HP

2.1 Connected Load

A part of the total load of specified region/area that is in operation for a particular instant is called connected load.

2.2 Demand Load

Total load either operational or non-operational of specified region or area for a particular instant is called demand load.

2.3 Base Load

In a specified area, the entire electrical appliance might be either operational or non-operational. The reason is that the connected loads are switched on and off regularly. Even for the loads those are switched on, they are not properly operating with their full ratings too. But every instant of time some load always remain presents and this load is called base load.

2.4 Peak Load

In a specified area, the entire electrical appliance might be either operational or non-operational. The reason is that the connected loads are switched on and off regularly. Even for the loads those are switched on, they are not properly operating with their full ratings too. But for a particular time interval, a maximum load can occur at any instant of time this load is called peak load.

2.5 Average Load

The average demand or average load of specified area for particular time interval is defined as the total energy delivered in a certain period divided by the time interval. Average load can be calculated as a daily average load, weekly average load, monthly average load and yearly (annual) average load.

These all terns are useful for statistical approach to forecast the load demand. But for time dependent study of loads dynamic behaviour, mathematical models for load are required.

3 Structure of Loads Model

For power system studies like load forecasting, planning for new power plant installation, load scheduling etc., definitions given in previous section are required. A demand can be planned for the upcoming days, weeks or even for the years using statistical approach but for dynamic studies, the load characteristics are very important in power system studies. This understanding of load characteristics is useful to formulate the load model which is required for real time mathematical based studies in power system. The loads may be classified in several groups but they have some important parameters that must be understood to define the load model.

1. Loads real and reactive power depends on the system voltage.
2. Loads real and reactive power depends on the system frequency.
3. Load characteristics affect the dynamic of power system.
4. Large number of loads of different characteristics.
5. Different loads clubbed together for individual area to develop an aggregate load.

Loads can be modeled for steady state analysis problems or transient state problems in power system. These all five parameters must be validated for developing the load model and therefore, achievement of loads characteristic means to know how the overall load real and reactive power behaves for the variation of voltage and frequency in power system.

Power system planners and operators attempt to accurately model the loads in order to analyze their systems. The information or knowledge about load model parameters, that properly depict load behaviour during electric power system disturbances, enables proper power system planning, reliable prediction of prospective operating scenarios and provides for adequate control actions to be chosen in order to prevent undesired system behaviour and ultimately system instability. With a poor load model we would need to operate the system with a higher safety margin [4] and therefore system becomes uneconomic.

A load model represents mathematical expression active and reactive power changes to power system voltage and frequency [7]. A load consists of several components that have very different dynamic characteristics [3]. However, it is very difficult to exactly describe the loads in a mathematical model due to several factors;

1. A large number of diverse load components.
2. Continuous changes in load demand.
3. Lack of precise information regarding composition of load.
4. Uncertainty regarding the characteristic of loads due to sudden disturbances.

Therefore, to identify actual load pattern for any power system study, a load model must be design including the factors as mentioned above. Influence of these factors may be accompanied with using static load, dynamic load and aggregate load model simultaneously. Such load models are called composite load model [8]. Traditionally loads are classified into two categories, static load and dynamic load. An aggregate load model can be developed by collecting all similar type loads together. Figure 1 shows a structure of composite load which include static load and number of induction motors connected in parallel to form a dynamic load by aggregate load modelling in power system.

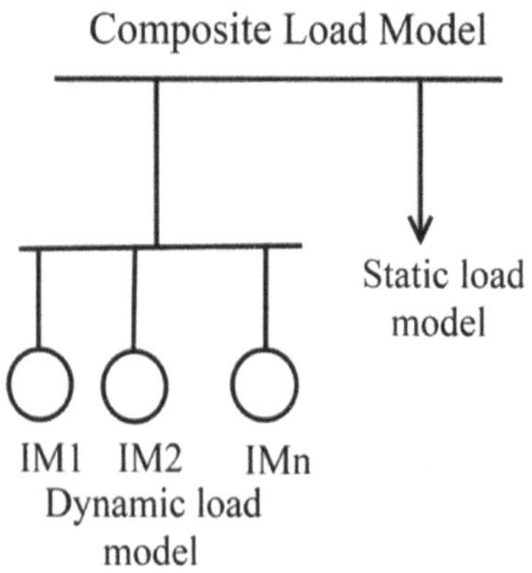

Fig. 1 Structure of composite load in power system

3.1 Structure for Static Load Model

In static load model, the load variation depends only on the instantaneous voltage or frequency input and is unrelated to the preceding inputs of voltage or frequency. Static load model are generally used for the calculation of steady state conditions and in steady state simulation study of power system.

Static load model is widely used to represent the load characteristics where real and reactive power depends on the present values of voltage and frequency. The active power, P and the reactive power, Q are being considered separately. The static load model can be represented in two ways, polynomial model or exponential model.

3.1.1 Polynomial Type Static Load Model

The load characteristics in terms of voltage are represented by the polynomial model [9];

$$P = P_0\left[k_1 + k_2\left(\frac{V}{V_0}\right) + k_3\left(\frac{V}{V_0}\right)^2\right] \tag{1}$$

$$Q = Q_0\left[k_4 + k_5\left(\frac{V}{V_0}\right) + k_6\left(\frac{V}{V_0}\right)^2\right] \tag{2}$$

where P and Q are the real and reactive power respectively at any instant when the bus voltage magnitude is V. The subscript 0 is used to represent the values of respective variables at initial operating conditions.

The real power expression given in Eq. (1) consists three terms as represented below in Eq. (3);

$$P = k_1 P_0 + k_2\left(\frac{V}{V_0}\right)P_0 + k_3\left(\frac{V}{V_0}\right)^2 P_0 \tag{3}$$

Considering the all three terms individually,

$$P = \begin{cases} k_1 P_0 \text{ Ist } term \\ + \\ k_2\left(\frac{V}{V_0}\right)P_0 \text{ IInd } term \\ + \\ k_3\left(\frac{V}{V_0}\right)^2 P_0 \text{ IIIrd } term \end{cases} \tag{4}$$

In Ist term, $P = constant$, it denotes constant power (P) load component.
In IInd term, $\frac{P}{V} = constant$, it denotes constant current (I) load component.

In IIIrd term, $\frac{P}{V^2} = constant$, it denotes constant impedance (Z) load component.

This polynomial type static load model, given in Eqs. (1) and (2) for real and reactive power respectively, consist three components namely constant impedance (*Z*) load component, constant current (*I*) load component and constant power (*P*) load component, therefore it is commonly known as *ZIP* model. The parameters k_1, k_2 and k_3 denote the proportional coefficients of real and reactive power. This static load can be generalized individually as constant power, constant current or constant impedance depending on the value of k_1, k_2 and k_3.

For constant impedance (*Z*) load model,

$$P = \begin{cases} k_1 P_0 = 0 \quad if \quad k_1 = 0 \\ + \\ k_2\left(\frac{V}{V_0}\right) P_0 = 0 \quad if \quad k_2 = 0 \\ + \\ k_3\left(\frac{V}{V_0}\right)^2 P_0 = \left(\frac{V}{V_0}\right)^2 P_0 \quad if \quad k_3 = 1 \end{cases} \tag{5}$$

So,

$$P = \left(\frac{V}{V_0}\right)^2 P_0 \tag{6}$$

For constant current (*I*) load model,

$$P = \begin{cases} k_1 P_0 = 0 \quad if \quad k_1 = 0 \\ + \\ k_2\left(\frac{V}{V_0}\right) P_0 = \left(\frac{V}{V_0}\right) P_0 \quad if \quad k_2 = 1 \\ + \\ k_3\left(\frac{V}{V_0}\right)^2 P_0 = 0 \quad if \quad k_3 = 0 \end{cases} \tag{7}$$

So,

$$P = \left(\frac{V}{V_0}\right) P_0 \tag{8}$$

For constant power (P) load model,

$$P = \begin{cases} k_1 P_0 = P_0 \quad if \quad k_1 = 1 \\ + \\ k_2\left(\frac{V}{V_0}\right) P_0 \quad if \quad k_2 = 0 \\ + \\ k_3\left(\frac{V}{V_0}\right)^2 P_0 = 0 \quad if \quad k_3 = 0 \end{cases} \tag{9}$$

So,

$$P = P_0 \tag{10}$$

All models explained above are derived for real power expressions only. The same concept can also be applied for getting the reactive power *ZIP* load model, *Z* load model, *I* load model and *P* load model.

The *ZIP* load model expression for reactive power,

$$Q = Q_0\left[k_4 + k_5\left(\frac{V}{V_0}\right) + k_6\left(\frac{V}{V_0}\right)^2\right] \tag{11}$$

The constant impedance (*Z*) load model expression for reactive power,

$$Q = \begin{cases} k_4 Q_0 = 0 \quad if \quad \mathrm{k}_4 = 0 \\ + \\ k_5\left(\frac{V}{V_0}\right)Q_0 = 0 \quad if \quad \mathrm{k}_5 = 0 \\ + \\ k_6\left(\frac{V}{V_0}\right)^2 Q_0 = \left(\frac{V}{V_0}\right)^2 Q_0 \quad if \quad \mathrm{k}_6 = 1 \end{cases} \tag{12}$$

So,

$$Q = \left(\frac{V}{V_0}\right)^2 Q_0 \tag{13}$$

The constant current (*I*) load model expression for reactive power,

$$Q = \begin{cases} k_4 Q_0 = 0 \quad if \quad \mathrm{k}_4 = 0 \\ + \\ k_5\left(\frac{V}{V_0}\right)Q_0 = \left(\frac{V}{V_0}\right)Q_0 \quad if \quad \mathrm{k}_5 = 1 \\ + \\ k_6\left(\frac{V}{V_0}\right)^2 Q_0 = 0 \quad if \quad \mathrm{k}_6 = 0 \end{cases} \tag{14}$$

So,

$$Q = \left(\frac{V}{V_0}\right)Q_0 \tag{15}$$

The constant power (*P*) load model expression for reactive power,

$$Q = \begin{cases} k_4 Q_0 = Q_0 \quad if \ \mathrm{k}_4 = 1 \\ + \\ k_5\left(\frac{V}{V_0}\right) Q_0 = 0 \quad if \ \mathrm{k}_5 = 0 \\ + \\ k_6\left(\frac{V}{V_0}\right)^2 Q_0 = 0 \quad if \ \mathrm{k}_6 = 0 \end{cases} \tag{16}$$

So,

$$Q = Q_0 \tag{17}$$

These all models given in Eqs. 1–17 are voltage dependent only. To make these models frequency dependent, expressions explained above can be multiplied with the factor denoting the frequency dependency of load real and reactive power. Since the permissible variation of frequency is very less so it can be considered as linear and therefore factor would be [10];

$$P = P(V) \times \{1 + k_7(f - f_0)\} \tag{18}$$

$$Q = Q(V) \times \{1 + k_8(f - f_0)\} \tag{19}$$

where $P(V)$ and $Q(V)$ are the expressions for voltage dependant real and reactive power respectively as explained in Eqs. 1–17 depending on the model type. k_7 and k_8 are the proportionality factors for real and reactive power expressions. P and Q are the real and reactive power respectively at any instant when the bus frequency is f. The subscript 0 is used to represent the values of respective variables at initial operating conditions.

3.1.2 Exponential Type Static Load Model

Polynomial load model explains the characteristic of load as *ZIP* model or individual load model like constant impedance (Z), constant power (P) or constant current (I) load only. But the actual load is more volatile in nature and even it cannot be possible to identify it in either of the form of *ZIP* load model. In such case, exponential type model contribute to identify the actual static load model. The load characteristics in terms of voltage are represented by the exponential model [11];

$$P = P_0\left(\frac{V}{V_0}\right)^{np} \tag{20}$$

$$Q = Q_0\left(\frac{V}{V_0}\right)^{nq} \tag{21}$$

where P and Q are the real and reactive power respectively at any instant when the bus voltage magnitude is V. The subscript 0 is used to represent the values of respective variables at initial operating conditions.

The parameters *np* and *nq* denote the exponent coefficients of real and reactive power. This static load can also be generalized as constant power, constant current or constant impedance depending on the value of *np* and *nq*.

Values of *np* and *nq* are considered to be zero for constant power load and this model is used in load flow studies.

Values of *np* and *nq* are considered to be 1 for constant current load and this model is used for system having large percentage of small rating induction motor.

Values of *np* and *nq* are considered to be 2 for constant impedance load and this model is used for transient studies.

Mathematically,

$$P = P_0\left(\frac{V}{V_0}\right)^{np}\begin{cases} np = 0\, constant\ power\ load \\ np = 1\, constant\ current\ load \\ np = 2\, constant\ impedance\ load \end{cases} \tag{22}$$

$$Q = Q_0\left(\frac{V}{V_0}\right)^{nq}\begin{cases} nq = 0\, constant\ power\ load \\ nq = 1\, constant\ current\ load \\ nq = 2\, constant\ impedance\ load \end{cases} \tag{23}$$

For frequency dependent models, expressions explained above can be multiplied with the factor denoting the frequency dependency of load real and reactive power. Since the permissible variation of frequency is very less so it can be considered as linear and therefore factor would be [10];

$$P = P_0\left(\frac{V}{V_0}\right)^{np} \times \{1 + k_9(f - f_0)\} \tag{24}$$

$$Q = Q_0\left(\frac{V}{V_0}\right)^{nq} \times \{1 + k_{10}(f - f_0)\} \tag{25}$$

where k_9 and k_{10} are the proportionality factors for real and reactive power expressions. P and Q are the real and reactive power respectively at any instant when the bus frequency is f. The subscript 0 is used to represent the values of respective variables at initial operating conditions.

The static load may have the aggregate characteristics that cannot be analyzed through either form of the *ZIP* load model. A composite load may have different exponent factors for steady state and transient sate analysis of load. References [12–17] gives a typical range for the Values of np and nq.

$$0 \le n_p \le 3$$
$$0 \le n_q \le 7$$

3.2 Structure of Dynamic Load

Dynamic load models are such for which load characteristics are affected by all of the voltage inputs over time used [18]. Motors consume 60–70% of energy from the power system; therefore, the dynamic characteristics of motors are critical for dynamic load modelling. In most of studies, induction motors as a load reduce system stability and is considered as the main dynamic load in their study system [19]. A basic paper for induction machine has been reported by H. C. Stanley in 1938 [20]. The analysis made by H. C. Stanley is based on the direct three-phase model using phase variables and its presentation in shifted reference axis. The most popular induction motor model is presented by P. C. Krause in 1986 and popularly known as Krause's model. This model is derived from direct-quadrature model or dynamic equivalent circuit model [21]. Reference [22] suggests that for most of the analysis, d–q model with currents as state variables is found most suitable, and the analysis can be carried out in any (stationary, rotor, synchronous or arbitrary) reference frame. Induction motor behaviour especially during transient conditions is investigated using MATLAB Simulink model in [23]. The simulink implementation based study for induction machine model appears to be black-boxes and therefore is not more suitable for researchers. While programming based approach for any system has much potential for the researchers to analyze the real time dependence of different parameters in system. In programming based approaches, mathematical expressions can be formulated for numbers of different system parameters, implemented depending on the individual real time problems for the individual system and analyzed for individual system conditions.

Normally, the induction motor parameters; mutual inductance, stator inductance, rotor inductance, stator resistance, rotor resistance, inertia of the rotor and load torque can be identified by experimental set up using no load and blocked rotor test. This experimental method is not suitable for simulation studies or on line testing of machines. Reference [24] suggests nonlinear least squares approach to identify the parameters of induction motor. In Ref. [17], the measured real and reactive power responses to voltage step are used as the input to parameter identification procedure based on curve fitting using least squares method and load model parameters are determined. Fuzzy logic controller, followed by initial reference parameters, is used for the parameter estimation of induction motor model [25].

Since, induction motors are the most vastly used dynamic load of inductive nature. Hence, knowledge of induction motor responses is essential for dynamic load modelling [26]. The induction motor is modeled using its dynamic equations. In ref. [8], the induction motor is modeled with the help of five dynamic equations but to simplify the induction motor's mathematical studies sometimes these five differential equations can be reduced to either three or one differential equations. On the basis of numbers of differential equations through which induction motor is being represented mathematically, model of induction motors are called fifth, third or first order model.

The fifth order model is very close to the real motor while third and first order model are the simple mathematical version of motor model. In third order model stator flux is considered constant while in first order model both stator and rotor fluxes are considered to be constant.

3.2.1 Manufacturer Data for Induction Motor

Commonly, the induction motor is defined by the output parameters, given in the catalogues as manufacturer data. For the mathematical analysis of induction motor like any other machine, an equivalent circuit is required. This equivalent circuit provides a platform for researchers for applying several fundamental theorems available in electrical engineering like network theorems, Kirchhoff's law, power flow equation etc. Once induction motor is ready to represent through its equivalent circuit by including all the induction motor parameters in its equivalent circuit, these parameters contribute in analyzing the induction motor's dynamics and performances in the system. Equivalent circuit showing the parameters for induction motor is given in Fig. 2. Table 2 shows the list of manufacturer data that are specified by the manufacturer with induction motor supply to buyers.

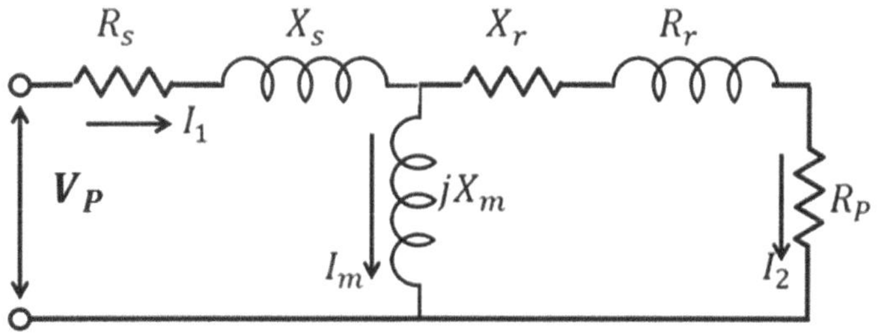

Fig. 2 Equivalent circuit for induction motor

Table 2 Manufacturer data for an induction motor

S. No.	Symbol	Description
1	P_{im}	Rating of induction motor in kW
2	V	Line voltage in Volt
3	f	Frequency of system in Hz
4	P	Number of poles
5	$\cos(\emptyset_{im})$	Power factor of induction motor
6	η_{im}	Efficiency of induction motor
7	s_{im}	Slip of induction motor

3.2.2 Estimation of Induction Motor Parameters Using Manufacture Data

As discussed in previous section, equivalent circuit parameters formulate a real time induction motor in the form of a network circuit. This network circuit helps researchers or engineers in performing several studies on motor with the help of its circuit parameters. Since the operating characteristics of induction motor is very dynamic in nature and therefore, most of the time induction motor performances, behaviors and dynamics are much influenced by these circuit parameters. Two types of studies can be followed for estimating these parameters; experimental type and analytical type. DC test, no load test and block rotor test are the popular experimental type methods to evaluate the electrical parameters of induction machine.

Analytical approach is most suitable for programming and simulation based studies and same is discussed in this chapter. To evaluate parameters, initial reference parameters are used as documented in Ref. [25]. Calculation of induction motor is being limited for squirrel cage induction motor in this chapter. In squirrel cage induction motor, rotor is short circuited and therefore $V_r = 0$. Induction motor parameters presented here can be estimated with the help manufacturer data given in Table 2 followed by the mathematical expressions as given in Eqs. (26)–(42).

If motor's pole pair is represented by pp, synchronous speed ω_s of induction motor can be expressed as in Eq. (26),

$$\omega_s = \frac{2\pi f}{pp} \tag{26}$$

Equation (27) represents expression of rotor speed for induction motor in terms of synchronous speed ω_s and slip s_{im}.

$$\omega_r = \omega_s(1 - s_{im}) \tag{27}$$

Figure 2 represents per phase equivalent circuit for induction motor. For line voltage V, per phase voltage is,

$$V_P = \frac{V}{\sqrt{3}} \tag{28}$$

Input current I_1 as in Fig. 2 can be formulated with P_{im} rating, $\cos(\emptyset_{im})$ power factor and η_{im} efficiency and it is given in Eq. (29).

$$I_1 = \frac{P_{im}}{\sqrt{3}V\cos(\emptyset_{im})\eta_{\mathrm{im}}} \tag{29}$$

For input terminal shown in Fig. 2, equivalent impedance, equivalent resistance and equivalent reactance can be expressed with the Eq. (30)–(32). All these values are referred to stator side as shown in Fig. 2.

$$Z = \frac{V_P}{I_1} \tag{30}$$

$$R^{'}_{eq} = Z\cos(\emptyset_{im}) \tag{31}$$

$$X^{'}_{eq} = Z\sin(\emptyset_{im}) \tag{32}$$

For calculation purpose, rotational losses in induction motor are being neglected. Since the rating of the induction motor is the shaft input power. This can be used equal to the mechanical power of induction motor. Neglecting the rotational losses in induction motor, mechanical torque T_m in terms of shaft power (rating of motor) P_{im} and rotor speed for induction motor ω_r is expressed in Eq. (33).

$$T_m = \frac{P_{im}}{\omega_r} \tag{33}$$

Resistance R_P in Fig. 2 represents mechanical power output or useful power in electrical terms. The power dissipated in that resistor is the useful power output or shaft power. Equation (34) shows the value of R_P in terms of rotor resistance and slip. Using this shaft power, value of rotor resistance R_r can be expressed as in Eq. (35).

$$R_P = \frac{1 - s_{im}}{s_{im}} R_r \tag{34}$$

$$R_r = \frac{s_{im}}{1 - s_{im}} \frac{(P_{im}/3)}{I_1^2} \tag{35}$$

To evaluate initial reference parameters, following expression available in literature are used for flux leakage coefficient σ, Mutual inductance of motor L_m and Stator self inductance L_s. It is also assumed that both stator self inductance L_s and rotor self inductance L_r are equal for calculation purposes. According to Ref. [25], Eq. (36)–(39) presents their mathematical expression.

$$\sigma = \frac{1 - \cos(\emptyset_{im})}{1 + \cos(\emptyset_{im})} \tag{36}$$

$$L_m = \frac{V_P}{2\pi f I_1 \sqrt{\sigma}} \tag{37}$$

$$L_s = 0.3\sigma L_m \tag{38}$$

$$L_r = L_s \tag{39}$$

All the parameters except stator resistance R_s for induction motor, as shown in Fig. 2, have been calculated above. To estimate the value of R_s, concept of equivalent circuit impedance is elaborated here. At input terminal equivalent impedance, equivalent resistance and equivalent reactance can be expressed as,

$$Z = R'_{eq} + jX'_{eq} \tag{40}$$

The value of impedance at input terminal can be expressed as,

$$Z = (R_s + jX_s) + \frac{(R_P + R_r + jX_r)(jX_m)}{(R_P + R_r + jX_r) + (jX_m)} \tag{41}$$

Putting the values of Z and R_P in Eq. (41),

$$R'_{eq} + jX'_{eq} = (R_s + jX_s) + \frac{\left(\frac{R_r}{s_{im}} + jX_r\right)(jX_m)}{\left(\frac{R_r}{s_{im}} + jX_r\right) + (jX_m)} \tag{42}$$

Using Eq. (42) and comparing the real parts of this expression in both the side, R_S can be calculated.

3.2.3 Voltage Estimation in d-q Model for Induction Motor

Equations (26)–(42) can be used to estimate the circuit parameters of induction motor with the help of manufacturer data as described in Table 2. Since the induction motor is a rotating machine and for any rotating machine, a reference frame is required to estimate the parameters. Reference [22] suggests that for most of the analysis, *d–q* model with currents as state variables is found most suitable, and the analysis can be carried out in any (stationary, rotor, synchronous or arbitrary) reference frame. Before carrying induction motor analysis in much detail, knowledge of stator and rotor voltage in d-q model is required and method to evaluate these values is presented in this section. Expressions for converting three phase voltage into d and q axis values are summarized here as in Ref. [27]. For the balanced operation of squirrel cage induction motor, the most widely used reference frame is synchronous rotating reference frame and same is used in this chapter [28].

For a balanced three phase system, phase values of stator voltages are

$$V_{sa} = \sqrt{2}V_P \cos(\omega_b t) \tag{43}$$

$$V_{sb} = \sqrt{2}V_P \cos\left(\omega_b t - \frac{2\pi}{3}\right) \tag{44}$$

$$V_{sc} = \sqrt{2}V_P \cos\left(\omega_b t + \frac{2\pi}{3}\right) \tag{45}$$

ω_b is a base frequency and is defined as represented in Eq. (46).

$$\omega_b = 2\pi f \tag{46}$$

Using Park Transformation, stator voltage in d-q model can be written for stator phase voltages given in Eqs. (43)–(45) with a constant $a = e^{-j\frac{2\pi}{3}}$.

$$V_s = \frac{2}{3}\{V_{sa} + aV_{sb} + a^2V_{sc}\} \tag{47}$$

The value of stator voltage in d-q model given in Eq. (47) is still in stator coordinates. This stator voltage in stator coordinated can be converted into synchronous rotating reference frame as expressed in Eq. (48).

$$V_{st} = V_s e^{-j\omega_s t} \tag{48}$$

The complex variables may be decomposed in plane along two orthogonal d and q axes rotating at speed ω_b to obtain the separate d–q (Park) model. So, direct axis stator voltage V_{ds} and quadrature axis stator voltage V_{qs} are,

$$V_{ds} = real(V_{st}) \tag{49}$$

$$V_{qs} = imag(V_{st}) \tag{50}$$

For squirrel cage induction motor the rotor side is short circuited so rotor direct and quadrature axis values V_{dr} and V_{qr} are zero.

$$V_{dr} = 0 \tag{51}$$

$$V_{qr} = 0 \tag{52}$$

3.2.4 Dynamic Equation of Induction Motor

In Ref. [8], mathematics for induction motor dynamic is explained in the detail. The induction motor can be expressed by five differential equations given in Eqs. (53)–(57). In these equations, the direct and quadrature axis voltages are the independent variables and fluxes and rotor speed are the dependent variables. For direct axis stator flux (φ_{qs}), quadrature axis stator flux (φ_{ds}), quadrature axis rotor flux (φ_{qr}), direct axis rotor flux (φ_{dr}) and rotor speed (ω_r) dynamic equations for induction motor are,

$$\frac{d\varphi_{qs}}{dt} = \omega_b\left[V_{qs} - \frac{\omega_s}{\omega_b}\varphi_{ds} - R_s I_{qs}\right] \tag{53}$$

$$\frac{d\varphi_{ds}}{dt} = \omega_b\left[V_{ds} + \frac{\omega_s}{\omega_b}\varphi_{qs} - R_s I_{ds}\right] \tag{54}$$

$$\frac{d\varphi_{qr}}{dt} = \omega_b\left[V_{qr} - \frac{\omega_s - \omega_r}{\omega_s}\varphi_{dr} - R_r I_{qr}\right] \quad (55)$$

$$\frac{d\varphi_{dr}}{dt} = \omega_b\left[V_{dr} + \frac{\omega_s - \omega_r}{\omega_s}\varphi_{qr} - R_r I_{dr}\right] \quad (56)$$

$$\frac{d\omega_r}{dt} = \frac{\omega_b}{2H}(T_e - T_L) \quad (57)$$

In these equations, dependent variable can be estimated using input supply voltage. The expression of direct and quadrature axis stator and rotor currents (i.e. I_{ds}, I_{qs} I_{dr} and I_{qr}) in terms of induction motor parameters and fluxes are given in Eqs. (58)–(61).

$$\varphi_{qs} = L_{ss} I_{qs} + L_m I_{qr} \quad (58)$$

$$\varphi_{ds} = L_{ss} I_{ds} + L_m I_{dr} \quad (59)$$

$$\varphi_{qr} = L_{rr} I_{qr} + L_m I_{qs} \quad (60)$$

$$\varphi_{dr} = L_{rr} I_{dr} + L_m I_{ds} \quad (61)$$

The term L_{ss} and L_{rr} are used to define the value as in Eqs. (62) and (63).

$$L_{ss} = L_s + L_m \quad (62)$$

$$L_{rr} = L_r + L_m \quad (63)$$

In Eq. (57), T_e and T_L denotes the electro-magnetic torque and load torque for induction motor. Expressions for T_e and T_L are elaborated in Eqs. (64) and (65).

$$T_e = \varphi_{qr} I_{dr} - \varphi_{dr} I_{qr} \quad (64)$$

Expression for per unit expression is;

$$T_L = B\frac{\omega_r}{\omega_b} \quad (65)$$

In Eq. (57), H is the machine inertia and in Eq. (65), B is the torque damping factor for induction motor. The induction motor dynamics are largely characterized by H and B along with J (moment of inertia for induction motor). These three parameters are popularly known as mechanical parameters of induction motor.

3.2.5 Method to Obtain Mechanical Parameters for Induction Motor

Equations (53)–(57) represents the dynamic model of induction motor. Fifth order model of induction motor can be represented by these equations while mathematical model can also be made simplified by making assumption of stator and rotor fluxes as discussed in starting of Sect. 3.2. To understand the operating characteristics of induction motor, one has to solve ordinary differential equations as given in last section. Electrical parameters can be estimated with the help of procedure as explained in Sect. 3.2.2 while method of estimating mechanical parameters (B, H and J) is not discussed yet in this chapter. The methods for calculating mechanical parameters are even not available much in literature. Some researchers have presented experimental procedure like retardation test [29–31] for estimating the mechanical parameters. But for programming and simulation based studies, parameters may directly be required in software coding using manufacturer data and so, analytical approach is more preferable to estimate these parameters directly by using manufacturer data as mentioned by the machine suppliers [32].

This section presents an analytical approach for estimating mechanical parameters of induction machine using induction motor dynamic response parameters. This section explains a software based approach for getting B, H and J values by run and trial method using rotor speed and slip responses. The value of B, H and J must be selected that satisfies the following operating constraints for rotor speed and slip.

(i) Slow varying speed with lower overshoots,
(ii) Slip should reach to its steady state value maintaining positive value of slip at all instants of response, and
(iii) Minimum value of J for which rotor speed should be equal to synchronous speed keeping continuous and differentiable step response for transfer function.

Example 1 For the manufacturer data given for induction motor as below, find its electrical parameters.

kW rating of induction motor, $P_{im} = 50\,\text{kW}$

Line voltage, $\text{V} = 400\,\text{V}$

Frequency of system, $\text{f} = 50\,\text{Hz}$

Number of poles $= 2$

Power factor of induction motor, $\cos(\emptyset_{im}) = 0.9$

Efficiency of induction motor, $\eta_{\text{im}} = 90\%$

slip of induction motor, $s_{im} = 4\%$

Solution For the data given above and using mathematical syntaxes in MATLAB command window for Eqs. 26–42, electrical parameters are estimated in this example. The MATLAB codes are also given just after this example. The variables selected in program are almost similar to the variables represented in chapter text. Still some parameters are changed due to the limitations of MATLAB parameter nomenclatures pattern and the parameters so chosen can easily be understand by the readers.

$P_{im} = 50 \times 10^3;$
$f = 50;$
$s = 0.04;$
$V = 400;$
$V_P = \frac{V}{\sqrt{3}} = 230.9401;$
$\cos(\emptyset_{im}) = 0.9;$
$\eta_{\text{im}} = 90\%;$
$I_1 = \frac{P_{im}}{\sqrt{3}V\cos(\emptyset_{im})\eta_{\text{im}}} = 89.0973;$
$S = \frac{P_{im}}{\cos(\emptyset_{im})} = 55.5556 \times 10^3;$
$\omega_b = 2 \times 3.14 \times f = 314;$
$\omega_s = \frac{2\times 3.14\times f}{2/2} = 314;$
$\omega_r = \omega_s(1 - s_{im}) = 301.44;$
$Z = \frac{V_P}{I_1} = \frac{230.9401}{89.0973} = 2.5920;$
$R'_{eq} = Z\cos(\emptyset_{im}) = 2.3328;$
$X'_{eq} = Z\sin(\emptyset_{im}) = 1.1298;$
$R_r = \frac{s_{im}}{1-s_{im}}\frac{(P_{im}/3)}{I_1^2} = 0.0875;$
$\sigma = \frac{1-\cos(\emptyset_{im})}{1+\cos(\emptyset_{im})} = 0.0526;$
$L_m = \frac{V_P}{2\pi f I_1\sqrt{\sigma}} = 0.0360;$
$L_s = 0.3\sigma L_m = 5.6808 \times 10^{-4};$
$L_r = L_s = 5.6808 \times 10^{-4};$
$X_m = 2\pi f L_m = 11.3097$
$X_s = 2\pi f L_s = 0.1785$
$X_r = 2\pi f L_r = 0.1785$

To find the value of R_s, compare the real part of Eq. (42) both the sides after rearranging the terms as,

$$R_s + jX_s = R'_{eq} + jX'_{eq} - \frac{\left(\frac{R_r}{s_{im}} + jX_r\right)(jX_m)}{\left(\frac{R_r}{s_{im}} + jX_r\right) + (jX_m)}$$

Hence, for the 50 kW induction motor electrical parameters are,

$R_s = 0.2875\ \Omega$
$R_r = 0.0875\ \Omega$
$X_s = 0.1785\ \Omega$
$X_r = 0.1785\ \Omega$
$X_m = 11.3097\ \Omega$

```
%%%% MATLAB codes for Example 1%%%%%
>> P_im = 50e3;
>> f = 50;
>> pp = 2/2;    %% pole pair which is pole divided by 2
>> s = 0.04;
```

```
≫ V = 400;
≫ powerfactor_im = 0.9;
≫ eff_im = 90/100;
≫ vp = v/sqrt(3);
≫ i1 = P_im/(sqrt(3)*v* Powerfactor_im *eff);
≫ ob = 2*pi*f; %base speed in rad per sec
≫ os = 2*pi*f/pp; % synchronous speed in rad per sec
≫ or = 2*pi*f*(1-s)/pp; % rotor speed in rad per sec
≫ z = vp/i1;
≫ Req_dash = z*powerfactor_im;
≫ Xeq_dash = z*sind(acosd(powerfactor_im));
≫ Rr = P_im*s/(3*(1-s)*i1^2);
≫ sigma = (1-igpf)/(1 + igpf);
≫ Lm = vp/(2*pi*f*i1*sqrt(sigma));
≫ Ls = 0.3*sigma*Lm;
≫ Lr = Ls;
≫ Xm = ob*Lm;
≫ Xs = ob*Ls;
≫ Xr = ob*Lr;
%% to find Rs, using the concept to equivalent circuit in next three steps
≫ a1 = ((Rr/s) + 1i*Xr)*(1i*Xm)/((Rr/s) + 1i*(Xr + Xm));
≫ b = Req_dash + 1i*Xeq_dash-a1;
≫ Rs = real(b);
%% Final results for example 1
Rs
Rr
Xs
Xr
Xm
```

Example 2 For the 50 kW induction motor given in last example, find mechanical parameters for induction motor.

Solution The MATLAB codes are not presented in this Example 2 because the same code will repeat again in Example 3. So only explanation is given here the program can be used for getting the solution of this Example 2 from the next Example 3.

Using the five differential equations given in Eqs. (53–57), induction motor rotor speed and slip responses can be plotted. Since the rotor speed or slip response depends on the selecting value of H and B while J can directly be calculated by the value of H. The values of H and B are selected by run and trial method here. Five differential equations of induction motor are solved using ODE45 in MATLAB program. The B and H values are estimated by analyzing rotor speed and slip responses with two constraints;

(i) Slow varying speed with lower overshoots, and
(ii) Slip should reach to its steady state value maintaining positive value of slip at all instants of its response.

This can be achieved by following these steps, (i) Set value of torque coefficient B such that the machine may run at ω_r, and (ii) for chosen value of B, set value of H keeping transients in specified range and limitations. It must also be noted that step response also helps to choose the best value of mechanical parameters and minimum value of H for which rotor speed should be equal to synchronous speed keeping continuous and differentiable step response for transfer function of reactive power to voltage change for induction motor load. The detail discussion about developing transfer function of reactive power to voltage change for induction motor load is explained in this chapter later on.

Since the above said method to define *B*, H and J use five differential equations of induction motor so, this is called fifth order model of induction motor. Once mechanical parameters are achieved using fifth order model, same electrical and mechanical parameters can be used for developing third order and first model of induction motor load.

For, 50 kW induction motor estimated mechanical parameters are,

$B = 0.5815$ N m s

$J = 0.09$ kg m^2

H = 0.0799

Table 3 represents the mechanical parameters for numbers of induction motor that are achieved by following the same procedure as explained in Example 2. It is assumed that all the manufacturer data for all the induction motors in Table 2 are same except their ratings in kW and equal to values as given in Example 1.

Example 3 For the induction motor of 50 kW rating given in previous example, draw the responses of output quantities (electro magnetic torque, slip and speed) and input quantities (active and reactive powers) for fifth, third and first order model of induction motor.

Solution After estimating the parameters for induction motor as discussed in previous examples, differential equations for induction motor are solved in MATLAB using ODE45 solver as explained below in this example. All the parameters used in MATLAB coding are same as discussed in this chapter during the theory discussion.

```
% Syntaxes for differential equation solver using ode45 is presented below
% state variables of induction motor five differential equations are.. %.....represented
by x
% all the parameters in MATLAB coding are written in per unit
% some parameters that can be written in MATLAB editor window directly
% therefore these are being changed with new variable
% but their equivalent parameters are also being mentioned
```

Table 3 Mechanical parameters estimated for different induction motor

S. No.	P_{im} (kW)	J (kg m^2)	H	B (N m s)
1	3	0.00104	0.0154	0.03489
2	7.5	0.0025	0.0148	0.08720
3	10	0.007	0.0311	0.1163
4	11	0.0072	0.0291	0.1279
5	18.5	0.02	0.0480	0.21515
6	50	0.0900	0.0799	0.58150
7	55	0.2	0.1615	0.6395
8	75	0.3500	0.2073	0.8720
9	90	0.40	0.1974	1.0465
10	100	0.47	0.2087	1.1625
11	110	0.61	0.2463	1.2790
12	150	0.65	0.1925	1.744
13	160	0.79	0.2193	1.86
14	200	1.1	0.2443	2.325
15	500	2.0	0.2576	5.8135
16	1000	6.2	0.2754	11.625
17	1500	8.2	0.2428	17.44
18	2000	10.6	0.2354	23.25

$\gg x00 = [0\ 0\ 0\ 0\ 0]$; % reset the initial conditions for 5 states variable
$\gg [t, x] = ode45(@order5, [0Tcheck], x00)$ % Syntax for ODE solver
$\gg function\, xdot = order5(t, x)$ % function for ode

% Estimation of d and q axis stator and rotor voltage
$\gg vo = vp/vbase$; % vbase is a base value chosen
$\gg Va = vo * (cosd(0) + 1i * sind(0))$;
$\gg Vb = vo * (cosd(-120) + 1i * sind(-120))$;
$\gg Vc = vo * (cosd(120) + 1i * sind(120))$;
$\gg Vsabc = [Va; Vb; Vc]$;
$\gg Vs1 = abs(Va) * sqrt(2)$;
$\gg Vs2 = abs(Vb) * sqrt(2)$;
$\gg Vs3 = abs(Vc) * sqrt(2)$;
$\gg Vsa = Vs1 * cos(ob * t)$;
$\gg Vsb = Vs2 * cos(ob * t - (2 * pi/3))$;
$\gg Vsc = Vs3 * cos(ob * t + (2 * pi/3))$;
$\gg a = \exp(1i * 2 * pi/3)$;
$\gg Vs = (2/3) * \left(Vsa + a * Vsb + a^\wedge 2 * Vsc\right)$; % in stator coordinates $\gg Vst = Vs * exp(-1i * (os) * t)$; % in synchronous reference frame
$\gg Vds = real(Vst)$;
$\gg Vqs = imag(Vst)$;

$\gg Vdr = 0;$
$\gg Vqr = 0;$
% Coding for ODE Solver
$\gg xdot = zeros(5, 1);$
$\gg Tl = B * x(5)/wbase;$ % wbase denotes ω_b $\gg xa = (Lm * Lm) - (Lss * Lrr);$
$\gg Iqs = ((Lm * x(3)) - (Lrr * x(1)))/xa;$
$\gg Ids = ((Lm * x(4)) - (Lrr * x(2)))/xa;$
$\gg Iqr = ((Lm * x(1)) - (Lss * x(3)))/xa;$
$\gg Idr = ((Lm * x(2)) - (Lss * x(4)))/xa;$
$\gg Te = (x(3) * Idr) - (x(4) * Iqr);$
%ob denotes ω_b
$\gg xdot(1) = ob * (Vqs - ((os/ob) * (x(2))) - (Rs * Iqs));$
$\gg xdot(2) = ob * (Vds + ((os/ob) * (x(1))) - (Rs * Ids));$
$\gg xdot(3) = ob * (Vqr - (((os - x(5))/ob) * (x(4))) - (Rr * Iqr));$
$\gg xdot(4) = ob * (Vdr + (((os - x(5))/ob) * (x(3))) - (Rr * Idr));$
$\gg xdot(5) = (ob/(2 * H)) * (Te - Tl);$
$\gg xdot = [xdot(1); xdot(2); xdot(3); xdot(4); xdot(5)];$
$\gg end$
$\gg t;$
$\gg x;$ % this command delivers an array of all five states

The array obtained for five state variables from five differential equations of induction motor is stored in ***x*** variable and time array is stored in ***t*** variable.

Parameters for fifth order induction motor model can be estimated by coding following MATLAB syntaxes,

% parameters estimation after getting the results of ODE45 solver
$\gg \varphi_{ds} = \mathrm{x}(:, 1);$
$\gg \varphi_{qs} = \mathrm{x}(:, 2);$
$\gg \varphi_{dr} = \mathrm{x}(:, 3);$
$\gg \varphi_{qr} = \mathrm{x}(:, 4);$
$\gg \omega_r = \mathrm{x}(:, 5);$
$\gg Iqs = ((Lm * x(:, 3)) - (Lrr * x(:, 1)))/xa;$
$\gg Ids = \frac{(Lm * x(:,4)) - (Lrr * x(:,2))}{xa};$
$\gg Iqr = ((Lm * x(:, 1)) - (Lss * x(:, 3)))/xa;$
$\gg Idr = \frac{(Lm * x(:,2)) - (Lss * x(:,4))}{xa};$
% tbase denotes base value of Te
$\gg Te = tbase * (x(:, 3). * Idr5) - (x(:, 4). * Iqr5);$
$\gg Tl = B. * x(:, 5)/wbase;$
$\gg slip = (os - x(:, 5))/os;$ % slip denotes s_{im}
% Syntax for estimating real and reactive power responses in kW and kVAR
% tbase denotes base value of Te
% sbase denotes the base power
$\gg P = sbase * (Vds. * Ids + Vqs. * Iqs)/1000;$
$\gg Q = sbase * (Vqs. * Ids - Vds. * Iqs)/1000;$

After getting the array of each parameter with respect to time individually, responses can be developed using syntax '***plot***'.

The above explained program can be reused to third and first order models of induction motor. Since, in third order model stator fluxes are assumed to be constant and in first order model both stator and rotor fluxes are assumed to be constant. The steady state value of these stator and rotor fluxes can be chosen as constant value from the array developed above for the same.

For third order model, three differential equations given in Eqs. (55)–(57) are used in ODE45 solver while for first order model, single differential equations given in Eq. 57 is only used in ODE45 solver. Rest explanations are same as for the fifth order model discussed above. Therefore, responses can be obtained for third and first order model of induction motor similar to its fifth order model.

The responses of output quantities (electromagnetic torque, slip and speed) and input quantities (active and reactive powers) are being compared for fifth, third and first order behaviour of induction motor in Figs. 3, 4, 5, 6 and 7.

Example 4 Conclude the responses obtained for 50 kW induction motor in Figs. 3, 4, 5, 6 and 7.

Solution Responses for fifth, third and first order model of 50 kW induction motor are shown above in Figs. 3, 4, 5, 6 and 7. Figures 3, 4 and 5 give rotor speed, slip and electromagnetic torque responses respectively and Figs. 6 and 7 give real and reactive responses for 50 kW induction motor. The steady state active and reactive power demand, shown in Figs. 6 and 7, is 41.93 kW and 13.34 kVAR respectively. The remaining power is being lost in the motor. Fifth order model of induction motor has more transients in its characteristics. It can be concluded that the behaviour of the third order model is quite similar to fifth order model while the first order model is

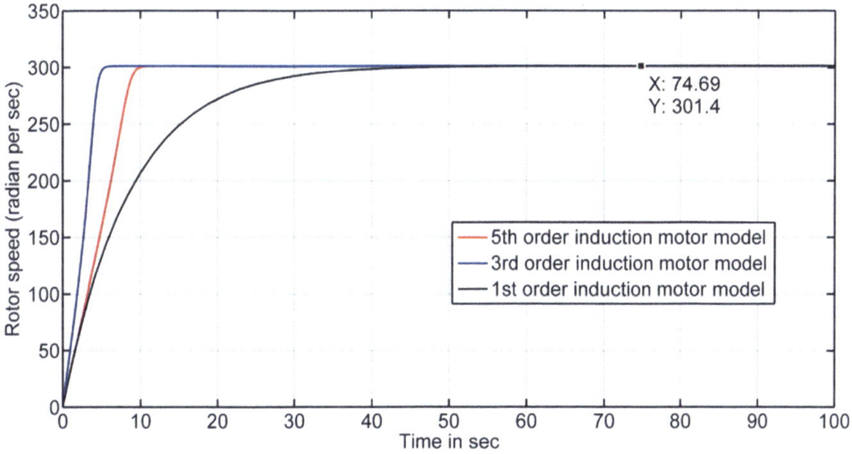

Fig. 3 Rotor speed for 50 kW induction motor

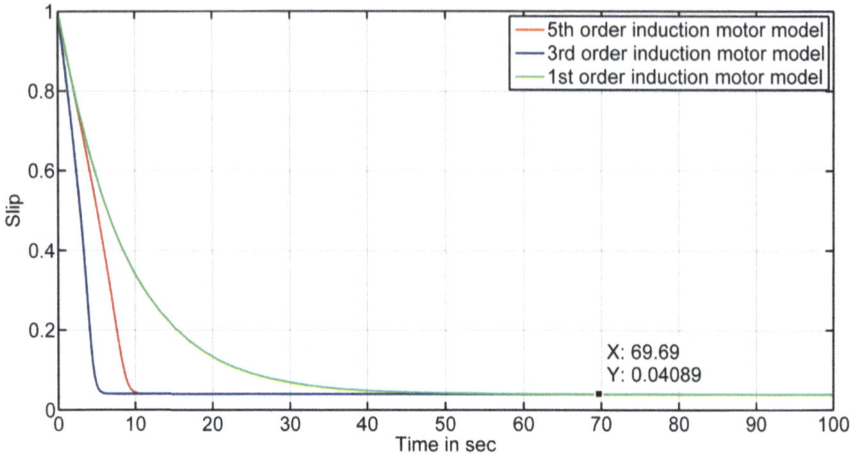

Fig. 4 Slip for 50 kW induction motor

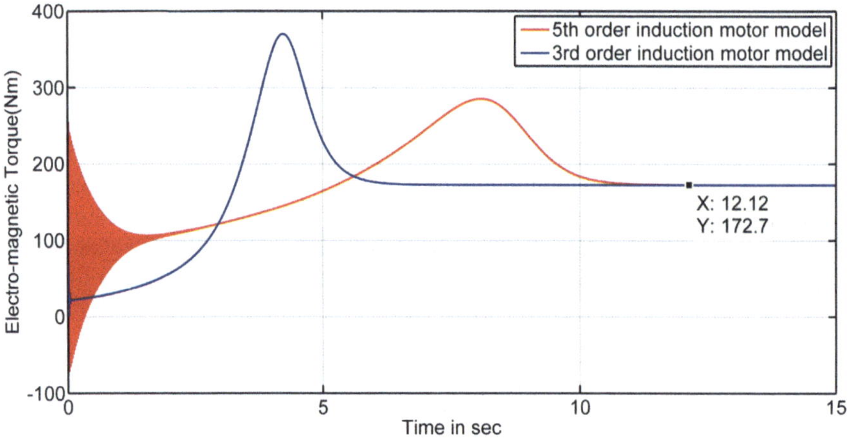

Fig. 5 Electro-magnetic torque for 50 kW induction motor

out of context. So it is obvious to use either fifth or third order model of the induction motor to get the equivalent mathematical model of induction motor.

3.3 Structure of Aggregate Load

Distribution system having large varieties of consumers is the most complex structure of the power system. For this study, we are considering only induction motor loads at distributing ends. Since the decentralized areas have many consumers which may

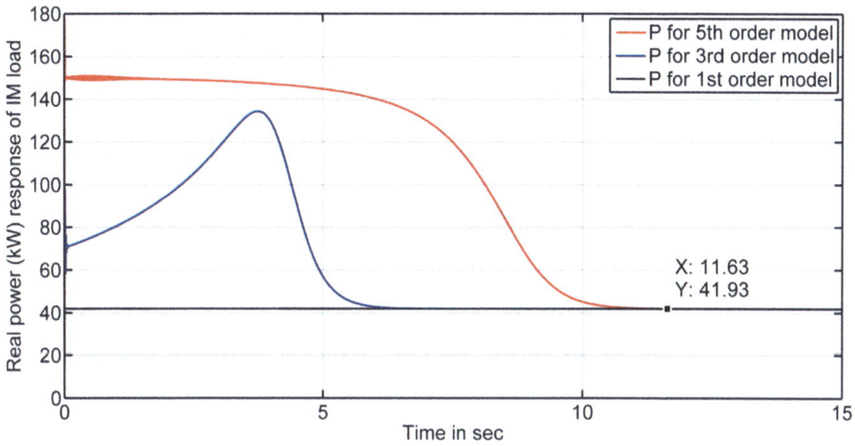

Fig. 6 Active power characteristics for 50 kW induction motor

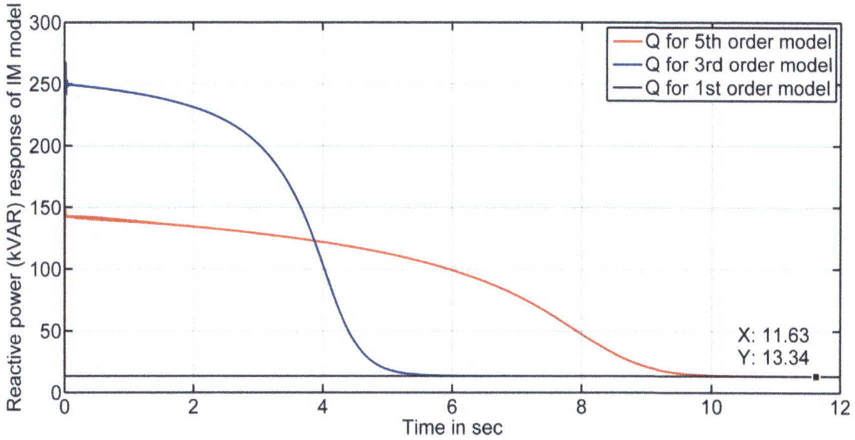

Fig. 7 Reactive power characteristics for 50 kW induction motor

consist large numbers of induction motors, it is not realistic to model every induction motor that is in the system because it is impractical to accurately represent each individual load due to the intense computation process involved. However, to analyze diversified load pattern in distribution systems, a group of motors can be considered in place of using a single large rating induction motor and then an aggregate load model of the induction motor can be developed for this distribution system.

The appropriate dynamic load model aggregation reduces the computation time and provides a faster and efficient model derivation and parameter identification. It is found that the small scale aggregation model gives acceptably accurate results than the large-scale aggregation model and is good for power system stability analysis [5]. Hence, aggregate models or single unit models with a minimum order of induction motor are needed to represent a group of motors [6]. Measurement based method to find aggregate model is presented in [33]. Reference [19] suggests two methods; weighted average of impedance and weighted average of admittance for finding aggregate model.

In this chapter, an aggregate model of the induction motor is developed using the law of energy conservation. According to energy conservation law, "Power absorbed by the aggregate motor model is equal to the sum of the individual motor's power absorbed".

For a system having n number of induction motor, if kVA rating of kth motor is denoted by S_k. For aggregate motor model, the aggregate kVA S^{agg} is,

$$S^{agg} = \sum_{k=1}^{n} S_k \tag{66}$$

Similarly, aggregate stator and rotor current can be defined as,

$$\overrightarrow{I_s^{agg}} = \sum_{k=1}^{n} \overrightarrow{I_{s,k}} \tag{67}$$

$$\overrightarrow{I_r^{agg}} = \sum_{k=1}^{n} \overrightarrow{I_{r,k}} \tag{68}$$

The aggregate motor load equivalent circuit parameters are;

$$R_s^{qgg} = \frac{\sum_{k=1}^{n} \left\{ \left| \overrightarrow{I_{s,k}} \right|^2 R_{s,k} \right\}}{\left| \overrightarrow{I_r^{qgg}} \right|^2} \tag{69}$$

$$R_r^{qgg} = \frac{\sum_{k=1}^{n} \left\{ \left| \overrightarrow{I_{r,k}} \right|^2 R_{r,k} \right\}}{\left| \overrightarrow{I_r^{qgg}} \right|^2} \tag{70}$$

$$X_s^{qgg} = \frac{\sum_{k=1}^{n} \left\{ \left| \overrightarrow{I_{s,k}} \right|^2 X_{s,k} \right\}}{\left| \overrightarrow{I_r^{qgg}} \right|^2} \tag{71}$$

$$X_r^{qgg} = \frac{\sum_{k=1}^{n} \left\{ \left| \overrightarrow{I_{r,k}} \right|^2 X_{r,k} \right\}}{\left| \overrightarrow{I_r^{qgg}} \right|^2} \tag{72}$$

$$X_m^{agg} = \frac{\sum_{k=1}^{n}\left\{\left|\overrightarrow{I_{s,k}} - \overrightarrow{I_{r,k}}\right|^2 X_{m,k}\right\}}{\left|\overrightarrow{I_s^{agg}} - \overrightarrow{I_r^{agg}}\right|^2} \tag{73}$$

The air gap power of the aggregate motor load is expressed as;

$$P_{airgap}^{agg} = \sum_{k=1}^{n}\left(Re\left(V\ I_{s,k}\right) - I_{s,k}^2 R_{sk}\right) \tag{74}$$

The slip and inertia of the aggregate motor can be then computed by,

$$slip^{agg} = \frac{\left|\overrightarrow{I_r^{agg}}\right|^2 R_r^{agg}}{P_{airgap}^{agg}} \tag{75}$$

$$H^{agg} = \frac{\sum_{k=1}^{n} H_k S_k}{S^{agg}} \tag{76}$$

The moment of inertia and inertia constant of the motor follows relation as;

$$H_k S_k = \frac{1}{2} J_k \omega_{sk}^2 \tag{77}$$

Therefore, the aggregate motor model can be estimated for the given set of several motors. The power system consist numbers of consumers with different rating induction motors, the overall dynamic load may be represented by an aggregate induction motor load.

Example 5 Manufacturer data for five induction motors of rating 3, 7.5, 10, 110 and 18.5 kW are given in Table 4. Find the aggregate motor model for these five induction motors and compare the results find for 50 kW induction motor as given in Examples 1 and 2.

Table 4 Manufacturer data for induction motors for 50 kW aggregate model

Motor specifications	Group of IMs for aggregate model				
	(IM1)	(IM2)	(IM3)	(IM4)	(IM5)
Induction motor rating (in kW)	18.5	11	10	7.5	3
Supply Voltage (in Volt)	400	400	400	400	400
Supply frequency (in Hz)	50	50	50	50	50
Power factor at full load (lagging)	0.9	0.9	0.9	0.9	0.9
Efficiency at full load	0.9	0.9	0.9	0.9	0.9
Slip at full load	0.04	0.04	0.04	0.04	0.04
Pole pair	1	1	1	1	1

Solution 50 kW aggregate model of induction motor is developed by clubbing five different ratings induction motor (IM1-IM5). All the specifications of induction motors are kept equal to 50 kW single unit of induction motor as in Table 4 except their rating, for simplifying the calculations. Equations 66–77 show the mathematical expressions for calculating the parameters of aggregate model of induction motor. Using mathematical syntaxes in MATLAB command window, electrical and mechanical parameters are estimated first for individual induction motors and then for aggregate model of 50 kW. The results found for 50 kW single unit of induction motor in Examples 2 and 3 are also listed in Table 5 for comparing with the results found in this example.

The circuit parameters of aggregate model are same as that of 50 kW single induction motor except the value of moment of inertia, inertia constant and torque-damping factor which denotes that the aggregate model will also behave like induction motor model but at different dynamics.

4 Modelling for Composite Load Model

In previous section structure of load is illustrated in detail. It has been explained that a load in distribution system may be any combination of static, dynamic and aggregate load. These loads can be classified according to the effect of the voltage on the load. Static loads are generally used for the calculation of steady state conditions and in steady state simulations of power system while dynamic loads are therefore necessary for analyzing power system behaviour following small or large disturbances [34]. Therefore a composite load can be developed for system by combining the entire available load in system together. Many papers have been published about composite load modelling which includes composite load as a combination of SLM and DLM.

Since, in distribution system, most of the connected load is inductive in nature and therefore the overall behaviour of the load will be similar to the induction motor behaviour. This nature of load will be discussed in this chapter later on. It can also be expressed here that induction motor is a machine that is more influenced by the action of reactive power compare to real power and this nature for induction motor can easily be illustrated with the support of following explanations;

(i) Reactive power influences are more dominated in induction machine because rotating magnetic field is the main process of transforming energy and this rotating magnetic field is developed by reactive magnetizing current in field winding.
(ii) Reactive magnetizing current id directly concerned with machine power factor and therefore, physical parameters (B, H and J) are highly dependent on reactive power structure of the induction motor.
(iii) Law of Decoupling explains the close relation of reactive power with voltage, moreover the response time of reactive power voltage control (Q-V loop) is

Table 5 List for evaluated parameters of induction motors for 50 kW aggregate model

	Rating (kW)	R_s (ohm)	R_r (ohm)	X_s (ohm)	X_r (ohm)	X_m (ohm)	J (kg m^2)	H	B (N m s)
IM1	18.5	0.7771	0.2364	0.4821	0.4821	30.5359	0.02	0.0480	0.21515
IM2	11	1.3070	0.3976	0.8109	0.8109	51.3558	0.0072	0.0291	0.1279
IM3	10	1.4377	0.4374	0.8920	0.8920	56.4913	0.007	0.0311	0.1163
IM4	7.5	1.9169	0.5832	1.1893	1.1893	75.3218	0.0025	0.0148	0.08720
IM5	3	4.7922	1.4580	2.9732	2.9732	188.3044	0.00104	0.0154	0.03489
Single unit of IM	50	0.2875	0.0875	0.1784	0.1784	11.2983	0.0900	0.0799	0.58150
Aggregate Model for IM1-IM6	50	0.2875	0.0875	0.1784	0.1784	11.3097	0.0377	0.0335	0.1462

very less compare to active power frequency control (P-f loop) in any rotating machines.

(iv) Voltage stability problems are the most prominent problems in power system that requires exact load modelling and suitable excitation system.

Hence, influence of reactive power on voltage control studies are more prominent compare to real power and frequency control. The reactive power voltage loop is prominently explained in this chapter for developing composite load model and so, the studies are limited to developing the transfer functions of change in reactive power with voltage change. This transfer function of change in reactive power with voltage change is developed for composite load model. This composite load model can be estimated by adding static and dynamic load model. Mathematically [8],

$$(D_v)_{CLM} = (D_v)_{SLM} + (D_v)_{DLM} \tag{78}$$

Parameters used in Eq. (78) are defined as,

$(D_v)_{CLM}$: Transfer function of reactive power change to voltage change for composite load
$(D_v)_{SLM}$: Transfer function of reactive power change to voltage change for static load
$(D_v)_{DLM}$: Transfer function of reactive power change to voltage change for dynamic load

Transfer function of reactive power change to voltage change for static and dynamic load are discussed in successive sub sections first and then transfer function of reactive power change to voltage change is also elaborated later on.

4.1 Development of Mathematical Model for Static Load

Exponential type load structure as explained in Sect. 3.1.2 is used for defining the static load model. This load can be represented by an exponential function of bus voltage magnitude as given in Eq. (21). In proportionality term, reactive power expression with load voltage can be represented as in Eq. (79).

$$Q_L^s \propto V^{nq} \tag{79}$$

For any instant of time; differentiating and solving Eq. (80),

$$\Delta Q = nq \frac{Q_L^s}{V} \Delta V \tag{80}$$

nq defines the exponential constant for reactive power and voltage relation. Q_L^s is defined as the reactive power for static load at the instant of load voltage V in system. Since, all the three parameters are constant so they can be replaced with a single parameter as in Eq. (81) and therefore,

$$nq\frac{Q_L^s}{V} = (D_v)_{SLM} \tag{81}$$

Therefore, transfer function for change in reactive power with voltage change for static load model can be represented as in Eq. (82).

$$(D_v)_{SLM} = \frac{\Delta Q}{\Delta V} = nq\frac{Q_L^s}{V} \tag{82}$$

Example 6 In a distribution system, an exponential type static load of 250 kW rating with load power factor is taken 0.9 lagging and exponential constant 3 is connected with a 400 V three phase system. For base voltage 400 V and base power 250 kW, find the expression for transfer function showing change in reactive power with voltage change.

Solution First mention all the given data,

Base power = 250 kW
Base voltage = 400 V
Exponential constant = $nq = 3$
Static load Real power at the instant of load voltage V, $P_L^s = 250$ kW
Power factor = 0.9 lagging
Reactive power for static load at the instant of load voltage V,
$Q_L^s = P_L^s \times \tan(\cos^{-1}(0.9)) = 121.0805$ kVAR
Per phase value, $Q_L^s(pu) = \frac{(121.0805/3)}{250} = 0.1614$
$\mathrm{V(pu)} = \frac{400}{400} = 1$(Assuming that load is delta connected)

Therefore,
$(D_v)_{SLM} = \frac{\Delta Q}{\Delta V} = nq\frac{Q_L^s}{V} = \frac{3\times 0.1614}{1} = 0.4842$

Example 7 Repeat the example 6 and draw the step response for $(D_v)_{SLM}$ using MATLAB for (i) 50 kW, (ii) 100 kW, (iii) 150 kW, (iv) 200 kW, and (v) 250 kW together.

Solution $(D_v)_{SLM}$ for 50, 100, 150, 200 and 250 kW can be calculated as in Example 6. The MATLAB command for getting step response is,

$\gg num = nq * Q_L^s$
$\gg den = V$
$\gg tf(num, den)$

These expressions in MATLAB command window execute a figure window for step response. All the step responses as asked in example can be presented on same figure window by writing syntax "≫ *hold on*" on command window. Therefore, the step responses as asked in this example are shown in Fig. 8.

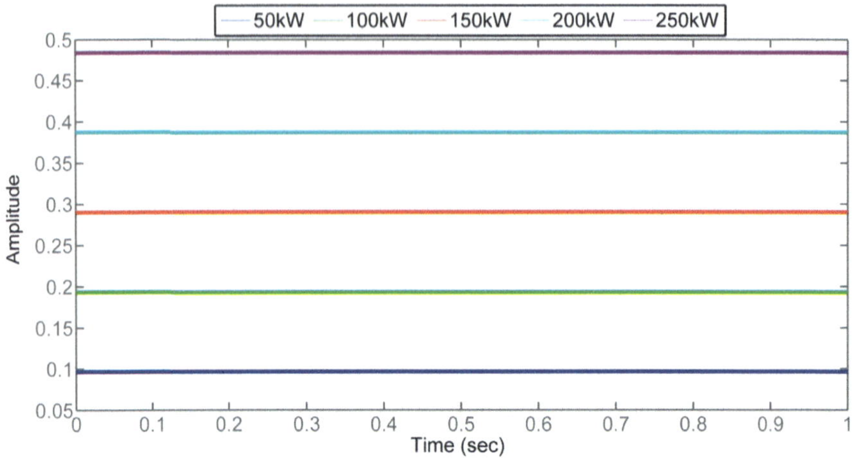

Fig. 8 Step response of $(D_v)_{SLM}$ for Example 7

4.2 Development of Mathematical Model for Dynamic Load

It has already been explained that induction motor is the most prominent dynamic load used in distribution system. There may be large numbers of induction motor available in system but they can be represented as aggregate model. For dynamic load (induction motor), $(D_v)_{DLM}$ is developed through its five differential equations using state space equations. The state space equations are being developed for fifth, third and first order models of induction motor.

To develop state space equations, control variable, disturbance variable and state variables are required. Since these studies are limited to developing the transfer functions of change in reactive power with voltage change, change in voltage ΔV is used as control vector, change in reactive power ΔQ is used as disturbance vector and five states of induction motor shown in five differential equations are used as state variable. Figure 9 represents a generalized block diagram for state space model of induction motor.

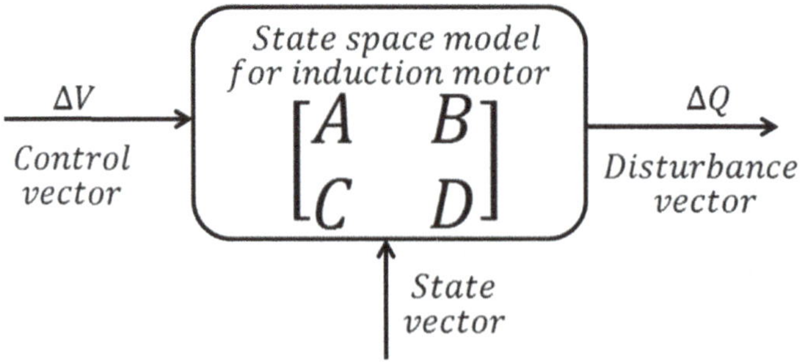

Fig. 9 State space model representation for induction motor

The solution can be obtained by estimating ABCD parameters from state space equations. Mathematically, generalized value of ABCD parameters can be represented as in Eqs. (82)–(86). If ε notation is used for state vectors and k notation is used for denoting the model order. For single input single output system the generalized values of matrices A B C and D can be formulated as;

$$Aq_k = \begin{bmatrix} \frac{\partial \dot{\varepsilon}_1}{\partial \varepsilon_1} & \frac{\partial \dot{\varepsilon}_1}{\partial \varepsilon_2} & \cdot\cdot & \frac{\partial \dot{\varepsilon}_1}{\partial \varepsilon_k} \\ \cdot & \cdot & \cdot\cdot & \cdot \\ \cdot & \cdot & \cdot\cdot & \cdot \\ \frac{\partial \dot{\varepsilon}_k}{\partial \varepsilon_1} & \frac{\partial \dot{\varepsilon}_k}{\partial \varepsilon_2} & \cdot\cdot & \frac{\partial \dot{\varepsilon}_k}{\partial \varepsilon_k} \end{bmatrix}_{k\times k} \tag{83}$$

$$Bq_k = \begin{bmatrix} \frac{\partial \dot{\varepsilon}_1}{\partial V} \\ \cdot \\ \cdot \\ \frac{\partial \dot{\varepsilon}_k}{\partial V} \end{bmatrix}_{k\times 1} \tag{84}$$

$$Cq_k = \begin{bmatrix} \frac{\partial Q}{\partial \varepsilon_1} & \cdot\cdot & \frac{\partial Q}{\partial \varepsilon_k} \end{bmatrix}_{1\times k} \tag{85}$$

$$Dq_k = \begin{bmatrix} \frac{\partial Q}{\partial V} \end{bmatrix}_{1\times 1} \tag{86}$$

Reactive power of Induction motor load Q_L^{IM} in terms of voltage can be represented as in Ref. [35],

$$Q_L^{IM} = V_{qs} I_{ds} - V_{ds} I_{qs} \tag{87}$$

Since V_{qs} is very close to zero (from the basic knowledge of three phase voltage), Q_L^{IM} can be rewritten as,

$$Q_L^{IM} = -V_{ds} I_{qs} \tag{88}$$

By using Eqs. 58–61, currents can be obtained as;

$$I_{qs} = \frac{L_m \varphi_{qr} - L_{rr}\varphi_{qs}}{L_m^2 - L_{ss}L_{rr}} \tag{89}$$

$$I_{ds} = \frac{L_m \varphi_{dr} - L_{rr}\varphi_{ds}}{L_m^2 - L_{ss}L_{rr}} \tag{90}$$

$$I_{qr} = \frac{L_m \varphi_{qs} - L_{ss}\varphi_{qr}}{L_m^2 - L_{ss}L_{rr}} \tag{91}$$

$$I_{dr} = \frac{L_m \varphi_{ds} - L_{ss}\varphi_{dr}}{L_m^2 - L_{ss}L_{rr}} \tag{92}$$

The used constants are denoted by new variables as;

$$x_a = L_m^2 - L_{ss}L_{rr} \tag{93}$$

$$X_1 = \frac{L_m}{x_a} \tag{94}$$

$$X_2 = \frac{L_{ss}}{x_a} \tag{95}$$

$$X_3 = \frac{L_{rr}}{x_a} \tag{96}$$

Substituting these constants value, equations for currents can be rewritten as,

$$I_{qs} = X_1\varphi_{qr} - X_3\varphi_{qs} \tag{97}$$

$$I_{ds} = X_1\varphi_{dr} - X_3\varphi_{ds} \tag{98}$$

$$I_{qr} = X_1\varphi_{qs} - X_2\varphi_{qr} \tag{99}$$

$$I_{dr} = X_1\varphi_{ds} - X_2\varphi_{dr} \tag{100}$$

And per unit electro-magnetic torque equation;

$$T_e = X_1\left(\varphi_{qr}\varphi_{ds} - \varphi_{dr}\varphi_{qs}\right) \tag{101}$$

The steady state values of fluxes, rotor speed and stator voltage are assumed to be;

$$\varphi_{qs}\ at\ steady\ state = x_{1ss} \tag{102}$$

$$\varphi_{ds}\ at\ steady\ state = x_{2ss} \tag{103}$$

$$\varphi_{qr}\ at\ steady\ state = x_{3ss} \tag{104}$$

$$\varphi_{dr}\ at\ steady\ state = x_{4ss} \tag{105}$$

$$\omega_r\ at\ steady\ state = \omega_r \tag{106}$$

$$V_{ds}\ at\ steady\ state = V_{ss} \tag{107}$$

With the help of five differential equation of induction motor and the equations given in this section, state space model of fifth, third and first order induction motor can be developed. Detail procedures to develop state space model (i.e. ABCD parameters) are discussed in successive subsections. These ABCD parameters are used for developing $(D_v)_{DLM}$ using the MATLAB syntax as,

$$\gg [num\_q, den\_q] = ss2tf(A, B, C, D); \tag{108}$$

$$\gg tfq5 = tf(num\_q, den\_q); \tag{109}$$

4.2.1 State Space Model for Fifth Order Model of Induction Motor

To develop state space model for fifth order model of induction motor, substitute all constant values of Sects. 4.2 and 3.2.4 in five differential equations of induction motor presented in Eqs. (53)–(57). Rewrite the expressions in state space equations format;

$$\begin{aligned}\dot{\varphi}_{qs} &= [R_s\omega_b X_3]\varphi_{qs} + [-\omega_s]\varphi_{ds} + [-R_s\omega_b X_1]\varphi_{qr} + [0]\varphi_{dr} \\ &\quad + [0]\omega_r + [0]V_{ds}\end{aligned} \tag{110}$$

$$\begin{aligned}\dot{\varphi}_{ds} &= [\omega_s]\varphi_{qs} + [R_s\omega_b X_3]\varphi_{ds} + [0]\varphi_{qr} + [-R_s\omega_b X_1]\varphi_{dr} \\ &\quad + [0]\omega_r + [\omega_b]V_{ds}\end{aligned} \tag{111}$$

$$\begin{aligned}\dot{\varphi}_{qr} &= [-R_r\omega_b X_1]\varphi_{qs} + [0]\varphi_{ds} + [R_r\omega_b X_2]\varphi_{qr} + [-\omega_b + \omega_r]\varphi_{dr} \\ &\quad + [x_{4ss}]\omega_r + [0]V_{ds}\end{aligned} \tag{112}$$

$$\begin{aligned}\dot{\varphi}_{dr} &= [0]\varphi_{qs} + [-R_r\omega_b X_1]\varphi_{ds} + [\omega_b - \omega_r]\varphi_{qr} + [R_r\omega_b X_2]\varphi_{dr} \\ &\quad + [-x_{3ss}]\omega_r + [0]V_{ds}\end{aligned} \tag{113}$$

$$\dot{\omega_r} = \left[\frac{-\omega_b X_1}{2H}\right]\varphi_{qs}\varphi_{dr} + \left[\frac{\omega_b X_1}{2H}\right]\varphi_{ds}\varphi_{qr} + \left[\frac{-B}{2H}\right]\omega_r + [0]V_{ds} \tag{114}$$

Reactive power expression in terms of these constants;

$$Q_L^{IM} = V_{ds}X_3\varphi_{qs} - V_{ds}X_1\varphi_{qr} \tag{115}$$

For small perturbations in system, Eqs. (110)–(115) can be rewritten in terms of incremental change in state, control and disturbance vectors. For five state vectors $\Delta\varphi_{qs}\Delta\varphi_{ds}\Delta\varphi_{qr}\Delta\varphi_{dr}$ and $\Delta\omega_r$ state space equations for fifth order model are;

$$\left[\Delta\dot{\varphi}_{qs}\ \Delta\dot{\varphi}_{ds}\ \Delta\dot{\varphi}_{qr}\ \Delta\dot{\varphi}_{dr}\ \Delta\dot{\omega}_r\right]^T = Aq_5\left[\Delta\varphi_{qs}\Delta\varphi_{ds}\Delta\varphi_{qr}\Delta\varphi_{dr}\Delta\omega_r\right]^T + Bq_5\Delta V \tag{116}$$

$$\Delta Q_L^{IM} = Cq_5\left[\Delta\varphi_{qs}\Delta\varphi_{ds}\Delta\varphi_{qr}\Delta\varphi_{dr}\Delta\omega_r\right]^T + Dq_5\Delta V \tag{117}$$

Aq_5, Bq_5, Cq_5 and Dq_5 are the constant matrices of the appropriate dimensions associated with the above control, state and disturbances vectors for fifth order model of induction motor.

Hence,

$$Aq_5 = \begin{bmatrix} R_s\omega_b X_3 & -\omega_s & -R_s\omega_b X_1 & 0 & 0 \\ \omega_s & R_s\omega_b X_3 & 0 & -R_s\omega_b X_1 & 0 \\ -R_r\omega_b X_1 & 0 & R_r\omega_b X_2 & -\omega_b + \omega_r & x_{4ss} \\ 0 & -R_r\omega_b X_1 & \omega_b - \omega_r & R_r\omega_b X_2 & -x_{3ss} \\ \frac{-\omega_b X_1}{2H}x_{4ss} & \frac{\omega_b X_1}{2H}x_{3ss} & \frac{\omega_b X_1}{2H}x_{2ss} & \frac{-\omega_b X_1}{2H}x_{1ss} & \frac{-B}{2H} \end{bmatrix} \tag{118}$$

$$Bq_5 = \begin{bmatrix} 0 \\ \omega_b \\ 0 \\ 0 \\ 0 \end{bmatrix} \tag{119}$$

$$Cq_5 = \left[\, V_{ss}X_3\ 0\ -V_{ss}X_1\ 0\ 0 \,\right] \tag{120}$$

$$Dq_5 = [-X_1x_{3ss} + X_3x_{1ss}] \tag{121}$$

4.2.2 State Space Model for Third Order Model of Induction Motor

To derive state space model for third order model of induction motor, stator direct and quadrature axis flux differential equations are considered to be zero. So, substituting $\dot{\varphi}_{qs} = 0$ and $\dot{\varphi}_{ds} = 0$ in Eqs. 110 and 111 and writing both equations in matrix form,

$$\begin{bmatrix} 0 \\ 0 \end{bmatrix} = \begin{bmatrix} R_s\omega_b X_3 & -\omega_s \\ \omega_s & R_s\omega_b X_3 \end{bmatrix}\begin{bmatrix} \varphi_{qs} \\ \varphi_{ds} \end{bmatrix} + \begin{bmatrix} -R_s\omega_b X_1 & 0 \\ 0 & -R_s\omega_b X_1 \end{bmatrix}\begin{bmatrix} \varphi_{qr} \\ \varphi_{dr} \end{bmatrix} + \begin{bmatrix} 0 \\ \omega_b \end{bmatrix}\begin{bmatrix} V_{ds} \\ V_{ds} \end{bmatrix} \tag{122}$$

Let the constant A_s and B_s such that,

$$A_s = J^{-1}\begin{bmatrix} R_s\omega_b X_1 & 0 \\ 0 & R_s\omega_b X_1 \end{bmatrix} \tag{123}$$

$$B_s = J^{-1}\begin{bmatrix} 0 \\ -\omega_b \end{bmatrix} \tag{124}$$

where,

$$J = \begin{bmatrix} R_s\omega_b X_3 & -\omega_s \\ \omega_s & R_s\omega_b X_3 \end{bmatrix} \tag{125}$$

So,

$$\begin{bmatrix} \varphi_{qs} \\ \varphi_{ds} \end{bmatrix} = A_s\begin{bmatrix} \varphi_{qr} \\ \varphi_{dr} \end{bmatrix} + B_s\begin{bmatrix} V_{ds} \\ V_{ds} \end{bmatrix} \tag{126}$$

$$\varphi_{qs} = A_s(1,1)\varphi_{qr} + A_s(1,2)\varphi_{dr} + B_s(1,1)V_{ds} \tag{127}$$

$$\varphi_{ds} = A_s(2,1)\varphi_{qr} + A_s(2,2)\varphi_{dr} + B_s(2,1)V_{ds} \tag{128}$$

Further, writing Eqs. 112 and 113 in matrix form,

$$\begin{bmatrix} \dot{\varphi}_{qr} \\ \dot{\varphi}_{dr} \end{bmatrix} = (A_r A_s + B_r)\begin{bmatrix} \varphi_{qr} \\ \varphi_{dr} \end{bmatrix} + A_r B_s\begin{bmatrix} V_{ds} \\ V_{ds} \end{bmatrix} + \begin{bmatrix} \varphi_{dr}\omega_r \\ -\varphi_{qr}\omega_r \end{bmatrix} \tag{129}$$

where,

$$A_r = \begin{bmatrix} -R_r\omega_b X_1 & 0 \\ 0 & -R_r\omega_b X_1 \end{bmatrix} \tag{130}$$

$$B_r = \begin{bmatrix} R_r\omega_b X_2 & -\omega_b \\ \omega_b & R_r\omega_b X_2 \end{bmatrix} \tag{131}$$

Let,

$$D_r = A_r A_s + B_r \tag{132}$$

$$E_r = A_r B_s \tag{133}$$

$$\begin{bmatrix} \dot{\varphi}_{qr} \\ \dot{\varphi}_{dr} \end{bmatrix} = D_r\begin{bmatrix} \varphi_{qr} \\ \varphi_{dr} \end{bmatrix} + E_r\begin{bmatrix} V_{ds} \\ V_{ds} \end{bmatrix} + \begin{bmatrix} \varphi_{dr}\omega_r \\ -\varphi_{qr}\omega_r \end{bmatrix} \tag{134}$$

Extracting $\dot{\varphi}_{qr}$ and $\dot{\varphi}_{dr}$ form Eq. (134),

$$\dot{\varphi}_{qr} = D_r(1,1)\varphi_{qr} + D_r(1,2)\varphi_{dr} + \varphi_{dr}\omega_r + E_r(1,1)V_{ds} \tag{135}$$

$$\dot{\varphi}_{dr} = D_r(2,1)\varphi_{qr} + D_r(2,2)\varphi_{dr} - \varphi_{qr}\omega_r + E_r(2,1)V_{ds} \tag{136}$$

Substituting values of Eqs. (127) and (128) in Eq. (114),

$$\begin{aligned}\dot{\omega_r} &= \left[F_r A_s(2,1)\varphi_{qr}^2 + F_r B_s(2,1)V_{ds}\varphi_{qr}\right] + \left[-F_r A_s(1,2)\varphi_{dr}^2 - F_r B_s(1,1)V_{ds}\varphi_{dr}\right] \\ &\quad - F_r A_s(1,1)\varphi_{dr}\varphi_{qr} + F_r A_s(2,2)\varphi_{dr}\varphi_{qr} + \left[\frac{-B}{2H}\right]\omega_r\end{aligned} \tag{137}$$

where,

$$F_r = \frac{\omega_b X_1}{2H} \tag{138}$$

Also, substituting values of Eqs. (127) and (128) in Eq. (16) of reactive power,

$$Q_L^{IM} = V_{ds}X_3\left[A_s(1,1)\varphi_{qr} + A_s(1,2)\varphi_{dr} + B_s(1,1)V_{ds}\right] - V_{ds}X_1\varphi_{qr} \tag{139}$$

For small perturbations in system, Eqs. (135), (136), (137) and (139) can be rewritten in terms of incremental change in state control and disturbance vectors. For three state vectors $\Delta\varphi_{qr}\Delta\varphi_{dr}$ and $\Delta\omega_r$, state space equations for third order model are;

$$\left[\Delta\dot{\varphi}_{qr}\Delta\dot{\varphi}_{dr}\Delta\dot{\omega}_r\right]^T = Aq_3\left[\Delta\varphi_{qr}\Delta\varphi_{dr}\Delta\omega_r\right]^T + Bq_3\Delta V \tag{140}$$

$$\Delta Q = Cq_3\left[\Delta\varphi_{qr}\Delta\varphi_{dr}\Delta\omega_r\right]^T + Dq_3\Delta V \tag{141}$$

Aq_3, Bq_3, Cq_3 and Dq_3 are the constant matrices of the appropriate dimensions associated with the above control, state and disturbances vectors for third order model of induction motor.

Hence,

$$Aq_3 = \begin{bmatrix} D_r(1,1) & D_r(1,2)+\omega_r & x_{4ss} \\ D_r(2,1)-\omega_r & D_r(2,2) & -x_{3ss} \\ Aq_3(3,1) & Aq_3(3,2) & \frac{-B}{2H}\end{bmatrix} \tag{142}$$

$$Bq_3 = \begin{bmatrix} E_r(1,1) \\ E_r(2,1) \\ F_r B_s(2,1)x_{3ss} - F_r B_s(1,1)x_{4ss}\end{bmatrix} \tag{143}$$

$$Cq_3 = \left[\{X_3 A_s(1,1) - X_1\}V_{ss}\ \ X_3 A_s(1,2)V_{ss}\ \ 0\right] \tag{144}$$

$$Dq_3 = X_3 A_s(1,1)x_{3ss} + X_3 A_s(1,2)x_{4ss} + 2X_3 B_s(1,1)V_{ss} - X_1 x_{3ss} \tag{145}$$

where,

$$Aq_3(3,1) = 2F_r A_s(2,1)x_{3ss} + F_r B_s(2,1)V_{ss} - F_r A_s(1,1)x_{4ss} + F_r A_s(2,2)x_{4ss} \tag{146}$$

$$Aq_3(3,2) = -2F_r A_s(1,2)x_{4ss} - F_r B_s(1,1)V_{ss} - F_r A_s(1,1)x_{3ss} + F_r A_s(2,2)x_{3ss} \tag{147}$$

4.2.3 State Space Model for First Order Model of Induction Motor

To derive state space model of first order model of induction motor, stator and rotor direct and quadrature axis flux differential equations are considered to be zero. Results obtained from third order model are used for obtaining first order model of induction by substituting $\dot{\varphi}_{qr} = 0$ and $\dot{\varphi}_{dr} = 0$ in Eq. 134,

$$D_r \begin{bmatrix} \varphi_{qr} \\ \varphi_{dr} \end{bmatrix} + E_r \begin{bmatrix} V_{ds} \\ V_{ds} \end{bmatrix} + \begin{bmatrix} \varphi_{dr}\omega_r \\ -\varphi_{qr}\omega_r \end{bmatrix} = 0 \tag{148}$$

Splitting matrix of Eq. (148) into equations,

$$D_r(1,1)\varphi_{qr} + \{D_r(1,2) + \omega_r\}\varphi_{dr} + E_r(1,1)V_{ds} = 0 \tag{149}$$

$$\{D_r(2,1) - \omega_r\}\varphi_{qr} + D_r(2,2)\varphi_{dr} + E_r(2,1)V_{ds} = 0 \tag{150}$$

Recollecting φ_{qr} and φ_{dr} again in terms of constants,

$$\begin{bmatrix} \varphi_{qr} \\ \varphi_{dr} \end{bmatrix} = H_r \begin{bmatrix} V_{ds} \\ V_{ds} \end{bmatrix} \tag{151}$$

where,

$$H_r = G_r^{-1}(-E_r) \tag{152}$$

$$G_r = \begin{bmatrix} D_r(1,1) & D_r(1,2) + \omega_r \\ D_r(2,1) - \omega_r & D_r(2,2) \end{bmatrix} \tag{153}$$

So, values of φ_{qr} and φ_{dr} are,

$$\varphi_{qr} = H_r(1, 1)V_{ds} \tag{154}$$

$$\varphi_{dr} = H_r(2, 1)V_{ds} \tag{155}$$

In Eq. (137), substituting values of φ_{qr} and φ_{dr},

$$\begin{aligned}\dot{\omega}_r = &\left[\left\{F_r A_s(2, 1)H_r^2(1, 1)\right\} + \{F_r B_s(2, 1)H_r(1, 1)\} + \left\{-F_r A_s(1, 2)H_r^2(2, 1)\right\}\right. \\ &+ \{-F_r B_s(1, 1)H_r(2, 1)\} + \{-F_r A_s(1, 1)H_r(2, 1)H_r(1, 1)\} \\ &+\{F_r A_s(2, 2)H_r(2, 1)H_r(1, 1)\}]V_{ds}^2 + \left[\frac{-B}{2H}\right]\omega_r\end{aligned} \tag{156}$$

Let,

$$\begin{aligned}&\left\{F_r A_s(2, 1)H_r^2(1, 1)\right\} + \{F_r B_s(2, 1)H_r(1, 1)\} + \left\{-F_r A_s(1, 2)H_r^2(2, 1)\right\} \\ &+ \{-F_r B_s(1, 1)H_r(2, 1)\} + \{-F_r A_s(1, 1)H_r(2, 1)H_r(1, 1)\} \\ &+ \{F_r A_s(2, 2)H_r(2, 1)H_r(1, 1)\} = J_r\end{aligned} \tag{157}$$

$$\dot{\omega}_r = J_r V_{ds}^2 + \left[\frac{-B}{2H}\right]\omega_r \tag{158}$$

Similarly, solving Eq. (139) for Q_L^{IM}

$$\begin{aligned}Q_L^{IM} = &X_3 A_s(1, 1)H_r(1, 1)V_{ds}^2 - X_1 H_r(1, 1)V_{ds}^2 + X_3 A_s(1, 2)H_r(2, 1)V_{ds}^2 \\ &+ X_3 B_s(1, 1)V_{ds}^2\end{aligned} \tag{159}$$

Let,

$$K_r = X_3 A_s(1, 1)H_r(1, 1) - X_1 H_r(1, 1) + X_3 A_s(1, 2)H_r(2, 1) + X_3 B_s(1, 1) \tag{160}$$

$$Q_L^{IM} = K_r V_{ds}^2 \tag{161}$$

For small perturbations in system, Eqs. (158) and (161) can be rewritten in terms of incremental change in state control and disturbance vectors. For single state vector $\Delta\omega_r$, state space equation for first order model;

$$\Delta\dot{\omega}_r = Aq_1\Delta\omega_r + Bq_1\Delta V \tag{162}$$

$$\Delta Q = Cq_1\Delta\omega_r + Dq_1\Delta V \tag{163}$$

Aq_1, Bq_1, Cq_1 and Dq_1 are the constant matrices of the appropriate dimensions associated with the above control, state and disturbances vectors for first order model of induction motor.

Hence,

$$Aq_1 = -\frac{B}{2H} \tag{164}$$

$$Bq_1 = 2J_r V_{ss} \tag{165}$$

$$Cq_1 = 0 \tag{166}$$

$$Dq_1 = 2K_r V_{ss} \tag{167}$$

Example 8 Draw the step responses for $(D_v)_{DLM}$ (fifth order type induction motor load) of ratings (i) 50 kW, (ii) 100 kW, (iii) 150 kW, and (iv) 200 kW using MATLAB coding. Use the same manufacturer data for all rating motors as in Example 1.

Solution To find the step response for $(D_v)_{DLM}$ following steps are followed;

1. For an induction motor manufacturer data first find electrical and mechanical parameters.
2. Develop the state space model for induction motor and find ABCD parameter.
3. Find transfer function $(D_v)_{DLM}$ from ABCD parameters.
4. Plot step response for $(D_v)_{DLM}$.
5. Repeat the same steps for all four ratings.

The results are shown in Fig. 10.

Example 9 Consider a composite load of 250 kW that includes 200 kW exponential type static load and 50 kW fifth order induction motor dynamic load participation. Draw the step response for $(D_v)_{CLM}$ for this load using MATLAB coding. All the required parameters for static and dynamic loads are same as in previous examples.

Solution To find the step response for $(D_v)_{CLM}$, find transfer function for 200 kW exponential type static load and 50 kW fifth order induction motor dynamic load as estimated in previous respective examples. Use Eq. (78) to find the $(D_v)_{CLM}$ through $(D_v)_{SLM}$ and $(D_v)_{DLM}$ expressions. Finally, plot step response for $(D_v)_{CLM}$. The results are shown in Fig. 11. Zoom view of plot is also shown in Fig. 12 for better understating.

For a composite load of 250 kW consisting 200 kW SLM and 50 kW DLM, transfer functions have been obtained. Actual view for this system shown in Fig. 11 represents that system will settle down at around 3 s. Zoom view of step responses for this composite load model, demonstrated in Fig. 12, show that the dynamic behaviour of load pattern is due to the participation of dynamic load. Static load will increase the magnitude of overall reactive power demand due to change in voltage.

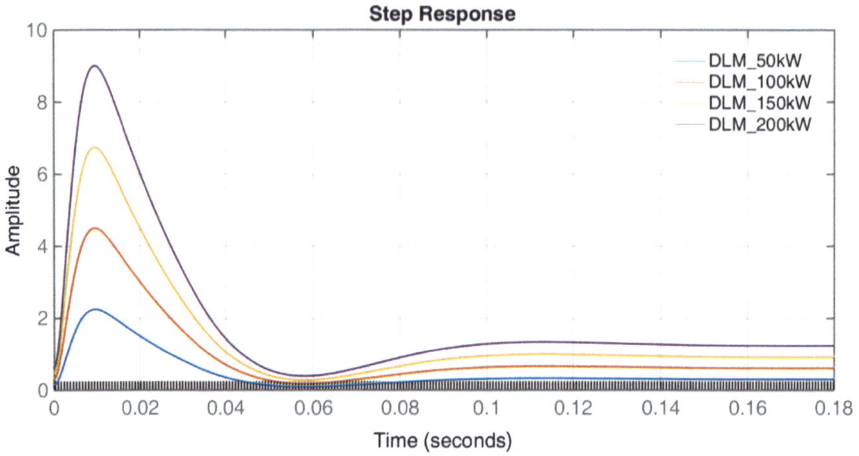

Fig. 10 Step responses for $(D_v)_{DLM}$ of fifth order model induction motor loads

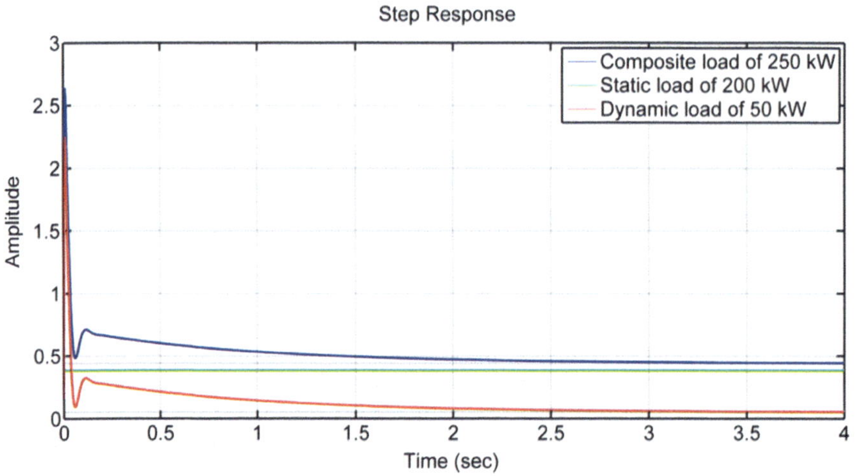

Fig. 11 Actual view of step responses for composite load of 250 kW having 200 kW static and 50 kW dynamic load

Example 10 For a composite load including participation of static and dynamic load as given in Table 6, draw the step response for $(D_v)_{CLM}$ using MATLAB coding. All the parameters are same as in previous examples.

Solution To find the step response for $(D_v)_{CLM}$, find transfer function for static load $(D_v)_{SLM}$ and dynamic load $(D_v)_{DLM}$ as estimated in Example 6 and 8. Use Eq. (78) to find the $(D_v)_{CLM}$ through $(D_v)_{SLM}$ and $(D_v)_{DLM}$ expressions. Finally, plot step response for $(D_v)_{CLM}$. The results are shown in Fig. 13.

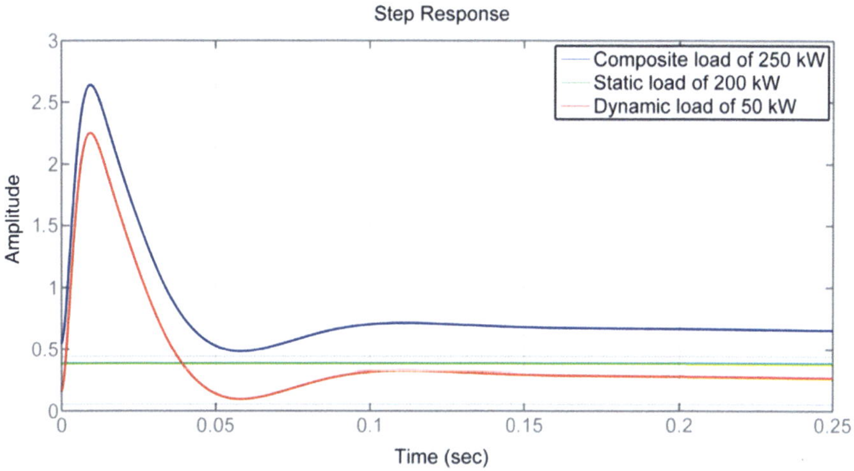

Fig. 12 Zoom view of step responses for composite load of 250 kW having 200 kW static and 50 kW dynamic load

Table 6 List of 250 kW composite load patterns based on participation of dynamic load

Title	Pattern 1 (kW)	Pattern 2 (kW)	Pattern 3 (kW)	Pattern 4 (kW)	Pattern 5 (kW)
Static load component	250	200	150	100	50
Dynamic load component	0	50	100	150	200
Total rating of composite load	250	250	250	250	250

Step responses for all five patterns given in Table 6 are compared in Fig. 13. This figure depicts that dynamic behaviour of composite load is due to the presence of induction motor as the dynamic load. It has also been observed that load with high participation of dynamic load attains stability in maximum time.

5 Conclusion

In this chapter, a detail discussion is focused on load model identification in power system. It has been concluded that variety of loads exist at load end due to which decision of choosing correct load model is very difficult for system. Static load model alone cannot be correctly quantified the load behaviour in system which was

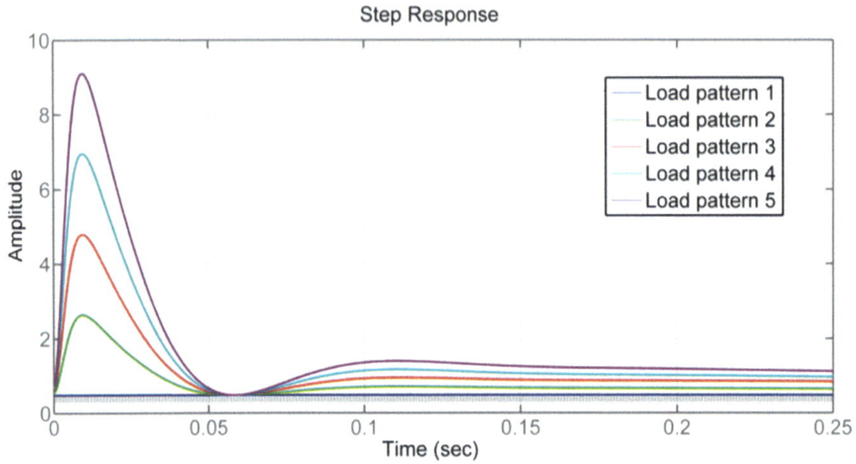

Fig. 13 Step responses for $(D_v)_{CLM}$ of fifth order model induction motor loads

reported by most of the authors in their work and therefore, a composite load is suitable for load modelling. This composite load comprises static as well as dynamic load compositions. Thus, a composite load model is designed by clubbing static and dynamic loads.

Static load can be modeled either as polynomial type or as exponential type. Exponential type model is the most generalized model to specify static load because it covers a wide range of load varieties depending on the exponential factor as described in the chapter.

Induction motor is the most versatile load and thus defines as the most commonly used dynamic load in power system. Induction motor's electrical and mechanical parameters are estimated in this chapter but more advance work can be done for investing these parameters in future. It has been observed that dynamic behaviour of composite load depends in proportion to participating factor of dynamic load.

Key Terms and Their Definitions

Static Load: A load that depends only on the instantaneous voltage input and is unrelated to the preceding voltage inputs.
Dynamic Load: A load that depends not only on the instantaneous voltage input but also it is related to the preceding voltage inputs.
Composite Load: A load that includes the participation of static and dynamic load.
Aggregate Load: A load that develops by collecting all loads together.
ZIP Load model: A polynomial type static load model having properties of constant power, constant impedance and constant current load individually or for any of their combinations.
Exponential Load model: A generalized static load model that can express any type of static load through its exponential factor.
Induction motor load: Most commonly used dynamic load in distribution system.

MATLAB Code

MATLAB Codes are given within the chapter with examples.

References

1. Saxena, N.K.: Investigation of static and dynamic reactive power compensation and cost analysis in isolated hybrid power system. Unpublished doctoral dissertation, Electrical Engineering Department, National Institute of Technology, Kurukshetra, India (2016)
2. Stojanovi, D.P., Korunovi, L.M., Milanovi, J.V.: Dynamic load modelling based on measurements in medium voltage distribution network. Electr. Power Syst. Res. **78**, 228–238 (2008)
3. Byoung, H.K., Kim, H., Lee, B.: Parameter estimation for the composite load model. J. Int. Council Electr. Eng. **2**(2), 215–218 (2012)
4. Parveen, T.: Composite Load Model Decomposition: Induction Motor Contribution. Doctoral dissertation, Faculty of Built Environment and Engineering, School of Engineering Systems, Queensland University of Technology (2009)
5. Muriuki, J.K., Muriithi, C.M.: Comparison of aggregation of small and large induction motors for power system stability study. Global Eng. Technol. Rev. **3**(2) (2013)
6. Aree, P.: Aggregating method of induction motor group using energy conservation law. ECTI Trans. Electr. Eng. Electr. Commun. **12**(1), 2014 (2014)
7. Performance, task force on load representation for dynamics. Load representation for dynamic performance analysis. IEEE Trans. Power Syst. **8**, 472–482 (1993)
8. Kundur, P.: Power System Stability and Control. Tata-Mcgraw-Hill, India (2006)
9. Murty, V.V.S.N., Kumar, A.: Comparison of optimal capacitor placement methods in radial distribution system with load growth and ZIP load model. Front. Energy **7**, 197–213 (2013)
10. Momoh, J.A.: Electrical Power System Application of Optimization. Marcel Dekker Inc, New York (2001)
11. Saxena, N.K., Kumar, A.: Estimation of composite load model with aggregate induction motor dynamic load for an isolated hybrid power system. Front. Energy **9**(4), 472–485 (2015)
12. Craven, R.H., Michael, M.R.: Load representations in the dynamic solution of the Queensland power system. J. Electr. Electron. Eng. **3**(1), 1–7 (1983)
13. Mauricio, W., Semiyen, A.: Effect of load characteristics on the dynamic stability of power system. IEEE Trans. Power Apparatus Syst. **91**, 2295–2304 (1972)
14. Milanovic, J.V., Hiskens, I.A.: Effect of load dynamics on power system damping. In: R. IEEE PEs Summer Meeting, vol. 94. SM 578-5 PWRS (1994), San Francisco (1994)
15. Karlsson, D., Hill, D.J.: Modelling and identification of non linear dynamics loads in power system. IEEE Trans. Power Syst. **9**(1), 157–166 (1994)
16. Fahmy, O.M., Attia, A.S., Badr, M.A.L.: A novel analytical model for electrical loads comprising static and dynamic components. Electr. Power Syst. Res. **77**, 1249–1256 (2007)
17. Son, S.E., Lee, S.H., Choi, D.H., Song, Y.B., Park, J.D., Kwon, Y.H., Hur, K., Park, J.W.: Improvement of composite load modelling based on parameter sensitivity and dependency analyses. IEEE Trans. Power Syst. **29**(1), 242–250 (2014)
18. Saxena, N.K., Kumar, A.: Reactive power control in decentralized hybrid power system with STATCOM Using GA, ANN and ANFIS methods. Int. J. Electr. Power Energy Syst. **83**, 175–187 (2016)
19. Parveen, T., Ledwich, G.: Decomposition of aggregated load: finding induction motor fraction in real load. In: Proceedings of Australasian Universities Power Engineering Conference, Sydney (2008)
20. Stanley, H.C.: An analysis of induction machine. AIEE Trans. **57** (1938)
21. Krause, P.C.: Analysis of Electric Machinery. McGraw-Hill Book Company, New York (1986)

22. Singh, G.K.: Self-excited induction generator research—a survey. Electr. Power Syst. Res. **69**, 107–114 (2004)
23. Rusnok, S., Sobota, P., Slivka, M., Svoboda, P.: Assessment transients during starting of induction motor in MATLAB Simulink and verification by measurement. Adv. Res. Sci. Areas (2012)
24. Wang, K., Chiasson, J., Bodson, M., Tolbert, L.M.: A nonlinear least-squares approach for identification of the induction motor parameters. In: Proceedings of 43rd IEEE Conference on Decision and Control, Atlantis, Paradise Island, Bahamas (2004)
25. Lehtla, T.: Parameter identification of an induction motor using fuzzy logic controller. In: Proceedings of PEMC, Budapest, part 3, pp. 292–296 (1996)
26. Hiskens, I.A., Milanovic, J.V.: Load modeling in studies of power system damping. IEEE Trans. Power Syst. **10**(4) (1995)
27. Boldea, I., Nasar, S.A.: The Induction Machine Handbook. CRC press (2002)
28. Saxena, N., Kumar, A.: Load modelling interaction on hybrid power system using STATCOM. In: Proceedings of IEEE INDICON 2010, Kolkata, India (2010)
29. Pedra, J., Sainz, L.: Parameter estimation of squirrel-cage induction motors without torque measurements. IEE Proc.-Electr. Power Appl. **153**(2) (2006)
30. Izosimov, D.B.: Experimental determination of parameters for an induction motor with a short circuited rotor. Russ. Electr. Eng. **84**(2), 81–88 (2013)
31. Ilina, I.D.: Experimental determination of moment of inertia and mechanical losses vs. speed, in electrical machines. In: Proceedings of The 7th International Symposium on Advanced Topics in Electrical Engineering, U.P.B Bucharest (2011)
32. Pedra, J.: Estimation of typical squirrel-cage induction motor parameters for dynamic performance simulation. IEE Proc.-Gener. Transm. Distrib. **153**(2), 137–146 (2006)
33. Zhang, Y., Zhang, W., Chu, X., Liu, Y.: Real-time optimal voltage control using measurement based. Electr. Power Syst. Res. **116**, 293–300 (2014)
34. Saxena, N.K., Kumar, A.: Analytical comparison of static and dynamic reactive power compensation in isolated wind diesel system using dynamic load interaction model. Electr. Power Compon. Syst. **53**(5), 508–519 (2015)
35. Choi, B.K., Chiang, H.D., Li, Y., Chen, Y.T., Huang, D.H., Lauby, M.G.: Development of composite load models of power systems using on line measurement data. J. Electr. Eng. Technol. **1**(2), 161–169 (2006)

A Novel Forward-Backward Sweep Based Optimal DG Placement Approach in Radial Distribution Systems

Farkhondeh Jabari, Somayeh Asadi and Sahar Seyed-barhagh

Abstract The huge value of the electricity consumption in different residential, commercial, industrial and agricultural sectors lead to the load-generation mismatch, voltage drops, cascading failures, and wide area blackouts. Therefore, the use of renewable energy resources based distributed generation (DG) units is rapidly growing in order to satisfy not-supplied electrical demand and reduce greenhouse gas emissions. Meanwhile, optimal placement of DG units in radial grids is crucial for minimization of the total active power losses and the voltage drops. This chapter proposes a novel backward-forward sweep (BFS) based methodology for optimal allocation of DG micro-plants in radial distribution systems aiming to minimize total real power losses of the whole system. Voltage permitted range limit and feeder capacity criterion are considered as optimization constraints. Simulation of BFS based DG placement method is conducted on the 33-bus distribution network to investigate its performance under different scenarios.

Keywords Forward-backward sweep (FBS) · Distributed generation (DG) · Optimal DG placement

---

F. Jabari · S. Seyed-barhagh
Faculty of Electrical and Computer Engineering, University of Tabriz, Tabriz, Iran
e-mail: f.jabari@tabrizu.ac.ir

S. Seyed-barhagh
e-mail: sbarhagh95@ms.tabrizu.ac.ir

S. Asadi (✉)
Department for Management of Science and Technology Development, Ton Duc Thang University, Ho Chi Minh City, Vietnam
e-mail: somayehasadi@tdtu.edu.vn

Faculty of Applied Sciences, Ton Duc Thang University, Ho Chi Minh City, Vietnam

M. Pesaran Hajiabbas and B. Mohammadi-Ivatloo (eds.),
Optimization of Power System Problems, Studies in Systems, Decision and Control 262,
https://doi.org/10.1007/978-3-030-34050-6_2

Nomenclature

$\dot{J}_{D,i}^{k+1}$	The current of the load i in scenario $(k + 1)$
$\dot{V}_i^k$	The voltage of the bus i in iteration k
$\dot{S}_i$, $\dot{S}_{D,i}$	Power injected to node/load i
$\dot{P}_{D,i}$, $\dot{Q}_{D,i}$	The active and reactive power utilizations in node i
$\dot{I}_{m,i}^{k+1}$	Current of branch m-i in scenario $(k + 1)$
$Z_{m,i}$	The impedance of the line m-i
F_{loss}	The real power losses as the objective function
$g_{i,j}$	The conductance of the branch i to j
θ_i, θ_j	The voltage angles of the buses i *and* j
$\dot{P}_{DG,i}$, $\dot{Q}_{DG,i}$	The active and reactive power productions of the DG unit in bus i
ϵ	Convergence coefficient
$V_i^{\min}$, $V_i^{\max}$	Lower and upper bounds of voltage magnitude for node i
I_b	The current of the branch b
$I_b^{\max}$	The current capacity of the line b

1 Motivation and Literature Review

Nowadays, the penetration level of distributed generators (DGs) in power systems is increasing due to power system restructuring, deregulation of electricity markets, global warming, and energy crisis [1]. Moreover, integration of DGs with power systems provides several benefits such as voltage profile improvement, ancillary services, power quality, and reliability enhancement, energy saving, loss and feeder congestion reduction [2].

Many types of research focused on the optimal allocation of DGs in distribution systems. For example, Gkaidatzis et al. [3] presented a particle swarm optimization (PSO) algorithm for siting and sizing of DGs considering load variations. In this study, total active power losses are minimized while satisfying the feeder capacity limit and the voltage permitted range constraint. In [4], simultaneous allocation of DGs and capacitors is optimized using a genetic algorithm to minimize their capital investment and maintenance costs, energy losses, and risk of not-supplied demand. In [5], sequential quadratic programming (SQP) and branch and bound method are integrated to solve a non-convex mixed integer non-linear programming problem for achieving better solutions in less calculation time than exhaustive load flow (ELF), improved analytical (IA) and PSO algorithms. Poornazaryan et al. [6] combined Cuckoo search method with a binary imperialistic competitive algorithm for minimization of real power losses and enhancement of voltage stability considering 50% variations in active and reactive loads. In [7], optimum places and capacities of DGs are determined by triangle number technique and multi-objective hybrid big-bang crunch to minimize the operation cost, power losses, pollutant emissions of greenhouse gases and maximize the voltage stability security margin. Reference [8] proposed a teaching learning algorithm for optimal placement of DGs in radial distribution systems in a way that voltage profile is improved in comparison with genetic and PSO algorithms. Reference [9] aims to mitigate feeder congestion and maximize energy saving by interrupting both active and reactive power consumptions of flexible loads considering their interruption costs using a genetic algorithm. In [10], implementing backward-forward sweep load flow algorithm coupled genetic algorithm, DGs are efficiently allocated and sized subject to voltage stability constraint. In [11], genetic, PSO and gravitational search algorithms are examined to find a good scenario with minimum DGs installation costs. In [12], optimal capacities of non-dispatchable photovoltaic (PV) power generation technology is determined to gain an interchange between minimum loss and maximum voltage stability by using a weighted rank sum ratio method. Kayal and Chanda [13] used a PSO algorithm for selection of optimum places and sizes of solar photovoltaic arrays and wind turbines in three 12, 15 and 33 bus radial distribution systems. In this research, reduction of grid power losses and enhancement of voltage stability index of the whole system are considered as optimization objectives. It is found that solar PV farms and wind turbines in lagging power factor operating mode lead to more voltage stability improvement in all buses. It is obvious that the voltage magnitude of all buses increases with the participation of DGs in active and reactive power compensations. In [14], non-dispatchable DGs such as solar PV panels and wind turbines and dispatchable energy sources such as biomass and biogas fueled gas turbine power generation cycle is optimally placed in the 51-bus radial distribution grid. Analytic hierarchy process (AHP) is employed in the PSO algorithm for solving a multi objective optimization problem including energy losses, feeder current capacity limit, voltage stability, and emission reduction aspects. In [15], it is revealed that symbiotic organisms search algorithm, which is based on the symbiotic relationship between different biological species, is more computationally efficient and fast than PSO, teaching-learning algorithm, cuckoo search optimization, artificial bee colony method, gravitational and stochastic fractal search approaches. Monte Carlo simulation (MCS) is developed by Sadeghi and Kalantar [16] to model variable outputs of solar and wind farms in dynamic planning of 9-bus radial distribution network. Covariance matrix adaptation evolutionary strategy determines the optimum planning scenario with maximum revenue using penalty and incentive factors. In [17], long-term forecasts of loads and yearly variations of renewable energy resources based power generation plants is incorporated in optimal reconfiguration and DG placement studies. Objectives of optimization problem includes the cost of line switching, power losses, investment and maintenance costs of DGs, and emission cost of DGs and upstream power system. Table 1 summarizes a taxonomy of different algorithms presented for optimal siting and sizing of DGs in distribution feeders.

As reviewed, different optimization algorithms have been implemented on distribution systems to find good places and optimal sizes of DGs and improve voltage stability and reduce system power losses. But, a search method with less calculation time and computational burden, no need to membership function of fuzzy logic, huge search space of MCS, cross over and mutation processes of genetic algorithm, and

Table 1 Comparison between different algorithms proposed for solving optimal DG placement problem

References	Search algorithm	Objective functions	Advantages
[18]	Point estimation method and genetic algorithm for probabilistic load flow and optimal allocation of DGs considering uncertainties of load, wind, electricity rate, solar, fuel price	Power loss minimization	Faster convergence than MCS based genetic algorithm
[19]	Cuckoo search algorithm	Power loss minimization and voltage profile improvement	Better objective functions than genetic and PSO algorithms
[20]	Non-dominated sorting genetic algorithm-II	Minimization of feeder losses, capital investment, and maintenance costs and voltage deviations	Improved cross over and mutation in comparison with genetic algorithm
[21]	Harmony search algorithm	Energy loss minimization and voltage stability enhancement	More accurate and faster than non-dominated Sorting Genetic Algorithm II
[22]	Plant growth simulation method	Minimum number of DGs, maximum voltage stability and minimum power losses	No need to cross over and mutation factors of genetic algorithm and membership function of fuzzy logic
[23]	Ant colony and artificial bee colony search algorithms	Loss and emission reduction	Less computational burden than point estimation method
[24]	Kalman filter model	Minimum power losses	More accurate than Gaussian linear optimizer
[25]	Continuous load flow analysis	Minimum losses and maximum loading margin	The accurate and computationally friendly approach in comparison with MCS

initial population of metaheuristic algorithms has not been proposed by scholars. This chapter aims to present a novel forward-backward sweep (BFS) based optimal DG placement strategy for radial distribution networks. In this method, the number and capacity of DGs are considered as known parameters. Total active power losses are considered as the objective function. Firstly, one of DGs is selected. Its active and reactive power generations are added to second (related to active power consumption) and third (related to reactive power consumption) columns of the bus data matrix. Then, BFS load flow is solved and total real power losses are calculated as a component of loss matrix in the 1st raw and 1st column. In loss matrix, a number of rows and columns are equal to the number of buses and DGs. Afterward, 1st DG is assumed to be installed on bus 2. A similar analysis is carried out and energy losses are computed as 2nd row and 1st column of loss matrix. When all buses are evaluated for placement of the 1st unit, 2nd DG is assumed to be located at buses 1 to *N*, respectively. where *N* refers to a number of nodes in the test distribution system. This process is repeated for all DGs and loss matrix is formed. Finally, the minimum values of columns are determined. If the minimum value of column *i* occurred in the *j*th row of loss matrix, bus *j* will be selected as a good place for installation of *i*th DG.

Other sections of this chapter are organized as follows: The BFS based DG allocation approach is mathematically modeled in Sect. 2. Simulations and results are then provided in Sect. 3. Afterward, Sect. 4 presents the conclusion.

2 Optimal DG allocation problem

2.1 FBS power flow

The single line diagram of the typical radial distribution system is depicted in Fig. 1. It is supposed that the power injection to the bus *i* is equal to $\dot{S}_i$. As obvious from Fig. 1 and given by (1), the value of the active/reactive power injected to the bus *i* is equal to the real/reactive load of this bus plus the sum of the power transmitted through the node *i* to other adjacent buses.

$$\dot{S}_i = \dot{S}_{D,i} + \sum_j \dot{S}_j \tag{1}$$

By considering the negative active ($-\dot{P}_{DG,i}$) and reactive ($\dot{Q}_{DG,i}$) power consumptions of the DG unit in bus *i*, its net demand can be calculated as (2).

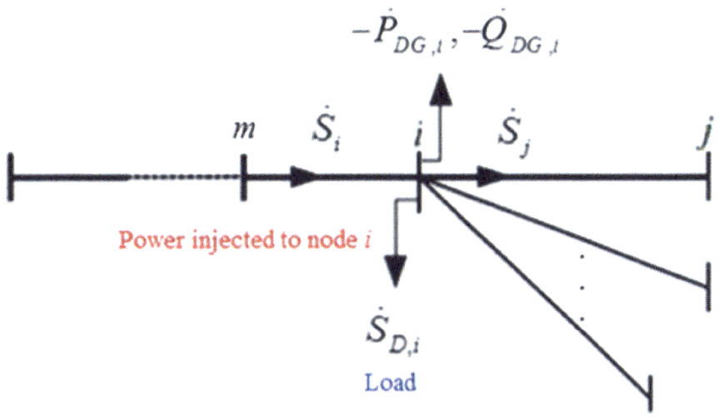

Fig. 1 The installation of the DGs in radial distribution systems

$$\dot{S}_{D,i} = \left(\dot{P}_{D,i} - \dot{P}_{DG,i}\right) + j\left(\dot{Q}_{D,i} - \dot{Q}_{DG,i}\right) \tag{2}$$

The current injected to the load i at iteration k+1 can be calculated based on its appearance power ($\dot{S}_{D,i}$) and voltage magnitude at iteration k ($\dot{V}_i^k$), as stated in equation (3). where, $\dot{J}_{D,i}^{k+1}$ is the current injected to the bus i in iteration k+1. The voltage of the node i at scenario k is stated as $\dot{V}_i^k$.

$$\dot{J}_{D,i}^{k+1} = \left(\frac{\dot{S}_{D,i}}{\dot{V}_i^k}\right)^* \tag{3}$$

In the backward sweep of the load flow analysis, the current in the line m-i at iteration k+1 is calculated as (20.4).

$$\dot{I}_{m,i}^{k+1} = \dot{J}_i^{k+1} + \sum_j \dot{I}_{i,j}^{k+1} \tag{4}$$

In the forward sweep, the voltage of the up-stream bus m at iteration k+1 depends on the value of the voltage drop in the transmission line, which connects the buses i and m to each other, as well as the voltage of the down-stream node i, as formulated by (5).

$$\dot{V}_m^{k+1} = \dot{V}_i^{k+1} - \dot{I}_{m,i}^{k+1} \times Z_{m,i} \tag{5}$$

If the convergence constraint (6) is satisfied for all buses, the backward and forward sweeps will be stopped. Therefore, the voltage of the bus j will be equal to V_j^{k+1} and the current of the branch i to j will be equal to $\dot{I}_{i,j}^{k+1}$.

$$\left|V_j^{k+1} - V_j^k\right| \le \varepsilon \tag{6}$$

2.2 Total active power loss, bus voltage limit, and feeder current capacity

The total real power loss of the distribution grid, F_{loss}, is calculated from (7). In which, $g_{i,j}$ is the conductance of branch i to j. In addition, n_l refers to the number of the transmission lines. The voltage angle of the bus i is defined as θ_i. According to (8), the lower ($V_i^{\min}$) and upper ($V_i^{\max}$) bounds of the bus voltage magnitude are considered as 0.9 and 1.05 per unit, respectively. Similarly, the current of the line b is limited by the maximum flow $I_b^{\max}$, as formulated by the inequality constraint (9).

$$F_{loss} = \text{Min} \sum_{\substack{i,j=1 \\ i \neq j}}^{n_l} g_{i,j}\left[V_i^2 + V_j^2 - 2V_i V_j \cos(\theta_i - \theta_j)\right] \tag{7}$$

$$V_i^{\min} \leq V_i \leq V_i^{\max} \tag{8}$$

$$I_b \leq I_b^{\max} \tag{9}$$

3 Proposed Algorithm and Illustrative Example

The BFS based strategy is used for optimal allocation of DGs in radial distribution systems and described based on MATLAB codes. The number and sizes of the DGs are known. As obvious from Fig. 2, "P_DG" and "Q_DG" refer to active and reactive capacities of DGs, respectively. The BFS based optimization algorithm finds the suitable places for installation of three distributed generation units and minimizes the real power loss, while satisfying the voltage limit and the feeder current capacity as (8) and (9). In the MATLAB codes, IEEE 33-bus radial distribution system [26] is tested to allocate three distribution generation units with active and reactive generation capacities of 70, 240, 545 kW and 36, 63, 250 kVAr, respectively. The single line diagram of the test network is illustrated in Fig. 3. The "bdata.not.per.unit" and "ldata.not.per.unit" are the bus and line data matrices, respectively. The 1st column of the node information matrix represents the bus number. The 2nd and 3rd ones refer to the real and reactive demands of the buses in terms of kW and kVAr, respectively. In the matrix "ldata.not.per.unit", the first and second columns show the starting and ending points of the branches. The 3rd and 4th columns report the resistance and reactance of the lines, respectively. In the first iteration of the proposed approach, the DG unit 1 is located at bus 1. The BFS power flow calculation is implemented on the updated node information matrix. The active power loss is then computed and saved as the 1st row and 1st column of the loss matrix. The loss matrix is defined as "Active_loss". In the 2nd iteration, the DG unit 1 is located at node 2 and the power flow calculation is performed. The active loss of the benchmark system is calculated and saved as the 2nd row and 1st column of the loss matrix. The same method is considered for the 2nd and 3rd DGs. As obvious from the loss matrix, which is shown in Fig. 4, if 1st DG unit is installed at bus 18, the total active power loss will be minimum and equal to 166.3765 kW. Moreover, 17th and 32nd buses are good choices for installation of 2nd and 3rd DGs. The voltage profile before and after installation of DGs under the best scenario and two other scenarios are depicted in Fig. 5. As expected, optimal placement of DGs using the forwardbackward sweep based search algorithm leads to a significant reduction in active power losses and improvement in bus voltage magnitude. Moreover, number of scenarios in search space of BFS based DG allocation method is reduced to 99 (number of buses × number of DGs).

Fig. 2 MATLAB codes of forward-backward sweep based optimal DG placement

```
P_DG=[70,240,545]; Q_DG=[36,63,250]; n_DG=3;
for i=1:33
    for j=1:3
bdata.not.per.unit=[%Bus    P(Kw)   Q(Kvar)
                    1       0       0
                    2       100     60
                    3       90      40
                    4       120     80
                    5       60      30
                    6       60      20
                    7       200     100
                    8       200     100
                    9       60      20
                    10      60      20
                    11      45      30
                    12      60      35
                    13      60      35
                    14      120     80
                    15      60      10
                    16      60      20
                    17      60      20
                    18      90      40
                    19      90      40
                    20      90      40
                    21      90      40
                    22      90      40
                    23      90      50
                    24      420     200
                    25      420     200
                    26      60      25
                    27      60      25
                    28      60      20
                    29      120     70
                    30      200     600
                    31      150     70
                    32      210     100
                    33      60      40];
ldata.not.per.unit=[
%         Inbus   Outbus  Resistance(ohm)  Reactance(ohm)
          1       2       0.0922           0.0470
          2       3       0.4930           0.2511
          3       4       0.3660           0.1864
          4       5       0.3811           0.1941
          5       6       0.8191           0.7070
          6       7       0.1872           0.6188
          7       8       0.7114           0.2351
          8       9       1.0300           0.7400
          9       10      1.0440           0.7400
          10      11      0.1966           0.0650
          11      12      0.3744           0.1238
          12      13      1.4680           1.1550
          13      14      0.5416           0.7129
          14      15      0.5910           0.5260
          15      16      0.7463           0.5450
          16      17      1.2890           1.7210
          17      18      0.7320           0.5740
          2       19      0.1640           0.1565
          19      20      1.5042           1.3554
          20      21      0.4095           0.4784
          21      22      0.7089           0.9373
          3       23      0.4512           0.3083
          23      24      0.8980           0.7091
          24      25      0.8960           0.7011
          6       26      0.2030           0.1034
          26      27      0.2842           0.1447
          27      28      1.0590           0.9377
          28      29      0.8042           0.7006
          29      30      0.5075           0.2585
          30      31      0.9744           0.9630
          31      32      0.3105           0.3619
          32      33      0.3410           0.5302];
bdata.not.per.unit(i,2)=bdata.not.per.unit(i,2)-P_DG(j);
bdata.not.per.unit(i,3)=bdata.not.per.unit(i,3)-Q_DG(j);
% Run FBS load flow algorithm presented in Chapter 14
        Active_loss(i,j)=Ploss;
        Reactive_loss(i,j)=Qloss;
    end
end
for i=1:busnum
    for j=1:n_DG
    if Active_loss(i,j)==min(Active_loss(:,j))
        best_place(j)=i;
end;end;end
```

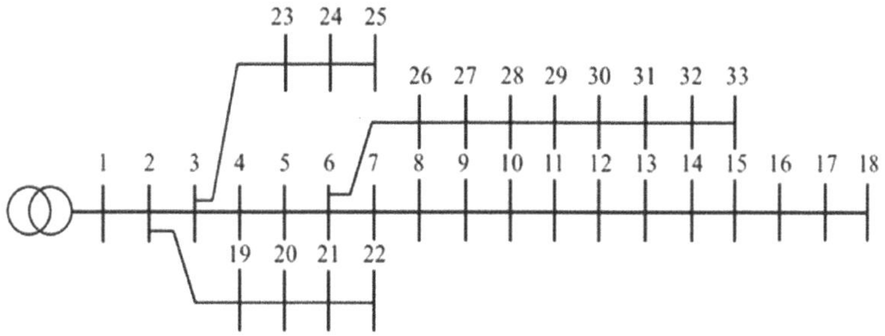

Fig. 3 Single line diagram of IEEE 33-bus radial distribution system

In other words, BFS based optimal DG placement strategy is a computationally friendly approach in achieving a global optimal solution in lower iterations and less calculation time. Total amounts of active power losses in three mentioned cases before installation of DGs, after optimal and non-optimal allocation of DGs can be summarized as Table 2. Figure 5 and Table 2 reveal that BFS search algorithm can find a global optimal solution vector after solving 99 (number of bus × number of DGs) load flow problems. The applicability of the proposed algorithm in finding the best DG places is compared with other recently published methods such as intersect mutation differential evolution (IMDE) [27], analytical [28], fuzzy genetic algorithm (FGA) [29], and bacterial foraging optimization algorithm (BFOA) [30]. Table 3 summarizes the optimal scenarios and the total real power losses obtained from the BFS load flow based DG placement approach and the other ones. It is obvious that the proposed methodology reduces the active power losses, significantly. Moreover, the minimum value of the bus voltage magnitude is more than that of other algorithms. In other words, if we consider that n and N respectively refer to the number of buses and DGs, BFS based DG allocation strategy not only reduces the number of scenarios from 2 to $n \times N$, but also improves the voltage profile and decreases the energy losses in comparison with other introduced methods.

```
Active_loss = [176.3658 176.3658 176.3658
 175.9749 175.2087 173.5816
 174.1316 169.7861 160.5752
 173.1698 167.0194 154.0177
 172.2220 164.2995 147.5957
 170.2387 158.6200 134.2038
 170.0273 158.0142 132.9573
 169.3800 156.1940 129.4099
 168.6691 154.2564 125.9960
 168.0126 152.4966 123.0269
 167.9011 152.2036 122.5602
 167.7084 151.7050 121.8212
 167.0529 150.0539 119.6827
 166.8480 149.5570 119.1739
 166.7076 149.2643 119.2485
 166.5727 149.0347 119.6705
 166.4189 148.8900 121.0057
 166.3765 148.9509 122.1083
 175.9179 175.0743 173.4661
 175.5402 174.2939 173.5152
 175.4769 174.2047 173.8303
 175.4359 174.2636 174.8981
 173.6913 168.5431 158.1002
 172.9058 166.3466 153.8640
 172.5312 165.4229 152.7556
 169.9968 157.9871 132.7875
 169.6762 157.1576 130.9427
 168.5490 154.2779 124.5832
 167.7423 152.2482 120.1322
 167.3024 151.1777 117.8484
 166.8907 150.1655 116.6126
 166.8099 149.9995 116.6033
 166.7991 150.0585 117.1865];
```

Fig. 4 4 Loss matrix obtained from solving backward forward sweep algorithm with 33×3 iterations (33=number of buses, 3=number of DGs)

Table 2 Comparison of different cases with and without DGs

Case study	Total real power losses (kW)
Without DGs	176.3658
With DGs (Best scenario)	86.7679
With DGs (1st DG at bus 12, 2nd DG at bus 20, 3rd DG at bus 4)	144.0289

Table 3 Comparison between the proposed approach and the other recently published algorithms

DG size (kW)	840, 1130	1000	600, 1100	633, 90, 947				
Algorithms	IMDE [27]	BFS	Analytical [28]	BFS	FGA [29]	BFS	BFOA [30]	BFS
Total energy losses (kW)	84.28	80.54	142.34	114.74	119.7	82.7	98.3	82
Minimum voltage magnitude in per unit (Bus number)	0.971 (33)	0.97 5(33)	0.931 (33)	0.933 (18)	0.935 (18)	0.963 (18)	0.964 (33)	0.965 (33)
Best places (Bus number)	14, 30	13, 30	18	30	7, 32	14, 30	7, 18, 33	14, 18, 30

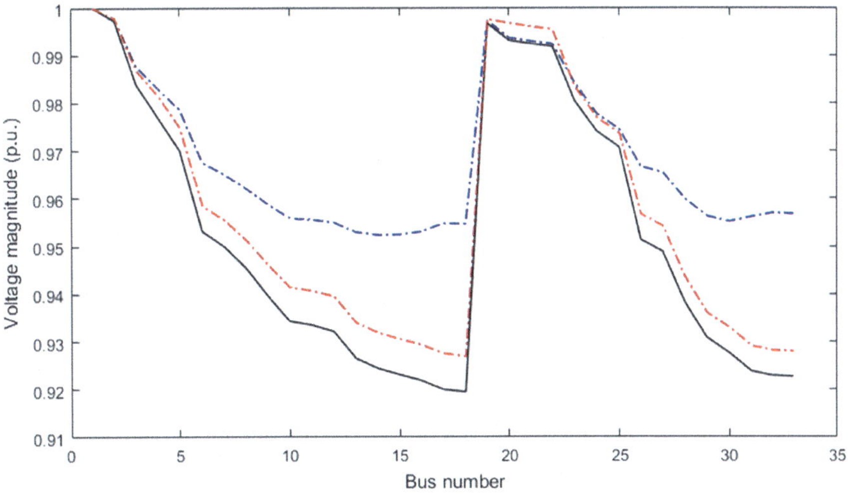

Fig. 5 Voltage profile in three cases: Without DGs (black), best scenario (blue), another scenario for installation of DGs (1st DG at bus 12, 2nd DG at bus 20, 3rd DG at bus 4, red)

4 Conclusions

In this chapter, the forward-backward load flow calculation was used for find-ing the optimum places for installing the DG units. At each iteration, one DG unit is considered to be located at one of buses. The BFS power flow algorithm is then run according to the line and updated bus data matrices. The active power loss is obtained and reported in the loss matrix. Finally, the good places are selected based on the loss matrix. For the nth DG unit, if the mth row of the loss matric is minimum, this DG unit should be installed at bus n. The numerical results revealed that the proposed DG allocating algorithm is more fast and accurate than other recently published methods because of its lower iterations and less ac-tive power losses.

References

1. Singh, B., Sharma, J.: A review on distributed generation planning. Renew. Sustain. Energy Rev. **76**, 529–544 (2017)
2. Sultana, U., Khairuddin, A.B., Aman, M.M., Mokhtar, A.S., Zareen, N.: A review of optimum DG placement based on minimization of power losses and voltage stability enhancement of distribution system. Renew. Sustain. Energy Rev. **63**, 363–378 (2016)
3. Gkaidatzis, P.A., Bouhouras, A.S., Doukas, D.I., Sgouras, K.I., Labridis, D.P.: Load variations impact on optimal DG placement problem concerning energy loss reduction. Electr. Power Syst. Res. **152**, 36–47 (2017)
4. Rahmani-andebili, M.: Simultaneous placement of DG and capacitor in distribution network. Electr. Power Syst. Res. **131**, 1–10 (2016)
5. Kaur, S., Kumbhar, G., Sharma, J.: A MINLP technique for optimal placement of multiple DG units in distribution systems. Int. J. Electri. Power Energy Syst. **63**, 609–617 (2014)
6. Poornazaryan, B., Karimyan, P., Gharehpetian, G.B., Abedi, M.: Optimal allocation and sizing of DG units considering voltage stability, losses and load variations. Int. J. Electr. Power Energy Syst. **79**, 42–52 (2016)
7. Esmaeili, M., Sedighizadeh, M., Esmaili, M.: Multi-objective optimal reconfiguration and DG (Distributed Generation) power allocation in distribution networks using Big Bang-Big Crunch algorithm considering load uncertainty. Energy **103**, 86–99 (2016)
8. Mohanty, B., Tripathy, S.: A teaching learning based optimization technique for optimal location and size of DG in distribution network. J. Electr. Syst. Inf. Technol. **3**(1), 33–44 (2016)
9. Shaaban, M.F., Atwa, Y., El-Saadany, E.: A multi-objective approach for optimal DG allocation. In: 2011 2nd International Conference on Electric Power and Energy Conversion Systems (EPECS), pp. 1–7. IEEE (2011)
10. Pradeepa, H., Ananthapadmanabha, T., Bandhavya, C.: Optimal allocation of combined DG and capacitor units for voltage stability enhancement. Procedia Technol. **21**, 216–223 (2015)
11. Alinezhad, P., Bakhoda, O.Z., Menhaj, M.B.: Optimal DG placement and capacity allocation using intelligent algorithms. In: 2015 4th Iranian Joint Congress on Fuzzy and Intelligent Systems (CFIS), pp. 1–8. IEEE (2015)
12. Fu, X., Chen, H., Cai, R., Yang, P.: Optimal allocation and adaptive VAR control of PV-DG in distribution networks. Appl. Energy **137**, 173–182 (2015)
13. Kayal, P., Chanda, C.K.: Placement of wind and solar based DGs in distribution system for power loss minimization and voltage stability improvement. Int. J. Electr. Power Energy Syst. **53**, 795–809 (2013)
14. Tanwar, S.S., Khatod, D.K.: Techno-economic and environmental approach for optimal placement and sizing of renewable DGs in distribution system. Energy **127**, 52–67 (2017)

15. Das, B., Mukherjee, V., Das, D.: DG placement in radial distribution network by symbiotic organisms search algorithm for real power loss minimization. Appl. Soft Comput. **49**, 920–936 (2016)
16. Sadeghi, M., Kalantar, M.: Multi types DG expansion dynamic planning in distribution system under stochastic conditions using Covariance Matrix Adaptation Evolutionary Strategy and Monte-Carlo simulation. Energy Convers. Manag. **87**, 455–471 (2014)
17. Zidan, A., Shaaban, M.F., El-Saadany, E.F.: Long-term multi-objective distribution network planning by DG allocation and feeders' reconfiguration. Electric Power Syst. Res. **105**, 95–104 (2013)
18. Evangelopoulos, V.A., Georgilakis, P.S.: Optimal distributed generation placement under uncertainties based on point estimate method embedded genetic algorithm. IET Gener. Transm. Distrib. **8**(3), 389–400 (2013). Available: http://digital-library.theiet.org/content/journals/10.1049/iet-gtd.2013.0442
19. Moravej, Z., Akhlaghi, A.: A novel approach based on cuckoo search for DG allocation in distribution network. Int. J. Electr. Power Energy Syst. **44**(1), 672–679 (2013)
20. Sheng, W., Liu, K.Y., Liu, Y., Meng, X., Li, Y.: Optimal placement and sizing of distributed generation via an improved nondominated sorting genetic algorithm II. IEEE Trans. Power Delivery **30**(2), 569–578 (2015)
21. Nekooei, K., Farsangi, M.M., Nezamabadi-Pour, H., Lee, K.Y.: An improved multi-objective harmony search for optimal placement of DGs in distribution systems. IEEE Trans. Smart Grid **4**(1), 557–567 (2013)
22. Esmaili, M.: Placement of minimum distributed generation units observing power losses and voltage stability with network constraints. IET Gener. Transm. Distrib. **7**(8), 813–821 (2013)
23. Kefayat, M., Lashkar, Ara A., Nabavi Niaki, S.A.: A hybrid of ant colony optimization and artificial bee colony algorithm for probabilistic optimal placement and sizing of distributed energy resources. Energy Convers. Manage. **92**, 149–161 (2015)
24. Lee, S.H., Park, J.W.: Optimal placement and sizing of multiple DGs in a practical distribution system by considering power loss. IEEE Trans. Ind. Appl. **49**(5), 2262–2270 (2013)
25. Hemdan, N.G.A., Kurrat, M.: Efficient integration of distributed generation for meeting the increased load demand. Int. J. Electr. Power Energy Syst. **33**(9), 1572–1583 (2011)
26. Venkatesh, B., Ranjan, R., Gooi, H.: Optimal reconfiguration of radial distribution systems to maximize loadability. IEEE Trans. Power Syst. **19**(1), 260–266 (2004)
27. Khodabakhshian, A., Andishgar, M.H.: Simultaneous placement and sizing of DGs and shunt capacitors in distribution systems by using IMDE algorithm. Int. J. Electr. Power. **82**, 599–607 (2016)
28. Naik, S.G., Khatod, D.K., Sharma, M.P.: Optimal allocation of combined DG and capacitor for real power loss minimization in distribution networks. Int. J. Electr. Power. **53**, 967–973 (2013)
29. Reddy, S.C., Prasad, P.V.N., Laxmi, A.J.: Placement of distributed generator, capacitor and DG and capacitor in distribution system for loss reduction and reliability improvement. Editors-in-Chief. **198** (2013)
30. Kowsalya, M.I.A.M.: Optimal distributed generation and capacitor placement in power distribution networks for power loss minimization. In: 2014 International Conference on Advances in Electrical Engineering (ICAEE), pp. 1–6 (2014)

Optimal Capacitor Placement in Distribution Systems Using a Backward-Forward Sweep Based Load Flow Method

Farkhondeh Jabari, Khezr Sanjani and Somayeh Asadi

Abstract Nowadays, the non-optimal placement of the shunt capacitors in distributed electricity systems may increase the total active power loss and lead to the voltage instability. Therefore, many researchers have recently focused on optimization of capacitor placement problem in radial and meshed distribution grids aiming to minimize transmission losses and improve the overall efficiency of the power delivery process. This chapter aims to present a backward-forward sweep (BFS) based algorithm for optimal allocation of shunt capacitors in distribution networks. The total real power loss of the whole system is minimized as the objective function. Moreover, the feeder current capacity and the bus voltage magnitude limits are considered as the optimization constraints. In addition, it is assumed that the sizes of capacitors are the known scalars. The 1st capacitor is considered to be located at the 1st bus of the test system. Then, the BFS load flow is run and the objective function is saved as 1st row and 1st column component of a loss matrix. Secondly, the 1st capacitor is assumed to be installed at bus 2 and the BFS load flow is run to obtain objective function as 2nd row and 1st column component of loss matrix. When all buses are assessed for installation of capacitor 1 and losses are calculated in each scenario, similar analyses are carried out for the 2nd capacitor bank and the values of the active power loss are saved as the 2nd column of the loss matrix. The same strategy is applied to other capacitors. Finally, a loss matrix is formed with number of rows and columns equal to the number of buses and shunt capacitors, respectively. The best places for installation of capacitors are determined based on the components of the loss matrix. Simulation of BFS based capacitor placement problem

F. Jabari · K. Sanjani
Faculty of Electrical and Computer Engineering, University of Tabriz, Tabriz, Iran
e-mail: f.jabari@tabrizu.ac.ir

K. Sanjani
e-mail: sanjani96@ms.tabrizu.ac.ir

S. Asadi (✉)
Department for Management of Science and Technology Development, Ton Duc Thang University, Ho Chi Minh City, Vietnam
e-mail: somayehasadi@tdtu.edu.vn

Faculty of Applied Sciences, Ton Duc Thang University, Ho Chi Minh City, Vietnam

M. Pesaran Hajiabbas and B. Mohammadi-Ivatloo (eds.),
Optimization of Power System Problems, Studies in Systems, Decision and Control 262,
https://doi.org/10.1007/978-3-030-34050-6_3

is conducted on the 33-bus distribution network to demonstrate its robustness and effectiveness in comparison with other procedures.

Keywords Optimal capacitor placement · Distribution system · Voltage improvement · Loss minimization · Feeder capacity

Nomenclature

$\dot{J}_i^{k+1}$	The current injected to the load i in iteration $(k + 1)$
$\dot{V}_i^k$	The voltage of the node i in iteration k
$\dot{S}_{D,i}$	The appearance power consumption of the load i
n_l	The number of lines
$\dot{I}_{h,i}^{k+1}$	The current of the feeder h-i in iteration $(k + 1)$
I_b	The current of the line b
$Z_{h,i}$	The impedance of the feeder h-i
$\dot{V}_h^{k+1}$	The voltage of the bus h in iteration $(k + 1)$
I_b^{max}	Maximum current of line b
F_{loss}	The real power loss of the distribution grid
$g_{i,j}$	The conductance of the line i-j
V_m	The voltage magnitude of the node m
θ_m	The voltage angle of the node m
$\dot{S}_i$	The appearance power injected to the bus i
$\dot{Q}_{C,i}$	The reactive power of the capacitor located at bus i
$\dot{P}_{D,i}$	The real power consumption at bus i
$V_i^{\min}$, $V_i^{\max}$	Minimum and maximum values of voltage magnitude for node i
$\dot{Q}_{D,i}$	The reactive power consumption at bus i

1 Introduction

Recently, optimization of capacitor placement problem in distribution systems has attracted more attention because of increased electricity demand and voltage drop, which may lead to load-generation mismatch and uncontrolled islanding of radial and meshed grids [1]. In [2], Gaussian and Cauchy probability distribution functions based particle swarm optimization (PSO) algorithm are employed for finding optimum places of capacitor banks, voltage profile improvement and energy loss reduction considering feeder loading capacity and voltage limits. Non-dominated sorting genetic algorithm (NSGA-II) is used in [3, 4] to investigate power losses, voltage stability and total harmonic distortion (THD). A clustering method is introduced in [5] for discrete optimization of capacitor places and sizes to minimize the sum of energy losses and capacitor costs. It is revealed that the clustering algorithm is computationally friendly and fast in comparison with fuzzy genetic algorithm [6, 7], direct search method [8], intersect mutation differential evolution strategy [9], teaching learning-based optimization [10], cuckoo search approach [11], self-adaptive harmony search algorithm [12] and artificial bee colony [13, 14]. In [15], cost of energy losses, capacitor installation cost, and voltage penalty factor are considered as main objectives of optimal capacitor allocation problem. In [16], a flower pollination algorithm (FPA) is presented for optimal allocating and sizing of capacitors in various distribution systems. Firstly, they suggested a set of candidate buses for installing capacitors using loss sensitivity factor. Secondly, FPA is employed to find the best scenario. In [17], artificial bee colony and artificial immune system are integrated for optimal co-placement of distributed generators and shunt capacitors. Authors of [18] proposed a shark smell optimization algorithm for determining suitable capacitor installation places using momentum gradient and rotational movement search strategies. Enhanced bacterial foraging optimization algorithm [19] is applied on sub-transmission systems to find the best sites and sizes of capacitors considering thermal loading of cables under the normal operating condition and different single line outage contingencies. Bacterial foraging optimization algorithm with loss sensitivity factor and voltage stability index is developed in [20] to find sizes and places of capacitor banks under all possible demand variations.

As reviewed, different optimization algorithms have been implemented on distribution systems to find good places and optimal sizes of shunt capacitors and improve voltage stability and reduce system power losses. But, a search method with less calculation time and computational burden, no need to membership function of fuzzy logic, huge search space of Monte Carlo simulations, cross over and mutation processes of genetic algorithm, and initial population of metaheuristic algorithms has not been proposed by scholars. This chapter aims to present a novel forward-backward sweep (BFS) based optimal capacitor placement strategy for radial distribution networks. In this method, the number and sizes of capacitors are considered as known parameters. Total active power losses are considered as the objective function. Firstly, one of the capacitors is selected. Its reactive power generation is added to third (related to reactive power consumption) column of the bus data matrix. Then, BFS load flow is solved and total real power losses are calculated as a component of loss matrix in the 1st raw and 1st column. In loss matrix, a number of rows and columns are equal to the number of buses and capacitors. Afterward, the 1st capacitor is assumed to be installed on bus 2. A similar analysis is carried out and energy losses are computed as 2nd row and 1st column of loss matrix. When all buses are evaluated for placement of the 1st unit, 2nd one is assumed to be located at buses 1 to N, respectively. where N refers to a number of nodes in the test distribution system. This process is repeated for all capacitors and loss matrix is formed. Finally, the minimum values of columns are determined. If the minimum value of column i occurred in the jth row of loss matrix, bus j will be selected as a good place for installation of the ith unit.

The remainder of the present chapter is organized as follows: The optimal capacitor placement strategy is formulated in Sect. 2. The illustrative example and discussions are provided in Sect. 3. Section 4 concludes the chapter.

2 Mathematical modeling of load flow based optimization problem

2.1 Forward-backward load flow

Figure 1 shows the sample radial distribution grid. As illustrated in this figure, the current injected to the load i in iteration k+1, $\dot{J}_i^{k+1}$, can be given by Eq. (1). In which, $\dot{V}_i^k$ represents the voltage magnitude of the node i in iteration k. Moreover, $\dot{S}_i$ refers to the appearance power of the node i. Equation (2) demonstrates the power balance criterion for each bus i. The appearance power injected to the bus i is equal to the power consumed by the load i plus the power transmitted from the node i to the adjacent bus j.

$$\dot{J}_i^{k+1} = \left(\frac{\dot{S}_{D,i}}{\dot{V}_i^k}\right)^* \tag{1}$$

$$\dot{S}_i = \dot{S}_{D,i} + \sum_j \dot{S}_j \tag{2}$$

where,

$\dot{S}_{D,i}$ The active and reactive power consumption in bus i
$\dot{S}_j$ The complex power flowing in node j

It is assumed that the reactive power injected by the capacitor units to the bus i is the negative reactive load, as expressed by Eq. (3). If the capacitor bank is installed in bus i, its reactive power, $\dot{Q}_{C,i}$, will be modeled as the negative reactive power consumption in this node. Note that $\dot{P}_{D,i}$ and $\dot{Q}_{D,i}$ are the active and reactive power consumptions of the node i, respectively.

$$\dot{S}_{D,i} = \dot{P}_{D,i} + j(\dot{Q}_{D,i} - \dot{Q}_{C,i}) \tag{3}$$

In the backward sweep, the current of the branch h-i in iteration k+1, $\dot{I}_{h,i}^{k+1}$, is calculated as (4).

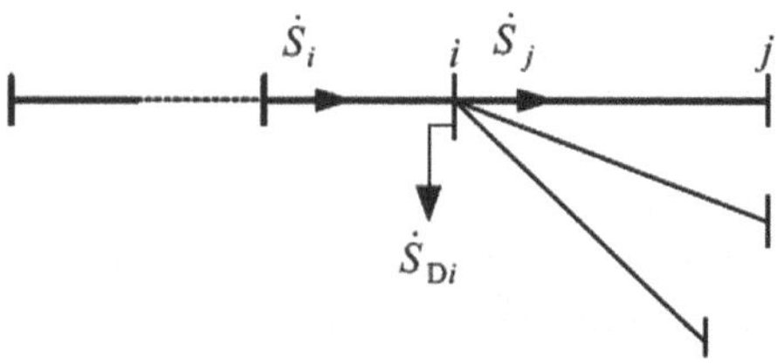

Fig. 1 A simple radial distribution network

$$\dot{I}_{h,i}^{k+1} = \dot{J}_i^{k+1} + \sum_j \dot{I}_j^{k+1} \tag{4}$$

In the forward sweep, the current of the branch h to i, $\dot{I}_{h,i}^{k+1}$, is used for calculating the voltage of the node h in iteration k+1, as fulfilled by (5). where, $Z_{h,i}$ is the impedance of the branch j.

$$\dot{V}_h^{k+1} = \dot{V}_i^{k+1} - \dot{I}_{h,i}^{k+1} \times Z_{h,i} \tag{5}$$

When the convergence criterion (6) is satisfied for all buses, the forward-backward sweep based power flow algorithm will be finished. The scalar ε is the convergence factor. If it is not satisfied for at least one bus, the equations (1)–(6) will be performed in the next iteration.

$$\left| V_j^{k+1} - V_j^k \right| \leq \varepsilon \tag{6}$$

2.2 *Optimal places for installation of capacitor banks*

Equation (7) demonstrates that the minimum total real power loss is considered for optimization of capacitor placement problem.

$$F_{loss} = \text{Min} \sum_{\substack{i,j=1 \\ i \neq j}}^{n_l} g_{i,j} \left[V_i^2 + V_j^2 - 2V_i V_j \cos(\theta_i - \theta_j) \right] \tag{7}$$

where,

F_{loss}	The active power loss of the whole system
$g_{i,j}$	The conductance of the line *i-j*
θ_i and θ_j	The voltage angle of the nodes i and j
n_l	The number of branches

Subject to:

- Voltage permitted range constraints

$$V_i^{\min} \leq V_i \leq V_i^{\max} \tag{8}$$

- Feeder capacity constraint

$$I_b \leq I_b^{\max} \tag{9}$$

In which,

$V_i^{\min}$ and $V_i^{\max}$	Minimum and maximum voltage magnitude of bus i, respectively
I_b	The current of the branch b
$I_b^{\max}$	The maximum current of the branch b

3 Illustrative Example

The backward-forward sweep based strategy is proposed for optimum allocation of shunt capacitors in radial distribution grids and comprehensively described based on MATLAB codes. In this research, it is assumed that the number and sizes of the capacitor banks are known parameters. As obvious from Fig. 2, "Q_cap" refer to the reactive capacities of the units. The BFS based optimization algorithm finds the good places for installation of three capacitor banks and minimizes the total active power losses while satisfying the voltage permitted range constraint and the feeder current limit as stated by (8) and (9), respectively. According to MATLAB codes, a 33-bus radial distribution system [21] is considered to allocate three capacitor units with reactive generation capacities of 50, 740, 260 kVAr, respectively. The single line diagram of IEEE 33-bus radial distribution system is illustrated in Fig. 3. The bus data matrix is defined as "bdata.not.per.unit". The first column of this matrix refers to the number of nodes. Active and reactive power consumptions in each bus are presented at the second and third columns of the bus data matrix in kW and kVAr, respectively. Similarly, "ldata.not.per.unit" is the line data matrix of the 33-bus radial benchmark network. In each row of the branch information matrix, the number of starting and ending points of each line is determined using the bus numbers. The third and fourth columns of the line data matrix represent the resistance and reactance of each branch in Ohm, respectively. Firstly, the capacitor unit 1 is considered to be installed at bus 1. Then, the backward-forward sweep algorithm is implemented on the updated bus data matrix. The total active power loss is then calculated and considered as the 1st row and 1st column of a loss matrix, which is defined as "Active_loss". In other words, the loss matrix has 33 rows (number of buses) and 3 columns (number of shunt capacitors). In the second iteration, the capacitor unit 1 is assumed to be located at bus 2 and the optimal power flow is run. The real power loss is obtained as the 2nd row and 1st column of the loss matrix. when the 1st capacitor bank is located at all buses, the 1st column of the loss matrix will be finished. The similar strategy will be repeated for the 2nd and 3rd units. Finally, the loss matrix will be formed as Fig. 4. According to this matrix, if the 1st reactive power bank is installed at bus 33, the total real power loss will be minimum and equal to 172.4784 kW. In the same manner, buses 30 and 32 are good choices for installation of the 2nd and 3rd units. The voltage before and after installation of capacitors are

Fig. 2 MATLAB codes of forward-backward sweep based optimal capacitor placement

```
clear all; close all; clc
Q_cap=[50,740,260];
n_cap=3;
for i=1:33
    for j=1:n_cap
bdata.not.per.unit=[%Bus   P(Kw)  Q(Kvar)
                     1     0      0
                     2     100    60
                     3     90     40
                     4     120    80
                     5     60     30
                     6     60     20
                     7     200    100
                     8     200    100
                     9     60     20
                     10    60     20
                     11    45     30
                     12    60     35
                     13    60     35
                     14    120    80
                     15    60     10
                     16    60     20
                     17    60     20
                     18    90     40
                     19    90     40
                     20    90     40
                     21    90     40
                     22    90     40
                     23    90     50
                     24    420    200
                     25    420    200
                     26    60     25
                     27    60     25
                     28    60     20
                     29    120    70
                     30    200    600
                     31    150    70
                     32    210    100
                     33    60     40 ];
ldata.not.per.unit=[
%         Inbus  Outbus Resistance(ohm) Reactance(ohm)
          1      2      0.0922          0.0470
          2      3      0.4930          0.2511
          3      4      0.3660          0.1864
          4      5      0.3811          0.1941
          5      6      0.8191          0.7070
          6      7      0.1872          0.6188
          7      8      0.7114          0.2351
          8      9      1.0300          0.7400
          9      10     1.0440          0.7400
          10     11     0.1966          0.0650
          11     12     0.3744          0.1238
          12     13     1.4680          1.1550
          13     14     0.5416          0.7129
          14     15     0.5910          0.5260
          15     16     0.7463          0.5450
          16     17     1.2890          1.7210
          17     18     0.7320          0.5740
          2      19     0.1640          0.1565
          19     20     1.5042          1.3554
          20     21     0.4095          0.4784
          21     22     0.7089          0.9373
          3      23     0.4512          0.3083
          23     24     0.8980          0.7091
          24     25     0.8960          0.7011
          6      26     0.2030          0.1034
          26     27     0.2842          0.1447
          27     28     1.0590          0.9377
          28     29     0.8042          0.7006
          29     30     0.5075          0.2585
          30     31     0.9744          0.9630
          31     32     0.3105          0.3619
          32     33     0.3410          0.5302];
bdata.not.per.unit(i,3)=bdata.not.per.unit(i,3)-Q_cap(j);
% Run FBS load flow algorithm presented in Chapter 14.
        Active_loss(i,j)=Ploss;
        Reactive_loss(i,j)=Qloss;
    end
end
for i=1:busnum
    for j=1:n_cap
        if Active_loss(i,j)==min(Active_loss(:,j))
            best_place(j)=i;
        end
    end
end
```

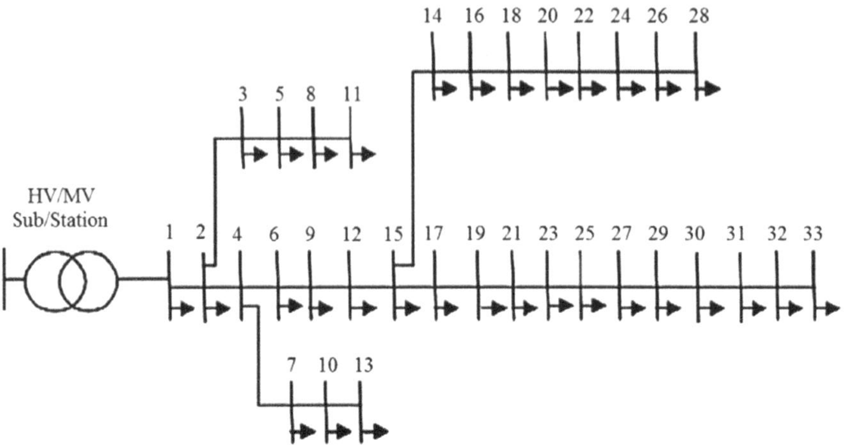

Fig. 3 Single line diagram of IEEE 33-bus radial distribution system

depicted in Fig. 5. As expected, the optimal placement of the capacitors using the forward-backward sweep based search algorithm leads to a significant reduction in active power losses and improvement in bus voltage magnitude. Moreover, the number of scenarios in search space of BFS based capacitor allocation method is reduced to 99 (number of buses × number of capacitors). In other words, the BFS based optimal capacitor placement strategy is a computationally efficient approach in achieving a global optimal solution in lower iterations and less calculation time. The total active power loss in two cases, before and after installation of capacitors, can be summarized as Table 1. Figure 5 and Table 1 reveal that the BFS search algorithm can find a global optimal solution vector after solving 99 (number of bus × number of capacitors) load flow problems.The applicability of the proposed algorithm in finding the best capacitor places is compared with other recently published methods such as intersect mutation differential evolution (IMDE) [9], analytical [22], fuzzy genetic algorithm (FGA) [23], and bacterial foraging optimization algorithm (BFOA) [24]. Table 2 summarizes the optimal scenarios and the total real power losses obtained from the BFS load flow based capacitor placement approach and the other ones. It is obvious that the proposed methodology reduces the active power losses, significantly. Moreover, the minimum value of the bus voltage magnitude is more than that of other algorithms. In other words, if we consider that n and N respectively refer to the number of buses and capacitors, BFS based capacitor allocation strategy not only reduces the number of scenarios from 2^n to $n \times N$, but also improves the voltage profile and decreases the energy losses in comparison with other introduced methods.

Fig. 4 Loss matrix obtained from solving backward forward sweep algorithm with 33×3 iterations (33=number of buses, 3=number of capacitors)

```
Active_loss =[176.3658 176.3658 176.3658
 176.2349 174.7226 175.7167
 175.6028 166.9380 172.5977
 175.2454 162.8148 170.8640
 174.8923 158.8030 169.1577
 174.1488 150.4074 165.5701
 174.0921 150.1654 165.3393
 173.9212 149.9026 164.6930
 173.7381 150.4733 164.0915
 173.5654 151.2445 163.5495
 173.5354 151.4260 163.4602
 173.4852 151.8755 163.3266
 173.3203 154.1122 162.9694
 173.2713 155.1124 162.8991
 173.2473 156.6405 162.9758
 173.2217 158.6390 163.0969
 173.1936 162.3288 163.3896
 173.1867 164.5594 163.6034
 176.2211 175.0407 175.7007
 176.1320 178.5131 175.7495
 176.1179 179.6097 175.8159
 176.1113 181.7699 176.0229
 175.4832 166.6047 172.1292
 175.2731 166.3559 171.3426
 175.1752 167.7625 171.1391
 174.0316 149.3201 165.0300
 173.8720 147.8636 164.2970
 173.2939 142.6808 161.6514
 172.8649 138.8935 159.6945
 172.6163 136.8316 158.5749
 172.5038 138.2712 158.3219
 172.4815 138.9306 158.3119
 172.4784 139.9698 158.4114];
```

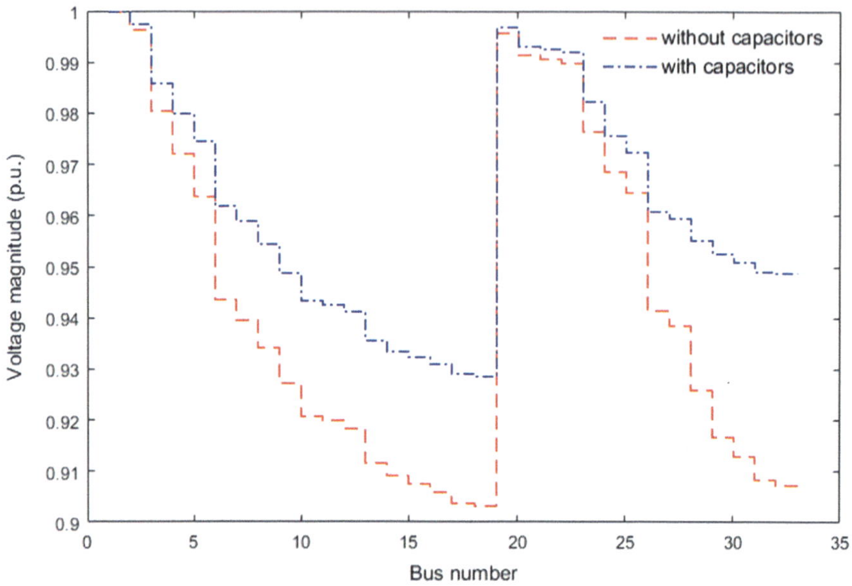

Fig. 5 Voltage profile in two cases: without capacitors (red), with capacitors (best scenario: blue)

Table 1 Comparison between two cases with and without installation of capacitor units

Case study	Total real power losses (kW)
Without capacitors	253.9667
With capacitors (Best scenario)	130.2507

Table 2 Comparison between the proposed approach and the other recently published algorithms

Capacitor sizes (kVAr)	475, 1037		1000		950, 700		350, 820, 277	
Algorithms	IMDE [9]	BFS	Analytical [22]	BFS	FGA [23]	BFS	BFOA [24]	BFS
Total energy losses (kW)	139.7	125.3	164.6	136.8	139.7 141.3	131.5	144.04	123.9
Minimum voltage magnitude in per unit (Bus number)	0.942 (18)	0.943 (18)	0.916 (18)	0.928 (18)	0.929 (18)	0.939 (18)	0.936 (18)	0.944 (18)
Best places (Bus number)	14, 30	9, 30	33	27	18, 30	8, 28	18, 30, 33	16, 30, 32

4 Conclusions

This chapter introduced a novel forward-backward sweep based capacitor placement strategy and used a loss matrix to determine the optimum places for installing the shunt capacitors. In this approach, each capacitor unit is considered to be located at one of the buses. Then, a backward-forward sweep based load flow analysis is implemented on test distribution system according to the line and updated bus data matrices. Total active power loss is calculated and reported as a component of loss matrix. When all buses are evaluated for installation of one capacitor, a similar strategy will be repeated for others. In summary, a loss matrix with a number of rows equals the number of buses and number of columns equals the number of capacitors is formed. Therefore, the optimum scenario for installation of the *i*th capacitor is a bus with a minimum value of power losses in the *i*th column. Robustness and effectiveness of BFS approach in finding global optimal places for installation of capacitors are proved using MATLAB codes and simulations on the 33-bus standard test system.

References

1. Aman, M.M., et al.: Optimum shunt capacitor placement in distribution system—a review and comparative study. Renew. Sustain. Energy Rev. **30**, 429–439 (2014)
2. Lee, C.-S., Ayala, H.V.H., Coelho, L.d.S.: Capacitor placement of distribution systems using particle swarm optimization approaches. Int. J. Electr. Power Energy Syst. **64**, 839–851 (2015)
3. Javadi, M.S., et al.: Shunt capacitor placement in radial distribution networks considering switching transients decision making approach. Int. J. Electr. Power Energy Syst. **92**, 167–180 (2017)
4. Cabral Leite, J., Pérez Abril, I., Santos Azevedo, M.S.: Capacitor and passive filter placement in distribution systems by nondominated sorting genetic algorithm-II. Electric Power Syst. Res. **143**, 482–489 (2017)
5. Vuletić, J., Todorovski, M.: Optimal capacitor placement in radial distribution systems using clustering based optimization. Int. J. Electr. Power Energy Syst. **62**, 229–236 (2014)
6. Das, D.: Optimal placement of capacitors in radial distribution system using a Fuzzy-GA method. Int. J. Electr. Power Energy Syst. **30**(6), 361–367 (2008)
7. Abul'Wafa, A.R.: Optimal capacitor placement for enhancing voltage stability in distribution systems using analytical algorithm and Fuzzy-Real Coded GA. Int. J. Electr. Power Energy Syst. **55**, 246–252 (2014)
8. Ramalinga Raju, M., Ramachandra Murthy, K.V.S., Ravindra, K.: Direct search algorithm for capacitive compensation in radial distribution systems. Int. J. Electr. Power Energy Syst. **42**(1), 24–30 (2012)
9. Khodabakhshian, A., Andishgar, M.H.: Simultaneous placement and sizing of DGs and shunt capacitors in distribution systems by using IMDE algorithm. Int. J. Electr. Power Energy Syst. **82**, 599–607 (2016)
10. Sultana, S., Roy, P.K.: Optimal capacitor placement in radial distribution systems using teaching learning based optimization. Int. J. Electr. Power Energy Syst. **54**, 387–398 (2014)
11. Arcanjo, D.N., et al.: Cuckoo Search Optimization technique applied to capacitor placement on distribution system problem. In: 2012 10th IEEE/IAS International Conference on Industry Applications (2012)

12. Rani, D.S., Subrahmanyam, N., Sydulu, M.: Self adaptive harmony search algorithm for optimal capacitor placement on radial distribution systems. In: 2013 International Conference on Energy Efficient Technologies for Sustainability (2013)
13. El-Fergany, A., Abdelaziz, A.: Multi-objective capacitor allocations in distribution networks using artificial bee colony algorithm. J. Electr. Eng. Technol. **9**(2), 441–451 (2014)
14. El-Fergany, A.A., Abdelaziz, A.Y.: Capacitor placement for net saving maximization and system stability enhancement in distribution networks using artificial bee colony-based approach. Int. J. Electr. Power Energy Syst. **54**, 235–243 (2014)
15. Azevedo, M.S.S., et al.: Capacitors placement by NSGA-II in distribution systems with non-linear loads. Int. J. Electr. Power Energy Syst. **82**, 281–287 (2016)
16. Vuletić, J., Todorovski, M.: Optimal capacitor placement in distorted distribution networks with different load models using Penalty Free Genetic Algorithm. Int. J. Electr. Power Energy Syst. **78**, 174–182 (2016)
17. Ng, H., Salama, M., Chikhani, A.: Classification of capacitor allocation techniques. IEEE Trans. Power Delivery **15**(1), 387–392 (2000)
18. Gnanasekaran, N., et al.: Optimal placement of capacitors in radial distribution system using shark smell optimization algorithm. Ain Shams Eng. J. **7**(2), 907–916 (2016)
19. Othman, A.M.: Optimal capacitor placement by Enhanced Bacterial Foraging Optimization (EBFO) with accurate thermal re-rating of critical cables. Electr. Power Syst. Res. **140**, 671–680 (2016)
20. Devabalaji, K.R., Ravi, K., Kothari, D.P.: Optimal location and sizing of capacitor placement in radial distribution system using Bacterial Foraging Optimization Algorithm. Int. J. Electr. Power Energy Syst. **71**, 383–390 (2015)
21. Venkatesh, B., Ranjan, R., Gooi, H.: Optimal reconfiguration of radial distribution systems to maximize loadability. IEEE Trans. Power Syst. **19**(1), 260–266 (2004)
22. Gopiya Naik, S., Khatod, D.K., Sharma, M.P.: Optimal allocation of combined DG and capacitor for real power loss minimization in distribution networks. Int. J. Elec. Power. **53**, 967–973 (2013)
23. Reddy, S.C., Prasad, P., Laxmi, A.J.: Placement of distributed generator, capacitor and DG and capacitor in distribution system for loss reduction and reliability improvement. Editors-in-Chief. **198** (2013)
24. Kowsalya, M.: Optimal distributed generation and capacitor placement in power distribution networks for power loss minimization. International Conference on Advances in Electrical Engineering (2014)

Optimal Capacitor Placement and Sizing in Distribution Networks

Arsalan Najafi, Ali Masoudian and Behnam Mohammadi-Ivatloo

Abstract Utilizing capacitor banks in order for local compensation of loads reactive power is common in distribution networks. Using capacitors has positive effects on networks such as power and energy loss reduction, voltage deviation and network harmonic reduction as well as improvement in network power factor. Capacitor placement is applied on the network in a form of single or multi-objective problems. Decreasing the total network loss is often the main reason for using capacitors in distribution networks. Capacitor placement approach involves the identification of location for capacitor placement and the size of the capacitor to be installed at the identified location. An optimization algorithm decides the location of the nodes where the capacitors should be placed. As we know, the capacitors are categorized in two main types of fixed and switchable capacitors. Selecting an appropriate type of capacitor is related to the topology of network, load value and economic situation. They are also different from coding point of view. In this section, the model of coding is presented at first, and then, the approach of applying is described based on optimization algorithm. The capacitors are often used for peak loads but they may be present in the network in off-peak due to the switching issues. The network voltage may be increased in off-peak with the presence of capacitors. Therefore, it is very important to consider both peak and off-peak in the capacitor sizing and placement problem. The proposed model is applied on IEEE 10 and 33-bus standard test cases in order to demonstrate the efficiency of the proposed model.

Keywords Capacitor placement · Teaching learning based optimization algorithm · Capacitor sizing

A. Najafi (✉)
Young Researchers and Elite Club, Sepidan Branch, Islamic Azad University, Spidan, Iran
e-mail: arsalan.najafi@iausepidan.ac.ir

A. Masoudian
Faculty of Engineering, University of Shiraz, Shiraz, Iran

B. Mohammadi-Ivatloo
Faculty of Electrical and Computer Engineering, University of Tabriz, Tabriz, Iran
e-mail: bmohammadi@tabrizu.ac.ir

M. Pesaran Hajiabbas and B. Mohammadi-Ivatloo (eds.),
Optimization of Power System Problems, Studies in Systems, Decision and Control 262,
https://doi.org/10.1007/978-3-030-34050-6_4

1 Introduction

Today, with the advent of science and technology, the use of electrical energy has grown dramatically. On the other hand, further flourishing needs a more powerful electricity grid. Power grid consists of three parts: production, transmission and distribution. Of course, distribution networks are important in the final section of the network. Due to lower voltage levels and higher currents, losses in distribution systems are higher than in transmission systems. The issue of reducing losses and improving the efficiency of electric energy supply to the power system is mainly addressed to the distribution network. Reducing the electric power losses is a way to increase the capacity of the production, transmission and distribution network without investing in production. Examples of such loss reduction solutions are reactive power control, cross-sectional variation of conductors, voltage level change, transformer load management, load management, over-distribution network topology change, and so on. The reactive power flow in the network increases the losses and reduces the useful capacity of the lines and transformers. The use of a capacitor as a reactive power generator is very common in order to regulate and control the voltage, preventing voltage fluctuations in the network and correcting the power factor due to the simplicity and low cost of the system. Installed capacitors reduce the network current and losses by reducing the reactive power flow of line from the main substation to the location of capacitor. The absorption and injection of reactive power should be carried out in such a way as to minimize the losses, and thus the capacitor optimal placement problem is discussed. The objective function of the capacitor optimal placement in distribution networks is the cost of installed capacitors, installation costs, etc., and the cost of power and energy losses. By minimizing the cost function along with the constraint, i.e., the permitted bus voltages and line currents, the optimal capacitor size and the location can be determined. Optimal capacitor placement problem can be formulated as a non-linear optimization problem with a series of equality and inequality constraints. Therefore, most of conventional optimization techniques are not able to solve this complex problem, thus evolutionary optimization methods need to be used to solve the problem.

Studies show that 13% of the total energy produced by power plants is dissipated as distribution losses [1], which caused by reactive power flow. However, losses due to reactive current can be reduced by shunt capacitor placement. In addition to reducing power and energy losses in load peak, optimal capacitor placement can free up distribution equipment capacity and improve the voltage profile. Hence, over the past decades, the optimal capacitor placement has been widely studied. Optimal capacitor placement involves determining the location, size and number of capacitors installed in the distribution system, so that the most benefit is obtained at different load levels.

2 Reactive Power Compensation

Reactive power compensation is known as a very important issue in a power system. Consuming load (residential, commercial, industrial, etc.) imposes active and reactive demands on the network. Active powers are converted into other forms of energies such as light, heat and rotational movement. Reactive power should be compensated for warranting the provision of active energies. Capacitor banks are used in a wide area in order to loss reduction, freeing up system capacity and improving the voltage profile. In the last 30 years, power capacitors have recovered greatly by improvement of dielectric materials and their manufacturing techniques. Capacitance sizes have increased from about 15 kVar to about 200 kVAR (Capacitor banks are in the range of about 300–1800 kVAR) [2]. Nowadays, power capacitors available to distribution companies are more efficient and less costly than 30 years ago. Under some conditions, even replacement of older capacitors is justified due to the lower losses of new ones. As a result, distribution companies can make their choices based on the economic evaluation of existing capacitor technology [2]. Shunt capacitors, i.e., capacitors connected in parallel to the grid, are used extensively in distribution systems. Shunt capacitors provide reactive power or reactive current to compensate for the out of phase component of the inductive load current. In addition, shunt capacitors correct lag characteristics of inductive loads by drowning the lead current that provides part or all of their lag component current. Therefore, a parallel capacitor has the same effect as a synchronous condenser, that is, an overexcited generator or synchronous motor. By using a shunt capacitor in the distribution feeder, the load current can be reduced and the line power factor can be improved. As a result, the voltage drop between the substation and the load decreases. The amount and quality of advantages are related to the number, type of shunt capacitors and their regulations. Therefore, an optimal way for capacitor placement is the main aspect in installing capacitors [3].

2.1 Benefit of Reactive Power Compensation

The installed shunt capacitors in the end of power system feeder for supplying reactive power have some advantages. In this section some of these advantages are investigated. Different methods are used by different companies to calculate the economic benefits of installing reactive power devices. In summary, the economic benefits of installing reactive compensators can be summarized as follows:

- Freeing up production capacity
- Reducing voltage drop, and consequently, obtaining an improved voltage profile
- Releasing feeder capacity and related equipment
- Delaying or removing investments for system reform and development
- Reducing power and energy losses.

2.1.1 Improving Voltage Profile

The feeders with large loads have a weak voltage profile and they face with voltage variations by loads changing. In a power system, voltage regulation in a small interval (5% of nominal voltage) and having a balance situation are proposed. By the way, the amounts of loads fluctuation and as a result, the voltage deviation will be more than allowable amount. The shunt capacitors are one of the main solutions to improve the voltage deviations. On the other hand, by keeping the voltage near to the nominal value, it is not required to use expensive regulators [3]. In addition, the revenue of distribution companies rises due to voltage increment by capacitors, which in turn increases energy consumption. This is especially true for domestic consumers. Increasing energy consumption depends on the nature of the equipment used; for example, the energy consumption of bulbs increases with the square of the voltage magnitude.

2.1.2 Loss Reduction by Capacitors

Delivering the reactive power at the load point leads to reduction in line current and losses. Within a determined study period, the amount of energy losses is also calculated. Now, taking into account the cost per kilowatt hour of energy production, the energy loss reduction benefit due to the capacitor placement can be calculated. Reducing losses at the peak load of the network has good benefits. By reducing the losses at the peak load, the power stations depart from their nominal values, thus reducing the need for production. Meanwhile, with the increase of new customers, the construction of new power plants is postponed. Modifying power factor can significantly reduce the network loss. This can lead to 15% rate of return in the network [4]. Modifying the power factor should be done near the customers in order to maximize the profit. Note that, installing capacitors in LV networks are more expensive in comparison with the MV and HV networks. In the many industrial places, the losses are about 2.5–7.5%, which is related to the states of operation, length of lines and feeder. Capacitors can only decrease a part of losses related to the reactive current [4].

2.1.3 Freeing up Power System Capacity

One of the other important advantages of capacitor placement in distribution network is to free up the capacity of feeders and related equipment, delaying or eliminating investment costs for improving or developing the system, and to free up the distribution transformers capacity. In addition, capacitor placement also frees up the capacity of production and transmission system. This leads to a better performance of operation and makes it possible for a larger number of customers to connect to the network and it does not require a new feeder to connect new customers.

In short, capacitors are very effective tools for reducing the costs of the electric power industry due to continuous increase in fuel and power costs. Power companies make profit whenever they are able to postpone or eliminate new power plant investments and reduce energy requirements. Therefore, capacitors help minimize operating costs and make it feasible for new consumers to invest as little as possible in the system. Today, American distribution companies have installed almost 1 kVAR capacitor per 2 kilowatt of installed power generation capacity to use from economic benefits of capacitor placement [5].

In addition, by using capacitors, a reactive current is supplied for transformers, motors and other devices. This action increases the power factor. It means, by a lower current (or apparent power) more active power usage is occurred. Therefore, capacitor banks can be utilized in order to decrease the load or give more flexibility to the network for increasing load.

2.1.4 Postponing Investment

By using the capacitors and freeing up the capacity, the cost of network expansion will be postponed. This snooze is started from distribution feeder to the substation and transmission networks [3]. It means more economic opportunities for network expansion planning.

2.2 *Disadvantages of Reactive Power Compensation*

Capacitor banks certainly have many benefits for the network. However, there are various states in which the capacitors make the system situation worse. In this section some bugs are investigated.

2.2.1 Resonance

Resonance is a situation in which capacitor and inductance reactance eliminates the effects of each other. As a result, the resistive impedance will be available in the network. The resulted frequency of this situation is called resonance frequency. The resonance increases extremely the current and voltage magnitude. This damages not only the capacitor but also the entire network.

2.2.2 Harmonic Resonance

If the resonance is occurred with the harmonic source (for example non-linear loads) simultaneously, the voltage and current will be increased extremely. Moreover, harmonic resonance will affect the performance of the capacitor.

2.2.3 Transient Switching of Capacitors

The transient mode of capacitor is occurred when a capacitor in high voltage is committed in the system.

2.2.4 Over-Voltage

The voltage of the system is varied in a predefined interval in the power system. Using capacitors can make over voltage in off-peak hours. It may exert unfavorable effects on the system [3].

3 Literature Review

Problem solving methods can be divided into four categories: analytical, numerical programming, evolutionary, and artificial intelligence. The next section summarizes the methods of each category and their advantages and disadvantages.

3.1 Analytical Approaches

In all of the early works on optimal capacitor placement, analytical methods have been used. These algorithms are used when powerful computing resources (high-capacity computers) are not available or expensive. Analytical methods include the use of algebra and calculus to determine the highest value of the saving function. This saving function is often provided as follows:

$$S = K_E \Delta E + K_P \Delta P - K_C C \tag{1}$$

where $K_E \Delta E$ and $K_P \Delta P$ are respectively the cost and energy reduction caused by capacitor placement, and $K_C C$ is the cost of capacitor placement.

Capacitor placement pioneers have used all analytical methods to solve this problem [6–9]. Although these methods can solve the problem in a simple form, they are based on unrealistic assumptions for feeders such as constant conductor size and uniform loading. From these studies, the famous two-thirds method is extracted. In the two-thirds method, for minimizing losses, a capacitor with a capacity of two-thirds of the reactive load of the feeder is placed at about two-thirds of the feeder length.

These early analytical methods were easy to understand and implement. Despite the disadvantages, some industries still use these methods for capacitor placement and some companies argue the rule as a guide. To improve the results, the feeder model is improved. References [8, 10, 11] have formulated the non-uniformity of the load and the different sizes of the conductor. Moreover, Refs. [12–14] have included the switching capacitors in the program, and further improved the situation by considering the location of the capacitor regulators. Another problem with analytical methods is to model the location and size of capacitors as continuous variables. Consequently, the calculated size capacitor may not correspond to the standard sizes, and also the location obtained does not match the allowed nodes for capacitor placement. Therefore, the results should be rounded to the nearest high or low standard; this results in over-voltage status or savings below the calculated value. Of course, most of the recent analytical methods are more accurate but require a lot of system information and longer time to run.

3.2 Numerical Computation Algorithms

Since access to computers became easier and computer memory was reduced, numerical programming algorithms were used to solve optimization problems. Numerical programming methods are repetitive techniques maximizing or minimizing the objective function of decision variables. The values of decision variables should also be constrained by a number of limits. The objective function is the cost saving for optimal location, size and number of capacitors. Voltage and currents can be decision variables that should satisfy all constraints. Numerical programming methods allow a more complex cost function to be optimized for the capacitor placement problem. The objective function can include all voltage constants, line loading, discrete capacitor sizes and physical locations of the nodes. Numerical programming can be used to formulate capacitor placement problem as follows:

$$MaxS = K_L \Delta L - K_C C \tag{2}$$

Subjected to:

$$\Delta V \leq \Delta V_{Max} \tag{3}$$

In this regard, $K_L \Delta L$ is cost savings that may include power and energy losses reduction at peak load as well as freeing up the system capacity. The parameter $K_C C$ is the cost of capacitor placement and V is the voltage variation that should not exceed ΔV_{Max}.

Reference [15] was the first to use dynamic programming to solve a capacitor placement problem, which considered only energy losses reduction with a discrete set of capacitor sizes. By examining all numerical programming methods, it can be seen that the level of growth and complexity of the models has progressed over time. This trend was due to increased computing capacity. Today, heavy calculations are relatively inexpensive, and many numerical optimization packages are available for each of the above algorithms. Some numerical programming methods consider the location of nodes and capacitor sizes as discrete variables; this has a good advantage over analytical methods. However, the preparation of data and the growth of the process for numerical techniques may require more time than analytical methods. To illustrate whether the answer obtained by numerical optimization planning methods are local or original, the convexity of capacitor placement problem should be determined. Considering the economic value of freed capacity and the effect of load growth in these methods may be very difficult.

3.3 Artificial Intelligent Algorithms

The recent popularity of artificial intelligence has led researchers to explore their uses in power engineering applications. In [16], a method based on GA is used for optimal capacitor placement. The size and location of the capacitors are encoded in the binary strings and the intersect operator is used to generate new populations. The problem formulation includes only the cost of capacitors and the reduction of peak power losses. References [17] and [18] are other studies that have used genetic algorithm to solve capacitor placement problem. In [19], the simulated annealing method is used to solve this problem. In recent years, the use of evolutionary algorithms has been increasing; some of these algorithms are: multi-objective algorithm of the immune system [20], differential evolution algorithm [21], firefly algorithm [22], inclusion and interchange of variables algorithm [23], particle swarm optimization [24], shark smell optimization algorithm [25], enhanced bacterial foraging optimization [26]. Moreover, [27] and [28], respectively, use neural networks and fuzzy logic to solve this problem.

4 Problem Formulation

The optimal capacitance problem has many variables and parameters, such as capacitor size and optimal capacitor location. In addition, constraints such as bus voltages are also involved. In this paper, objectives and constraints are considered as follows:

4.1 Objective Function

Different objectives in the case of capacitor placement can be considered. The following objective is considered here.

The objective function f shows the total cost of the loss and the cost of the capacitor [29]:

$$f = K_P P_{loss}^{peak} + \sum_{j=1}^{Nc} K_C Q_C^j + \sum_{l=1}^{Nl} K_E T_l P_{loss}^l \tag{4}$$

where P_{loss}^{peak} represents the losses in the peak, Q_C the cost of the capacitor j, Nc the number of capacitor points, K_P the power loss into cost conversion factor, K_C the cost of the capacitor per kilovar and K_E the cost per kilowatt of energy losses.

It should be noted that losses in the distribution network include both power and energy loss. Power loss is related to peak loading and energy loss is related to loading during the year and can be calculated from power loss according to the loss factor.

4.2 Constraints

In addition to minimizing voltage deviation as an objective, the voltage deviation of individual buses should not exceed the limits and must be between the maximum and minimum values [29].

$$V_{Min} \leq V_i \leq V_{Max} \tag{5}$$

Furthermore, due to economic and technical considerations, the capacitor placement in distribution networks is usually done in such a way that the total capacitance in the network does not exceed a certain limit of

$$\sum_{j=1}^{Nc} Q_C^j \leq Q_{Max} \tag{6}$$

5 Modeling and Optimization Algorithm

5.1 Teaching and Learning Based Optimization Algorithm

Teaching and learning based optimization (TLBO) is an algorithm inspired by the teacher's influence on the students. This algorithm is based on the transfer of knowledge from the teacher to the students. This algorithm has two phases of knowledge transfer; the first phase is teacher phase in which the knowledge is transferred from teacher to class and the second phase is the student phase. In the student phase, information is exchanged between students themselves. In this phase, knowledge transfer is from the side of students with the higher knowledge to their cohorts [30].

- Teacher phase

The first part of the TLBO algorithm is teacher training. The main goal of the teacher is to transfer knowledge and increase the positive output of knowledge in the class. Efficiency in this area dates back to the teacher. Mathematically, this phase is expressed as [30]:

$$X_{i,new} = X_i + r_1.(X_{Teacher} + T_F X_{Mean}) \tag{7}$$

In this regard, X_i is a solution (a student) of the set of problem solutions, $X_{Teacher}$ the best answer to the problem that plays the role of the teacher, X_{Mean} the mean value of the answers to the problem and T_F the teacher factor, which is obtained from relation 8:

$$T_F = \text{round}(1 + r_2) \tag{8}$$

In this case, r_2 is a random number between zero and one. Using *round*, the factor value is rounded.

In the process of answer generation, if the generated answer is better than the previous one, it replaces the previous answer.

- Student phase

In this phase, students increase their knowledge through the exchange of information, in which there is no stable process, and as a result, each student can exchange knowledge with another student. These cases are mathematically motivated. Two random answers i and j are selected. Note that the two answers are not the same. Then, the student phase is completed using the following formula:

If the ith answer is better than the jth answer

$$X_{i,new} = X_i + r_3.\left(X_i - X_j\right) \tag{9}$$

If the ith answer is worse than the jth answer

$$X_{i,new} = X_i + r_3.\left(X_j - X_i\right) \tag{10}$$

where r_3 is a vector of random numbers between zero and one. In this case, r_3 is a vector of random numbers between zero and one.

After the answers are made, the new answer would replace the previous answer if the newly generated random answer is better than the previous one. In summary, the TLBO algorithm can be seen in the flowchart of Fig. 1.

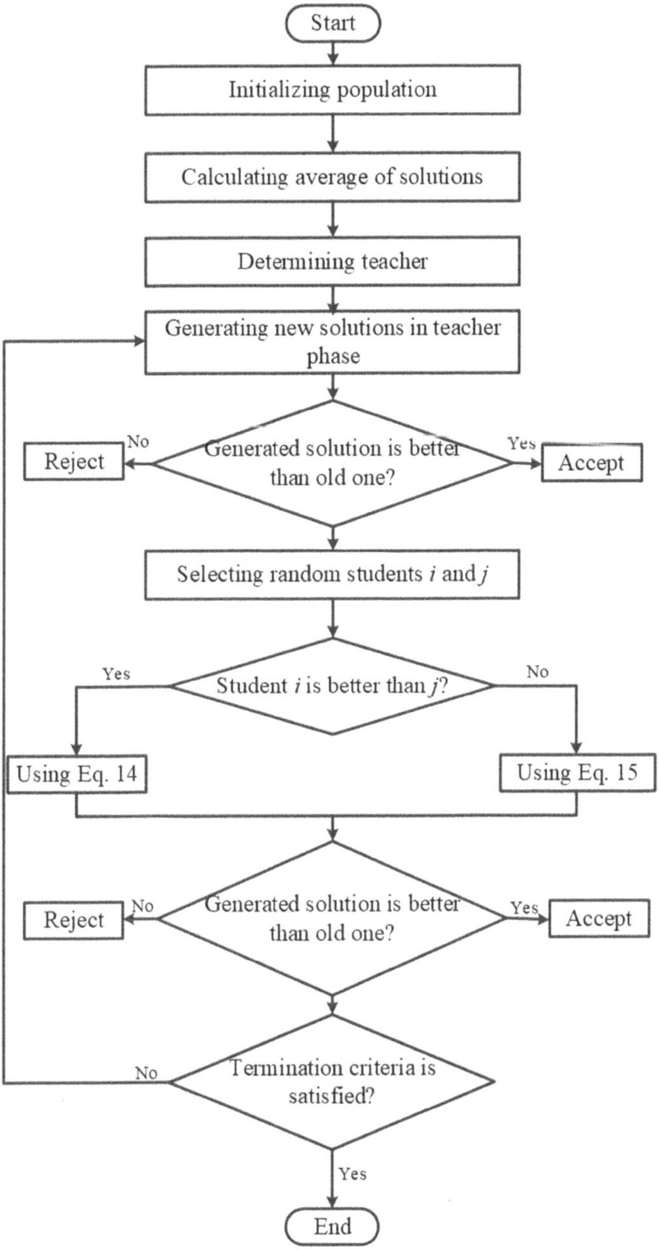

Fig. 1 Flowchart of TLBO algorithm

Fig. 2 Proposed model for codding

Bus 1	Bus 2	Bus 3	....	Bus n
x_1	x_2	x_3	....	x_n

$$x_i \in \{0,1,2,3,\ldots,N_c\}$$

5.2 Matching TLBO with Capacitor Placement Problem

In this case, each answer represents the optimal location and size of the capacitor. Figure 2 shows a coding of an answer to the algorithm. This code is such that each of the elements of this code is assigned to a bus. A number between zero and the maximum number of types of capacitors can be placed in these elements. The value zero means that the bus is not allocated to the capacitor and the value 1 means type 1 capacitor and the rest numbers obey the same order.

5.3 Load Model

Choosing a suitable location for capacitors is highly dependent on the system load. Therefore, the loading information of all load points is required to be known. On the other hand, in order to reduce the calculation, the total load of the system is estimated as a step. Figure 3 shows the load model used in this problem [31]. These load levels are usually expressed as a percentage of peak load. Additionally, the number of surfaces considered for the load is not limited and the capacitor placement problem can be solved easily without any need for modification of the model for several load levels as well as different loading levels for different load points.

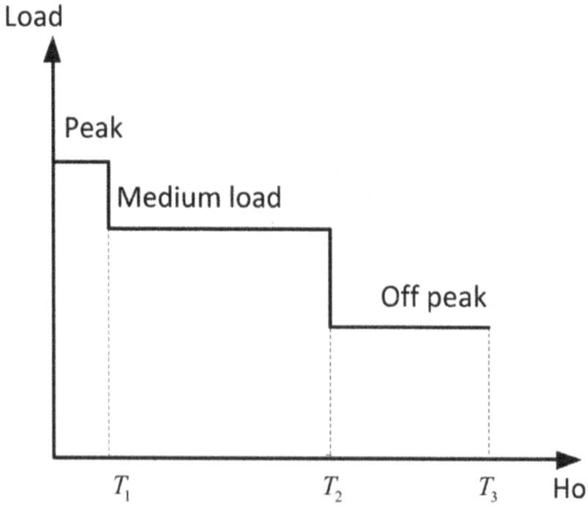

Fig. 3 Proposed load model

6 Numerical Results

6.1 Test Cases

In order to perform simulations, two IEEE 10 [32] and 33-bus test cases have been used [33]. In the 10-bus system, the primary power and energy losses are 7070.16 kW and 6,371,219 kWh, respectively. The initial active and reactive power of the system are also 12,368 and 4186 kW, respectively. In the 33-bus system, the power and energy losses are 40.15 kW and 387,058 kWh, respectively. In both systems the objective is to reduce the losses and capacitor placement cost. The schematic of these two systems is shown in Figs. 4 and 5, respectively. The load and network lines specifications are given in Tables 1 and 2, respectively. We will continue to introduce each of these test cases individually.

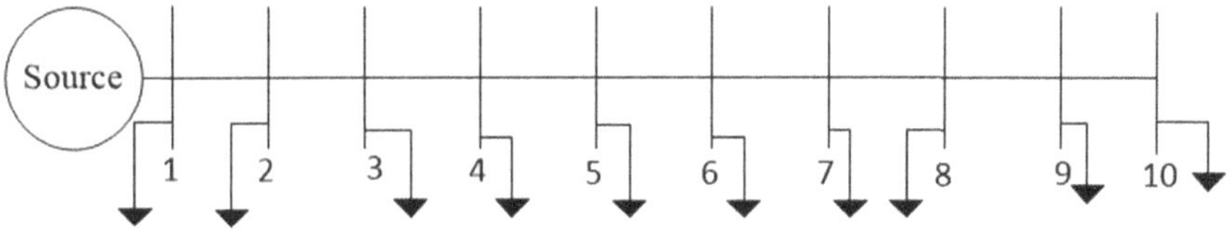

Fig. 4 Schematic of 10-bus test case

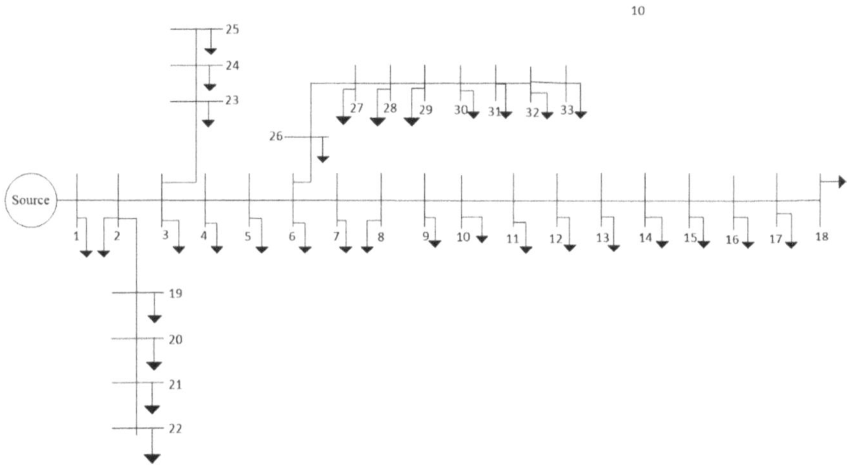

Fig. 5 Schematic of 33-bus test case

Table 1 Characteristics of 10-bus test case

From bus i	To bus j	R (Ohm)	X (Ohm)	P (kW)	Q (kVar)
1	2	0.1233	0.4127	1840	460
2	3	0.014	0.6051	980	340
3	4	0.7463	1.2050	1790	446
4	5	0.6984	0.6084	1598	1840
5	6	1.9831	1.7276	1610	600
6	7	0.9053	0.7886	780	110
7	8	2.0552	1.1640	1150	60
8	9	4.7953	2.716	980	130
9	10	5.3434	3.0264	1640	200

6.2 10-Bus Test Case

In this system, in all three modes, low, medium and peak load, voltages at the ending buses are lower than the limit. The acceptable range of voltage is between 0.9 and 4.1, but the voltage drop in initial conditions is about 0.84, which is less than the limit. After optimization, according to Fig. 6, the worst voltage that is again at the end of the network is about 0.91, which is above the lower limit. Moreover, the power loss is up to 704.64, which is dropped to acceptable levels of casualties. The energy loss reaches 6,100,342.29 (see Table 3). The optimum locations are buses 3, 5, 7, and 9. Figure 7 shows rapid convergence of the algorithm with a acceptable rate of 200 iterations (Table 4).

6.3 33-Bus Test Case

In this system, the power and energy losses are 57.48 kW and 387,058.9721 kWh, respectively. The permitted voltage range is also 0.95 to 1.05. After performing the simulations, Table 5 shows the power loss rate decreased to 43.41. Meanwhile, the energy losses have dropped to 348,408.4 kWh. Voltage profiles are shown in Fig. 8 before and after simulation. In cases, low, medium and peak load, the voltage profile is improved and voltage deviation is reduced. However, while capacitors where allowed to be assigned to all buses, only a capacitor of 450 kV is assigned to bus 29, and all positive effects are only due to this capacitor. This indicates the importance of selecting the correct location for capacitor placement. Figure 9 also depicts the convergence graph of the TLBO algorithm. The final cost of the case is 3421.6.

Table 2 Characteristics of 33-bus test case

From bus i	To bus j	R (Ohm)	X (Ohm)	P (kW)	Q (kVar)
1	2	0.0922	0.0477	100	60
2	3	0.493	0.2511	90	40
3	4	0.366	0.1864	120	80
4	5	0.3811	0.1941	60	30
5	6	0.819	0.707	60	20
6	7	0.1872	0.6188	200	100
7	8	1.7114	1.2351	200	100
8	9	1.03	0.74	60	20
9	10	1.04	0.74	60	20
10	11	0.1966	0.065	45	30
11	12	0.3744	0.1238	60	35
12	13	1.468	1.155	60	35
13	14	0.5416	0.7129	120	80
14	15	0.591	0.526	60	10
15	16	0.7463	0.545	60	20
16	17	1.289	1.721	60	20
17	18	0.732	0.574	90	40
2	19	0.164	0.1565	90	40
19	20	1.5042	1.3554	90	40
20	21	0.4095	0.4784	90	40
21	22	0.7089	0.9373	90	40
3	23	0.4512	0.3083	90	50
23	24	0.898	0.7091	420	200
24	25	0.896	0.7011	420	200
6	26	0.203	0.1034	60	25
26	27	0.2842	0.1447	60	25
27	28	1.059	0.9337	60	20
28	29	0.8042	0.7006	120	70
29	30	0.5075	0.2585	200	600
30	31	0.9744	0.963	150	70
31	32	0.3105	0.3619	210	100
32	33	0.341	0.5302	60	40

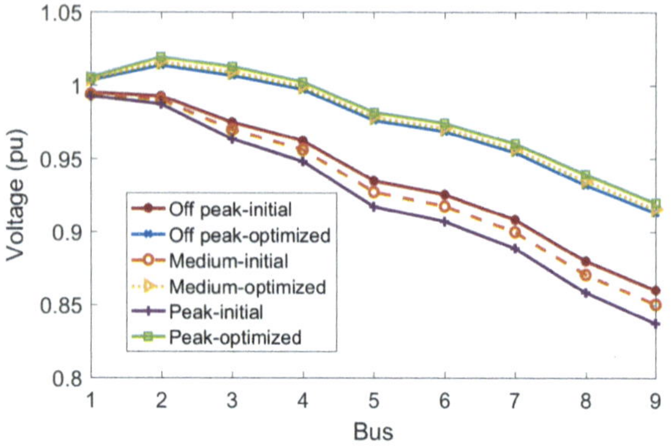

Fig. 6 Voltage profile of 10-bus test case

Table 3 Results of 10-bus test case

Output	Initial	Optimized
Loss (kW)	783.7763	704.64
Energy Loss (kwh)	6,371,219.877	6,100,342.29

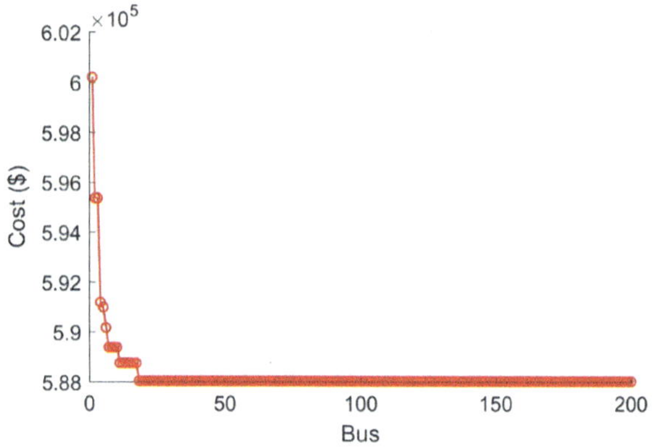

Fig. 7 Convergence of 10-bus test case

Table 4 Place and size of optimal capacitors

Bus	Bus 2	Bus 3	Bus 4	Bus 5	Bus 6	Bus 7	Bus 8	Bus 9	Bus 10
Value	0	4800	0	1800	0	600	0	600	0

Table 5 Results of 10-bus test case

Output	Initial	Optimized
Loss (kw)	57.489	43.41
Energy Loss (kwh)	387,058.972	348,408.4

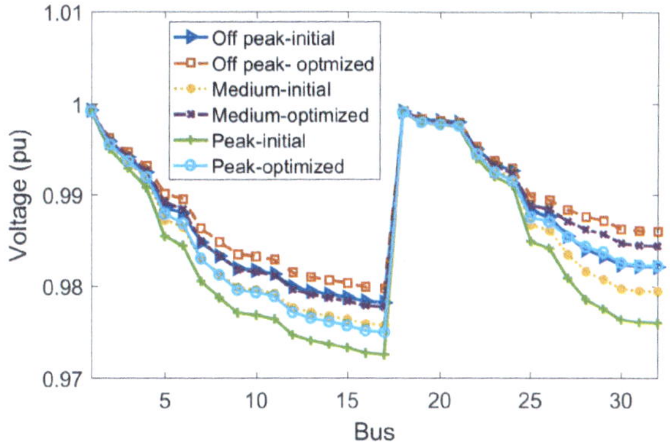

Fig. 8 Voltage profile of 33-bus test case

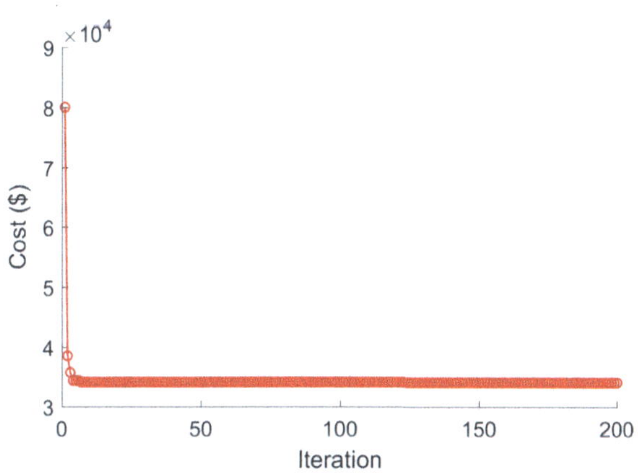

Fig. 9 Convergence of 33-bus test case

7 Conclusion

In this chapter, the optimal location and sizing of capacitors in a distribution network were investigated using the TLBO optimization algorithm. Each answer (student) in the TLBO algorithm was considered to be the location and optimal size of the capacitors. The goal was to reduce the power and energy losses and the cost of locating the capacitors. In order to consider the energy losses, a three-level model of loads, including off-peak, medium and peak load was used. Simulations were implemented in two standard 10 and 33-bus systems. The results showed that there is a voltage drop problem at the end of the system in the 10-bus system, and this voltage drop can be improved by capacitor placement. In addition, network losses can be reduced. In the 33-bus system, network loss reduction and voltage profile improvement can be seen.

Acknowledgements This book chapter is gratefully dedicated to my beloved wife, Samira, who taught me how to be a better man.

MATLAB Code

In this part the MATLAB codes of optimal capacitor placement for the 10-bus test case is presented. The code of 33-bus test case is similar. Therefore only one of test cases is presented here. The code is separated to the some functions. Each function should be copied in a separate MATLAB m-file and then the first code should be run.

```
%% Main mfile should be run
clc
clear all
definParameters();
global No Cap Type NBus No pop Iter Cap Price Ke
Loaddata Strdata T OffPeak T Medium T Peak NLoadLevel
Kp
tic
PLoss = zeros(No pop,1);f =
zeros(No pop,1);LoadDataBase = Loaddata(:,3);
LoadOffPeak = 0.3*LoadDataBase;LoadMedium =
0.6*LoadDataBase;LoadPeak = LoadDataBase;
Loaddata(:,3) = LoadOffPeak; %%%% Evaluating Initial
conditions
[PLossOutOffPeak0,VbusOutOffPeak0,IsecOut0]=DLF(Strdata
,Loaddata); Loaddata(:,3) = LoadMedium;
[PLossOutMedium0,VbusOutMedium0,IsecOut0]=DLF(Strdata,L
oaddata); Loaddata(:,3) = LoadPeak;
[PLossOutPeak0,VbusOutPeak0,Isec0]=DLF(Strdata,Loaddata
);
EnergyLossIni = T OffPeak*PLossOutOffPeak0 +
T Medium*PLossOutMedium0 +...
T Peak*PLossOutPeak0; Loaddata(:,3) = LoadDataBase;

%%%
p = ceil(rand(No pop,NBus-1)*No Cap Type);%%% Initial
popoulation
pop = Cap Mvar determine(p); %%% Allocation MVAr to the
generated population
for i = 1:size(p,1)
      pop(i,:) = Cap Mvar determine(p(i,:));
      Load(:,1) = LoadOffPeak - (pop(i,:))';Load(:,2) =
      LoadMedium - (pop(i,:))';
```

```
        Load(:,3) = LoadPeak -
        (pop(i,:))';Total Cap Price
        =sum(Cap Price((p(i,:))));
        for il=1:NLoadLevel
              Loaddata(:,3) = Load(:,il);
              [PLoss(i,il),Vbus,Isec(i,il,:)]=DLF(Strdata
              ,Loaddata);%%% Running load flow
              PenaltyVoltageL(i,il)= PenV(Vbus); %%%
              Calculating amount of penalties
        end
        PenaltyVoltage(i) = sum(PenaltyVoltageL(i,:),2);
        f(i) = Ke*(T OffPeak*PLoss(i,1) +
        T Medium*PLoss(i,2) + T Peak*PLoss(i,3)) +
        Kp*PLoss(i,1) + Total Cap Price;%%% Calculating
        objective function
        f(i) = f(i) + PenaltyVoltage(i);
end
PBest = p;PBestValue = f;[GTeacherValue, index] =
min(f); GTeacher = PBest(index,:); %%% The best
solution
Xmean = mean(p);
for k = 1:Iter
        k
        [f,p,GTeacher, GTeacherValue, Xmean,
        PenaltyVoltage, PenaltyVoltageBest] =
        UpdateSolutions(GTeacher, p, Xmean, f,
        PenaltyVoltage, LoadOffPeak, LoadMedium,
        LoadPeak);
        %%% Generating new solutions
        fff(k) = GTeacherValue;
end
toc
ij = 1:Iter;
hold on
plot(ij,fff,'r')

%% Function of defining input parameters
function definParameters()
global No Cap Type Cap MVar NBus No pop Iter Cap Price
VLoadMax VLoadMin PF Loaddata Strdata pMax pMin Ke Kp T
Kl T OffPeak T Medium T Peak NLoadLevel
No Cap Type = 7; %%% Number of capacitor types
Cap MVar = 4*[0 150 300 450 600 900 1200];%%% MVar of
capacitors
```

```
Cap Price =4*[0 750 975 1140 1320 1650 2040];%%% Price
of capacitors
No pop =100; %%% Number of population
Iter = 200; %%% Iteration number
VLoadMax = 1.1; %%% Upper voltage bound
VLoadMin = 0.9; %%% Lower voltage bound
PF = 5000; %%% Penalty factor
%%%          Bus       P        Q
Loaddata= [2         1840     460
          3         980      340
          4         1790     446
          5         1598     1840
          6         1610     600
          7         780      110
          8         1150     60
          9         980      130
          10        1640     200
          ];
%%%      From Bus  To Bus   Length      R          X    Imax
       Cap
Strdata = [1        2          1    0.1233    0.4126   0    0
          2        3          1    0.014     0.6051   0    0
          3        4          1    0.7463    1.205    0    0
          4        5          1    0.6984    0.6084   0    0
          5        6          1    1.9831    1.7276   0    0
          6        7          1    0.9053    0.7886   0    0
          7        8          1    2.0552    1.164    0    0
          8        9          1    4.7953    2.716    0    0
          9        10         1    5.3434    3.0264   0    0
          ];
NBus = size(Loaddata,1) + 1; %%% Number of buses
pMax = No_Cap_Type; %%% Maxiumum bound of populations
pMin = 1; %%% Minimum bound of populations
Ke = 0.06; %%% Coeffcent of energy loss
Kp = 300;  %%% coeffcent of power loss
T = 8760; %%% time period
Kl = 168;           %%%
T OffPeak = 3000; %%% Off peak hours
T Medium = 5300;        %%% Medium load hours
T Peak = 460;           %%% Peak hours
NLoadLevel = 3;         %%% Number of load levels

%% Function of updating solutions
```

```
function [f,p,GTeacher, GTeacherValue, Xmean,
PenaltyVoltage, PenaltyVoltageBest] =
UpdateSolutions(GTeacher, p, Xmean, f,
PenaltyVoltage, LoadOffPeak, LoadMedium, LoadPeak)
global Cap Price Ke Kp Strdata Loaddata T OffPeak
T Medium T Peak NLoadLevel No Cap Type
for i = 1:size(p,1) %%%%%%%%%%%%%%% Teacher phase
%%%%%%%%%%%%%%%
      TF = round(1+rand);pnew(i,:) = p(i,:) +
      rand(1,size(p,2)) .* (GTeacher - TF*Xmean);
      pnew(i,:) = round(pnew(i,:));
      for k = 1:size(p,2)
            if pnew(i,k)>No Cap Type
                  pnew(i,k) = No_Cap_Type;
                  elseif pnew(i,k)<1
                        pnew(i,k)= 1;
            end
      end
      pop(i,:) = Cap Mvar determine(pnew(i,:));
      Load(:,1) = LoadOffPeak - (pop(i,:))';
      Load(:,2) = LoadMedium - (pop(i,:))';
      Load(:,3) = LoadPeak - (pop(i,:))';
      Total Cap Price
      =sum(Cap Price((pnew(i,:))));
      for il=1:NLoadLevel
            Loaddata(:,3) = Load(:,il);
            [PLoss(i,il),Vbus,Isec(i,il,:)] =
            DLF(Strdata,Loaddata);
            PenaltyVoltageL(i,il)= PenV(Vbus);
      end
      PenaltyVoltageNew(i) =
      sum(PenaltyVoltageL(i,:),2);
      fnew(i) = Ke*(T OffPeak*PLoss(i,1) +
      T Medium*PLoss(i,2) + T Peak*PLoss(i,3)) +
      Kp*PLoss(i,1) + Total Cap Price;
      fnew(i) = fnew(i) + PenaltyVoltageNew(i);
      if fnew(i)<f(i)
            p(i,:) = pnew(i,:);f(i)=fnew(i);
            PenaltyVoltage(i) =
            PenaltyVoltageNew(i);
      end
      j = round(1 + rand*(i-1)); %%%%%%%%%%%
      Student phase %%%%%%%%%%%%%%%%%%%%%
      if j~=i
```

```
if f(i)<f(j)
      pnew(i,:) = p(i,:) +
      rand(1,size(p,2)).*(p(i,:) -
      p(j,:));
else
      pnew(i,:) = p(i,:) +
      rand(1,size(p,2)).*(p(j,:) -
      p(i,:));
end
pnew(i,:) = round(pnew(i,:));
for k = 1:size(p,2)
      if pnew(i,k)>No Cap Type
            pnew(i,k)= No Cap Type;
      elseif pnew(i,k)<1
            pnew(i,k)= 1;
      end
end
pop(i,:) =
Cap Mvar determine(pnew(i,:));
Load(:,1) = LoadOffPeak - (pop(i,:))';
Load(:,2) = LoadMedium - (pop(i,:))';
Load(:,3) = LoadPeak - (pop(i,:))';
Total Cap Price
=sum(Cap Price((pnew(i,:))));
for il=1:NLoadLevel
      Loaddata(:,3) = Load(:,il);
      [PLoss(i,il),Vbus,Isec(i,il,:)] =
      DLF(Strdata,Loaddata);
      PenaltyVoltageL(i,il)=
      PenV(Vbus);
end
PenaltyVoltageNew(i) =
sum(PenaltyVoltageL(i,:),2);
fnew(i) = Ke*(T OffPeak*PLoss(i,1) +
T Medium*PLoss(i,2) +
T Peak*PLoss(i,3)) + Kp*PLoss(i,1) +
Total Cap Price;
fnew(i) = fnew(i) +
PenaltyVoltageNew(i);
if fnew(i)<f(i)
      p(i,:) = pnew(i,:);
      f(i)=fnew(i);
      PenaltyVoltage(i) =
      PenaltyVoltageNew(i);
```

```
            end
        end
    end
    [GTeacherValue, index] = min(f);
    GTeacher = p(index,:);
    PenaltyVoltageBest = PenaltyVoltage(index(1));
    Xmean = mean(p);

    %% Function of backward forward load flow
    function [PLoss,Vbus,Isec]=DLF(Strdata,Loaddata)
    % Strdata->> 1-from/2-to/3-Length(km)/4-R(ohm/km)/5-
    X(ohm/km)/6-Imax(Amp)/7-Capacitor (kvar)
    % Loaddata->> 1-bus/2-P(kw)/3-Q(kw)
    PLoss = [];

    Nsec=length(Strdata(:,1)); %Number of sections (or to
    buses)
    Vbase=23000;                %V base of the system (v)
    Isec=zeros(Nsec,1);
    Vbus=Vbase*ones(Nsec,1);
    Cbus=zeros(Nsec,1);
    Sbus=zeros(Nsec,1);
    Rsec=Strdata(:,4).*Strdata(:,3);
    Xsec=Strdata(:,5).*Strdata(:,3);
    Zsec= Rsec + i*Xsec;
    %=============================Algorithm===============
    ===
    BI=zeros(Nsec,Nsec+1);
    BI(1,1)=1;
    BV=BI;
    for k=1:Nsec
        BI(:,Strdata(k,2))=BI(:,Strdata(k,1));
        BI(k,Strdata(k,2))=1;
        BV(:,Strdata(k,2))=BV(:,Strdata(k,1));
        BV(k,Strdata(k,2))=Zsec(k);
    end
    BI(:,1)=[];
    BV(:,1)=[];
    BV=BV.';
    Cbus(Strdata(:,2))=Strdata(:,7)*1000;
    Sbus(Loaddata(:,1))=(Loaddata(:,2)+i*Loaddata(:,3))*100
    0; Cbus(1,:)=[];
    Sbus(1,:)=[];
    Iter=0;
```

```
NERROR=1;
S_bus=Sbus-i*(Cbus.*(Vbus/Vbase).^2);      %for P
constant Ibus=conj(S bus./(sqrt(3)*Vbus));
while ((Iter<100)&&(NERROR>1e-5))
      Iter=Iter+1;
      OldIbus=Ibus;
      VD=sqrt(3)*(BV*BI)*Ibus;
      Isec=BI*Ibus;
      Vbus=Vbase-VD;
      S bus=Sbus-i*(Cbus.*(Vbus/Vbase).^2);      %for P
      constant Ibus=conj(S bus./(sqrt(3)*Vbus));
      NERROR=max(max(abs(Ibus-OldIbus)));
end
%=======================================================
===
LossSec=3*abs(Isec).^2.*(Rsec)/1000;
PLoss=sum(LossSec);
Vbus=abs(Vbus)/Vbase; % voltage of to buses
return

%% Function of allocating MVAr to the generated
population function pop = Cap Mvar determine(p)
global Cap MVar NBus
for i = 1:size(p,1)
      pop row = p(i,:);
      pop row MVar = zeros(1,NBus-1);
      for j=1:NBus-1
            pop row MVar(j) = Cap MVar(pop row(j));
      end
      pop(i,:) = pop row MVar;
end

%% Function of applying upper and lower bounds of
population
function p = ApplyingConstraint(p)
global No Cap Type
for i=1:size(p,1)
      for j=1:size(p,2)
            if p(i,j)>No Cap Type
                  p(i,j) = No Cap Type;
            elseif p(i,j)<1
                  p(i,j) = 1;
            end
      end
```

```
end

%% Function of penalizing infeasible solutions
function PenaltyVoltage = PenV(Vbus)
global VLoadMax VLoadMin PF

for i=1:size(Vbus,1)
      if (Vbus(i)>VLoadMax) || (Vbus(i)<VLoadMin)
            Penalty(i) = PF;
      else
            Penalty(i) = 0;
      end
end
PenaltyVoltage = sum(Penalty);

%% Function of initializing population
function p= Initialazation()
global   No_pop   VgMin   VgMax   No_generator   NTrans
NTransStep TransTap  NQComp  QCompMin QCompMax

V = VgMin + rand(No_pop,No_generator)*(VgMax-VgMin);
TT = ceil(NTransStep*rand(No_pop,NTrans));
T = TransTap(TT);
tic
for ii = 1:No_pop
      for jj=1:NQComp
            QComp(ii,jj) = QCompMin(jj) + rand*
            (QCompMax(jj) - QCompMin(jj));
      end
end

QCompValue = ceil(1 + rand(No_pop,NQComp)*
(length(QComp) - 1));
p = [V TT QComp];
```

References

1. Bunch, J.B., Miller, R.D., Wheeler, J.E.: Distribution system integrated voltage and reactive power control. IEEE Trans. Power Apparatus Syst. **101**(2), 284–289 (1982)
2. Arcanjo, D.N., Pereira, J.L.R., Oliveira, E.J., Peres, W., de Oliveira, L.W., da Silva Junior, I.C.: Cuckoo search optimization technique applied to capacitor placement on distribution system problem. In: Proceedings of 10th IEEE/IAS International Conference on Industry Applications (2012)

3. Fuchs, E., Masoum, M.: Power Quality in Power Systems and Electrical Machines, pp. 261–300. Elsevier (2008)
4. Saric, A.T., Calovic, M.S., Djukanovic, M.B.: Fuzzy optimization of capacitors in distribution systems. In: IEE Proceedings on Generation, Transmission and Distribution (1997)
5. Wu, D., Tang, F., Guerrero, J.M., Vasquez, J.C.: Autonomous control of distributed generation and storage to coordinate P/Q sharing in islanded microgrids-an approach beyond droop control. In: Proceeding of International Energy Conference (ENERGYCON) (2014)
6. Neagle, N.M., Samson, D.R.: Loss reduction from capacitors installed on primary feeders. Trans. Am. Instit. Electr. Eng. Part III. IEEE Trans. Power Apparatus Syst. **75**(3), 1 (1956)
7. Cook, R.F.: Optimizing the application of shunt capacitors for reactive-volt-ampere control and loss reduction. Trans. Am. Instit. Electr. Eng. Part III. IEEE Trans. Power Apparatus Syst. **80**(3), 430–441 (1961)
8. Grainger, J.J., Lee, S.H.: Optimum size and location of shunt capacitors for reduction of losses on distribution feeders. IEEE Trans. Power Apparatus Syst. **100**(3), 1105–1118 (1981)
9. Chang, N.E.: Locating shunt capacitors on primary feeder for voltage control and loss reduction. IEEE Trans. Power Apparatus Syst. **88**(10), 1574–1577 (1969)
10. Lee, S.H., Grainger, J.J.: Optimum placement of fixed and switched capacitors on primary distribution feeders. IEEE Trans. Power Apparatus Syst. **100**(1), 345–352 (1981)
11. Salama, M.M.A., Chikhani, A.Y., Hackam, R.: Control of reactive power in distribution systems with an end-load and fixed load condition. IEEE Power Eng. Rev. **5**(10), 39–46 (1985)
12. Civanlar, S., Grainger, J.J.: Volt/Var control on distribution systems with lateral branches using shunt capacitors and voltage regulators part II: the solution method. IEEE Trans. Power Apparatus Syst. **104**(11), 3284–3290 (1985)
13. Civanlar, S., Grainger, J.J.: Volt/Var control on distribution systems with lateral branches using shunt capacitors and voltage regulators part III: the numerical results. IEEE Trans. Power Apparatus Syst. **104**(11), 3291–3297 (1985)
14. Grainger, J.J., Civanlar, S.: Volt/Var control on distribution systems with lateral branches using shunt capacitors and voltage regulators part I: the overall problem. IEEE Trans. Power Apparatus Syst. **104**(11), 3278–3283 (1985)
15. Dura, H.: Optimum number, location, and size of shunt capacitors in radial distribution feeders a dynamic programming approach. IEEE Trans. Power Apparatus Syst. **87**(9), 1769–1774 (1968)
16. Boone, G., Chiang, H.D.: Optimal capacitor placement in distribution systems by genetic algorithm. Int. J. Electr. Power Energy Syst. **15**(3), 155–161 (1993)
17. Sundhararajan, S., Pahwa, A.: Optimal selection of capacitors for radial distribution systems using a genetic algorithm. IEEE Trans. Power Syst. **9**(3), 1499–1507 (1994)
18. Miu, K.N., Chiang, H.D., Darling, G.: Capacitor placement, replacement and control in large-scale distribution systems by a GA-based two-stage algorithm. IEEE Trans. Power Syst. **12**(3), 1160–1166 (1997)
19. Ananthapadmanabha, T., Kulkarni, A.D., Gopala Rao, A.S., Raghavendra Rao, K., Parthasarathy, K.: Knowledge-based expert system for optimal reactive power control in distribution system. Int. J. Electr. Power Energy Syst. **18**(1), 27–31 (1996)
20. Huang, T.L., Hsiao, Y.T., Chang, C.H., Jiang, J.A.: Optimal placement of capacitors in distribution systems using an immune multi-objective algorithm. Int. J. Electr. Power Energy Syst. **30**(3), 184–192 (2008)
21. Chiou, J.P., Chang, C.F., Su, C.T.: Capacitor placement in large-scale distribution systems using variable scaling hybrid differential evolution. Int. J. Electr. Power Energy Syst. **28**(10), 739–745 (2006)
22. Kavousi, Fard A., Niknam, T.: Optimal stochastic capacitor placement problem from the reliability and cost views using firefly algorithm. IET Sci. Meas. Technol. **8**(5), 260–269 (2014)
23. Perez, Abril I.: Algorithm of inclusion and interchange of variables for capacitors placement. Electr. Power Syst. Res. **148**, 117–126 (2017)
24. Lee, C.-S., Ayala, H.V.H., Coelho, L.S.: Capacitor placement of distribution systems using particle swarm optimization approaches. Int. J. Electr. Power Energy Syst. **64**, 839–851 (2015)

25. Gnanasekaran, N., Chandramohan, S., Kumar, P.S., Mohamed Imran, A.: Optimal placement of capacitors in radial distribution system using shark smell optimization algorithm. Ain Shams Eng. J. **7**(2), 907–916 (2016)
26. Othman, A.M.: Optimal capacitor placement by Enhanced Bacterial Foraging Optimization (EBFO) with accurate thermal re-rating of critical cables. Electr. Power Syst. Res. **140**, 671–680 (2016)
27. Gu, Z., Rizy, D.T.: Neural networks for combined control of capacitor banks and voltage regulators in distribution systems. IEEE Trans. Power Deliv. **11**(4), 1921–1928 (1996)
28. Chin, H.C.: Optimal shunt capacitor allocation by fuzzy dynamic programming. Electr. Power Syst. Res. **35**(2), 133–139 (1995)
29. Das, D.: Optimal placement of capacitors in radial distribution system using a Fuzzy-GA method. Int. J. Electr. Power Energy Syst. **30**(6), 361–367 (2008)
30. Gautam, G.D., Pandey, A.K.: Teaching learning algorithm based optimization of Kerf deviations in pulsed Nd: YAG laser cutting of Kevlar-29 composite laminates. Infrared Phys. Technol. **89**, 203–217 (2017)
31. Falaghi, H., Singh, C.: Optimal conductor size selection in distribution systems with wind power generation. In: Wang, L., Singh, C., Kusiak, A. (eds.) Wind Power Systems: Applications of Computational Intelligence, pp. 25–51. Springer, Heidelberg (2010)
32. Chung, T.S., Leung, H.C.: A genetic algorithm approach in optimal capacitor selection with harmonic distortion considerations. Int. J. Electr. Power Energy Syst. **21**, 561569 (1999)
33. Afzalan, E., Taghikhani, M.A., Sedighizadeh, M.: Optimal placement and sizing of DG in radial distribution networks using SFLA. Int. J. Energy Eng. **2**(3), 73–77 (2012)

Binary Group Search Optimization for Distribution Network Reconfiguration

Hamid Teimourzadeh, Behnam Mohammadi-Ivatloo and Somayeh Asadi

Abstract In this chapter, the binary group search optimization algorithm (BGSO) is proposed to tackle with optimal network reconfiguration problem in distribution systems. Here, total loss minimization is considered as the objective which is solved subject to system radial operation and power flow constraints. Here, the basics of GSO algorithm is presented first and then, necessary modification for developing BGSO is discussed. The main part of this chapter deals with a source code, which expresses step by step implementation of BGSO method to optimal network reconfiguration problem. Needless to emphasize that the BGSO and associated source code presented in this chapter is a general engine that can be easily adjusted to any optimization problem with binary variables. In addition, the source code associated with the developed forward-backward sweep-based load flow study is also provided. The simulation studies are performed on different distribution networks to examine the scheme at various conditions and problem complexities. Comprehensive simulation studies conducted in this chapter verifies effectiveness of the BGSO and developed source code for solving optimal distribution network reconfiguration problem.

Keywords Group search optimization · Distribution network reconfiguration · Source code · Forward-backward load flow

---

H. Teimourzadeh · B. Mohammadi-Ivatloo
Faculty of Electrical and Computer Engineering, University of Tabriz, Tabriz, Iran
e-mail: hteimour96@ms.tabrizu.ac.ir

B. Mohammadi-Ivatloo
e-mail: bmohammadi@tabrizu.ac.ir

S. Asadi (✉)
Department for Management of Science and Technology Development, Ton Duc Thang University, Ho Chi Minh City, Vietnam
e-mail: somayehasadi@tdtu.edu.vn

Faculty of Applied Sciences, Ton Duc Thang University, Ho Chi Minh City, Vietnam

M. Pesaran Hajiabbas and B. Mohammadi-Ivatloo (eds.),
Optimization of Power System Problems, Studies in Systems, Decision and Control 262,
https://doi.org/10.1007/978-3-030-34050-6_5

1 Literature Review

Distribution network reconfiguration (DNR) is the manner of system topology adjustment through varying on/off state of switches and while preserving radial operating structure of the network. In distribution systems, DNR is a common practice for reducing system losses, satisfying operating constrictions, balancing the load, improving voltage quality and augmenting system security [1]. DNs are commonly running in a radial arrangement. These systems also implemented by the large quantity of sectionalizing switches and some of tie switches [2].

Various researchers discussed the DNR problems employing different methods [2–4]. The DNR problem has been solved in [2], implementing the algorithm which is based on the blending of a novel fuzzy adaptive PSO and Nelder–mead simplex search algorithm named NFAPSO–NM. In [4], DNR problem has been solved implementing genetic algorithm (GA) with the objective of power losses reduction. In [1], the DNR problem has been defined for load balancing and loss reduction as an integer programming problem. A heuristic method for reconfiguration has been introduced in [5] that presents a subsequent switch opening based on the branch power flow. Fuzzy multi-objective approach along with heuristic based method has been presented in [6], in order to optimize network configuration. In [7], a novel non-revisiting genetic algorithm has been introduced for solving the reconfiguration problem. The binary group search optimization algorithm (BGSO) is presented in [8] for solving the optimal DNR problem for loss minimization. In [9] , harmony search algorithm has been presented in order to solve the reconfiguration of the unbalanced distribution network problem.

In this chapter, at first, a small summary of the general GSO is displayed. The searching space mode of the general GSO method is in continues. However, a binary searching device is needed to solve the DNR problem. The main part of this chapter deals with a source code which expresses step by step implementation of BGSO method to optimal network reconfiguration problem. To this end, the source code for each step is also provided after each part of the implementation discussed. Then, the BGSO algorithm is performed for the distribution system re-configuration problem. Simulations are conducted on 69-node and 119-node distribution test systems to prove the performance of the BGSO algorithm and developed source code in comparison with other procedures.

2 Group Search Optimization Algorithm (GSO)

In this part, at first, a summary on the basics of the GSO algorithm is presented. Thereafter, the BGSO algorithm is discussed in details that is efficient for handling problems containing binary variables.

2.1 Basics of GSO

The principal design of the GSO is encouraged by animal group-living principles. Group living is a general event in the animal behavioral environment which has been severely analyzed. One result of living collectively is that group searching lets each of the members to increase spot gaining speeds as well as to decrease the difference of hunt success. The authors in [10] observed that the searching area of white crappie might be cones or set of wedges, which were identified by most hunt height, most hunt distance, and most hunt angle. The top of any cone is the spot at which the fish stops and looks for hunt. Deployment of natural phenomena and animal behavior is common practice to devise heuristic algorithms [11, 12]. This has regularly attended to the selection of two tactics for foraging in groups: 1) producing, e.g., food seeking; and 2) scrounging, e.g., meeting sources unsealed by other members [13]. Joining is a universal feature observed in most social animals such as spiders, lions, fish, and birds. People in a community that are successful at sources hunting give sources at their cost to less successful ones [14]. Producer–scrounger (PS) [15] and information sharing (IS) [16] are two types that have been introduced to investigate the optimal strategy for joining. Foragers in the PS type are supposed to do joining or producing tactics individually. On the other hand, the IS type supposes that foragers seek synchronously for their individual source while seeking for chances to join. At least for the joining procedure of ground-feeding birds, current researches recommend that the PS type is more credible than IS type [14]. The GSO is a heuristic algorithm and based on population which named the group [17]. In this theory, the members of the group are categorized as follows:

- Producer: The one which maintains the most suitable location in comparison to other ones and implemented by vision ability.
- Scroungers: Members which follow the producer to join it.
- Rangers: Rangers perform random walking in the search space.

The producer examines the nearness of its existing location to find the best one. This searching technique is named vision ability and relates to a process of testing some specific point in the vicinity of the producer member. Scroungers define their searching path with respect to the producer member. Eventually, scrounger members try to be in the closest location relative to the producer member. Moreover, in order to avoid getting stuck in local minima, rangers are committed as random walkers to perform random search. The continuous, binary and integer searching spaces have different features. Therefore, the mathematical formulation for modeling the behavior of GSO group members should be different in each space. In the following sections, appropriate mathematical formulation for modeling GSO group members in continuous and binary searching space, are presented. The population of GSO algorithm is called a group and each individual in the population is called a member. In n-dimensional search space, the ith member in kth searching iteration has a current position $X_i^k \in R^n$ and a head angle $\varphi_i^k = \left(\varphi_{i1}^k, \ldots, \varphi_{i(n-1)}^k\right) \in R^{n-1}$. Where, R is the set of real numbers and φ_{ik} is polar angle of ith member relative to the kth dimension.

The search direction of ith member, $D_i^k(\varphi_i^k)$ could be calculated from φ_i^k via a polar to Cartesian coordinate transformation [18]:

$$d_{i1}^k = \prod_{q=1}^{n-1} \cos(\varphi_{iq}^k) \tag{1}$$

$$d_{ij}^k = \sin\left(\varphi_{i(j-1)}^k\right) \prod_{q=j}^{n-1} \cos\left(\varphi_{iq}^k\right) \quad j = 2, \ldots, n-1 \tag{2}$$

$$d_{in}^k = \sin\left(\varphi_{i(n-1)}^k\right) \tag{3}$$

$$D_i^k(\varphi_i^k) = \left(d_{i1}^k, d_{i2}^k, \ldots, d_{in}^k\right) \tag{4}$$

Scroungers define their searching path with respect to the location of producer member and try to reach the producer member as much as possible. To avoid getting stuck in local minima in the optimization process, rangers are deployed to perform random walk in the search space.

At the k-th iteration the producer X_p behaves as follows:

1. The producer will scan at zero degree and then scan laterally by randomly sampling three points in the scanning field as follow:

$$X_z = X_p^k + r_1 l_{\max} D_p^k(\varphi^k) \tag{5}$$

$$X_r = X_p^k + r_1 l_{\max} D_p^k(\varphi^k + r_2\theta_{\max}/2) \tag{6}$$

$$X_l = X_p^k + r_1 l_{\max} D_p^k(\varphi^k - r_2\theta_{\max}/2) \tag{7}$$

where, $r_1 \in R^1$ is a normally distributed random number with zero mean and one standard deviation and $r_2 \in R^{n-1}$ is a uniformly distributed random number sequence in the range of (0, 1).

2. Once the zero, right and left points are defined, these points should be evaluated. The producer will then find the new point. If the new point has a better value in comparison with its current position, producer flies to the new point. If not, it will stay in its current position and turn its head using Eq. (8)

$$\varphi^{k+1} = \varphi^k + r_2 a_{\max} \tag{8}$$

where, $a_{\max} \in R^1$ is the maximum turning angle.

If the producer cannot find a better area after a iterations, it will turn its head back to zero degree as follows:

$$\varphi^{k+a} = \varphi^{k} \tag{9}$$

where, a is a constant.

During each searching bout, a number of group members are selected as scroungers. The scroungers will keep searching for opportunities to join the resources found by the producer. At the kth iteration, the area copying behavior of the ith scrounger can be modeled as a random walk toward the producer.

$$X_i^{k+1} = X_i^k + r_3 \circ (X_p^k - X_i^k) \tag{10}$$

where, $r_3 \in R^n$ is a uniform random sequence in the range (0, 1). Operator "$\circ$" is the Hadamard product or the Schur product, which calculates the entry-wise product of two vectors. During scrounging, the ith scrounger will keep searching for other opportunities to join the producer. The rest of the group members will be dispersed from their current positions. Random walks, which are thought to be the most efficient searching method for randomly distributed resources are employed by the rangers. At the kth iteration, a ranger generates a random head angle φ_i using (8), and then it chooses a random distance from (11) and moves to the new point using (12).

$$l_i = ar_1 l_{\max} \tag{11}$$

$$X_i^{k+1} = X_i^k + l_i D_i^k(\varphi^{k+1}) \tag{12}$$

More details on GSO could be found in [19].

2.2 *Binary Group Search Optimization (BGSO)*

2.2.1 Producer

In binary searching space, all the members of GSO group are either 0 or 1. According to (5–7), the zero, right and left points are defined using the term $r_1 l_{\max}$ as a random length. To simulate the producer searching ability, a random part of producer array should be chosen. In BGSO, random length selection is performed using two random pointers. The sub array between the two pointers is the search space that producing process could be performed on the selected sub array. In Fig. 1, the random length selection for the producer member is illustrated. According to Fig. 1 for $(1 \times n)$ array, r_4 is a random pointer in the range of (1, n) and r_5 is also a random pointer in the range (r_4, n).

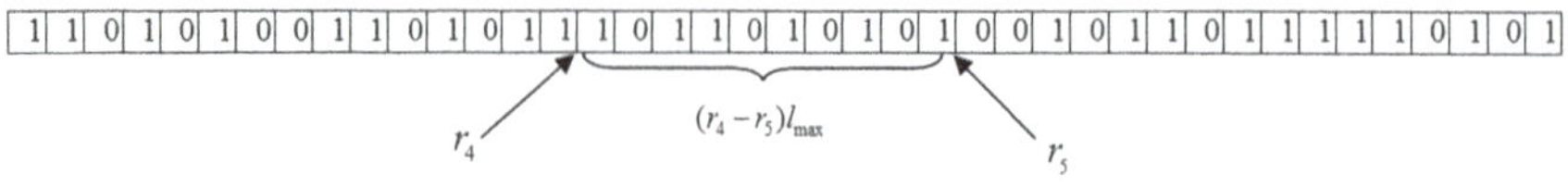

Fig. 1 Searching process of producer in binary space

Fig. 2 **a** Choosing three new test points for sample sub-array, **b** Producing at second step of head angle

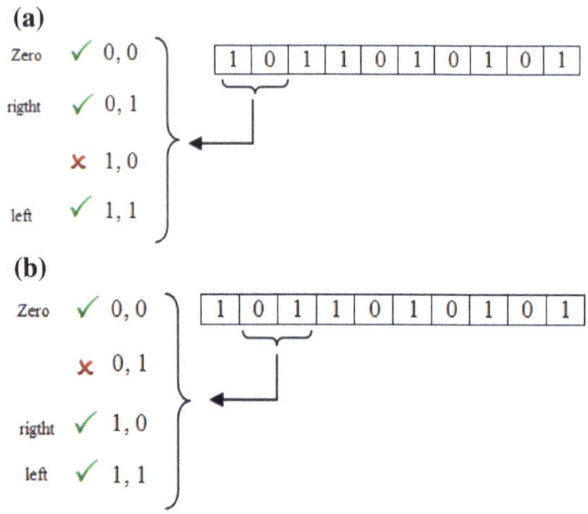

After simulating the random length, zero, right and left points should be discriminated similar to (5–7). For the selected sub array, each two sequential columns could be defined as one step of head angle revising procedure. Since there are 2 binary variables in each step, four states are possible. One of them is the initial condition and other three states should be checked by the producer member as zero, right and left points. Each of these test points forms a new sub-array and the new sub-array should be inserted to the conventional producer array. Therefore, (5–7) are simulated by generating three new arrays. If the generated new sub-array has better fitness value in comparison with that of the older producer, the new array is chosen as a producer member. On the contrary, if better solution is not achieved, producer should change its head angle and performs producing. The scanning process for a sample sub-array along with producing action at second step of head angle are depicted in Fig. 2a and b, respectively.

The source code associated with producer at binary search space is developed as follows:

```
%% Producer
    producer=pupMAT(1,:);
    k=1;
    for k=1:teta-1
        PMAT=producer(1,length(Sec)+1:end);
        testmat=PMAT(k:k+1);
        counter=0;
        stateMAT=[0 0;0 1;1 0;1 1];
        for kk=1:4
            if (stateMAT(kk,1)==testmat(1,1) &&
stateMAT(kk,2)==testmat(1,2))
            else
                counter=counter+1;
                newPMAT(counter,:)=stateMAT(kk,:);
            end
        end
        PMATr=PMAT;
        PMATr(k:k+1)=newPMAT(1,:);
        PMATz=PMAT;
        PMATz(k:k+1)=newPMAT(2,:);
        PMATl=PMAT;
        PMATl(k:k+1)=newPMAT(3,:);
        Xr=[producer(1,1:length(Sec)),PMATr];
        Xz=[producer(1,1:length(Sec)),PMATz];
        Psign=0;
        Pcount=0;
        while (Psign~=1 && Pcount<10)
            producer2=producer(1,1:length(Sec));
            OsecP=ceil(brch.*rand+1);
            if (producer2(1,OsecP)==0)
                OsecP=ceil(brch.*rand+1);
            end
            if (producer2(1,OsecP)==0)
                OsecP=ceil(brch.*rand+1);
            end
            if (producer2(1,OsecP)==0)
                OsecP=ceil(brch.*rand+1);
            end
            producer2(1,OsecP)=0;
            Xl=[producer2,PMATl];
            InMAT=Xl;
            [newlinedata]=NewDataMake(InMAT,linedata);
            [radial]=radialChek(newlinedata,nbus);
           if (radial == 1)
```

```
                [Tloss V
IL]=powerflow(newlinedata,busdata,nbus,Vbase);
                Psign=1;
                if (Tloss < LossMAT(1,1))
                    Psign=1;
                    pupMAT(1,:)=Xl;
                    LossMAT(1,1)=Tloss;
                    producer=Xl;
                end
            end
            Pcount=Pcount+1;
        end
        InMAT=Xz;
        [newlinedata]=NewDataMake(InMAT,linedata);
        [radial]=radialChek(newlinedata,nbus);
        if (radial == 1)
            [Tloss V
IL]=powerflow(newlinedata,busdata,nbus,Vbase);
            if (Tloss < LossMAT(1,1))
                pupMAT(1,:)=Xz;
                LossMAT(1,1)=Tloss;
                producer=Xz;
            end
        end
        InMAT=Xr;
        [newlinedata]=NewDataMake(InMAT,linedata);
        [radial]=radialChek(newlinedata,nbus);
        if (radial == 1)
            [Tloss V
IL]=powerflow(newlinedata,busdata,nbus,Vbase);
            if (Tloss < LossMAT(1,1))
                pupMAT(1,:)=Xr;
                LossMAT(1,1)=Tloss;
                producer=Xr;
            end
        end
    end
    [LossMAT index]=sort(LossMAT);
    pupMAT=pupMAT(index,:);
    Fin=[Fin LossMAT(1,1)];
```

2.2.2 Scrounger

In order to simulate area copying behavior of scroungers, the position of *i*th scrounger should be subtracted from producer member position. However, for the binary array,

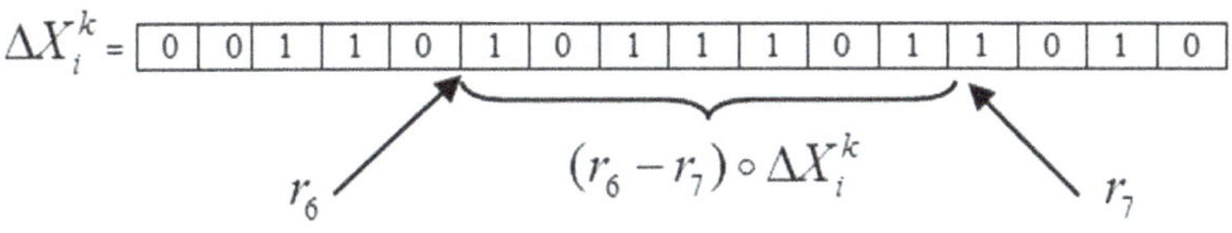

Fig. 3 The scrounging process in the binary space

this function is not possible. Hence, the *XOR* function is proposed, which can be efficient in binary search space. In binary space, random length selection is performed using two random pointers and the sub array between the two pointers is considered as the search space. Scrounging at binary space can be formulated as follow:

$$\Delta X_i^k = XOR(X_p^k, X_i^k) \tag{13}$$

where, X_p^k is producer member position at kth iteration and X_i^k is ith scrounger member at kth iteration. After computing the term ΔX_i^k, a random length using pointers r_6 and r_7 should be selected (see Fig. 3). In Fig. 3, for $(1 \times n)$ array, r_6 is a random number in the range of (1, n) and r_7 is another random number in the range of (r_6, n). Components of the sub-array which are equal to 1, represent the difference. Therefore, the state of corresponding components in the X_i^k should be changed. Doing so, a new scrounger member is generated and scrounging process is realized.

The source code associated with scrounger at binary search space is developed as follows:

```
%% Scroungers
ScrPercent=0.4;
ScrSize=round(ScrPercent*size(pupMAT,1));
Randpop=randperm(size(pupMAT,1));
pro=find(Randpop==1);
Randpop(pro)=[];
Xp=pupMAT(1,:);
for k=1:ScrSize
    ScrMAT(k,:)=pupMAT(Randpop(k),:);
end
Scounter=0;
for k=1:ScrSize
    Xsc=ScrMAT(k,:);
    DEl=xor(Xp,Xsc);
    r3=ceil((brch/2).*rand+1);
    r4=r3+ceil((brch/2).*rand+1);
    m=Xsc;
```

```
    for kk=r3:r4
       if (DEl(kk)==1)
          m(kk)=Xp(kk);
       end
    end
    radial=0;
    [newlinedata]=NewDataMake(m,linedata);
    [radial]=radialChek(newlinedata,nbus);
    mm=m;
    kkk=0;
    while (radial~=1 && kkk<length(m))
        mm=m;
        kkk=kkk+1;
        if (mm(kkk)==0)
            mm(kkk)=1;
            [newlinedata]=NewDataMake(m,linedata);
            [radial]=radialChek(newlinedata,nbus);
        end
    end
    if (radial==1)
        Scounter=Scounter+1;
        ScrPop(Scounter,:)=mm(1,:);
    end
end
```

2.2.3 Ranger

In (12), ranging process is performed using a random length and head angle. Similar to the producer and scroungers, random length is accomplished by using two random pointers for binary ranger members. First, a random length using pointers r_8 and r_9 should be generated in which, r_8 is a random number in the range of (1, n) and r_9 is a random number in the range of (r_9, n). Next, the random direction is generated as:

$$\Delta X_i = rand\,\mathrm{int}\,(1, l) \tag{14}$$

where, ΔX_i is the selected sub-array using r_8 and. r_9 $rand\mathrm{int}\,(1, x)$ is an operator which provides a random binary array with length of x, and l is the generated random length. The source code associated with ranger at binary search space is developed as follows:

```
%% Ranger
RangPercent=0.6;
RangSize=round(RangPercent*popsize)-1;
RangCounter=0;
for k=ScrSize+1:length(Randpop)
    RangCounter=RangCounter+1;
    RangMAT(RangCounter,:)=pupMAT(Randpop(k),:);
end
RangRand=randint(RangSize,length(tie));
Rcounter=0;
for k=1:RangSize
    RangMAT(k,length(Sec)+1:end)=RangRand(k,:);
    Zcounter=0;
    for kk=1:length(tie)
        if (RangMAT(k,kk)==1)
            Zcounter=Zcounter+1;
        end
    end
    Ycounter=0;
    for kk=1:length(Sec)
        if (RangMAT(k,kk)==0)
            Ycounter=Ycounter+1;
        end
    end
    if (Zcounter>Ycounter)
        radial=0;
        xo=0;
        while (radial~=1 && xo<25)
            RangMAT2=RangMAT;
            for kk=1:(Zcounter-Ycounter)
                uu(kk)=ceil(brch.*rand+1);
            end
            RangMAT2(k,uu)=0;
            InMAT=RangMAT2(k,:);
            [newlinedata]=NewDataMake(InMAT,linedata);
            [radial]=radialChek(newlinedata,nbus);
            xo=xo+1;
        end
        if (radial==1)
            Rcounter=Rcounter+1;
            Rangpop(Rcounter,:)=RangMAT2(k,:);
        end
    end
end
```

3 Problem Formulation

The reconfiguration of distribution network is the process of changing the topology of distribution systems by altering the open/closed status of tie and sectionalizing switches. This maneuver is efficient to fulfill operational requirements such as minimization of system total loss. In this chapter, the objective is to minimize total system active power loss while satisfying load flow and other operating constraints of distribution network. The problem could be formulated as follows:

$$Min : Z = \sum_{i=1}^{L} r_i \frac{P_i^2 + Q_i^2}{V_i^2} \tag{15}$$

where, Z is the objective function, r_i is the resistance of ith branch and L is the total number of branches. The P_i and Q_i are active and reactive power at the end bus of ith branch, respectively.

Moreover, the power flow analysis should be derived. For each configuration, the power flow analysis should be carried out to compute the nodal voltage, system total loss and current of each branch. It is clear that, for the proposed configuration, the computed voltages and currents should be in their premising range. Additionally, the proposed configuration should be a radial network and all load points should be supplied.

Radial Check: Distribution system should operate in radial configuration and all loads should be supplied. These constraints are considered using Kirchhoff algebraic method based on bus incidence matrix [20]. For the proposed configuration, the incidence matrix, $\hat{\mathbf{A}}$, should be set up as follow:

$$a_{ij} = 0 \quad \text{if branch } i \text{ is not conneced to node } j \tag{16}$$

$$a_{ij} = -1 \quad \text{if branch } i \text{ is directed toward node } j \tag{17}$$

$$a_{ij} = 1 \quad \text{if branch } i \text{ is directed away node } j \tag{18}$$

After setting the incidence matrix up, the reference node should be eliminated. Afterwards, the reference node should be eliminated. The obtained sub-matrix is called $\mathbf{A}$. If $|\det(\mathbf{A})| = 1$, the system is radial and all loads are being supplied. The source code associated with radial check is developed as follows:

```
%%radialChek
function [radial]=radialChek(newlinedata,nbus)
B(:,1)=newlinedata(:,2);
B(:,2)=newlinedata(:,3);
nedge = size(B,1);
A = zeros(nedge,nbus);
for i=1:nedge,
  A(i,B(i,1)) = 1;
  A(i,B(i,2)) = -1;
end



sizeA=size(A);
sutun=sizeA(2);
for k=2:sutun
    AA(:,(k-1))=A(:,k);
end
AAsize=size(AA);
if (AAsize(1)~=AAsize(2))
    radial=0;
    return
end
if (AAsize(1)==AAsize(2))
    if (abs(det(AA))==1)
       radial=1;
        return
    elseif (det(AA)==0)
        radial=0;
        return
    end
end
```

The source code associated with power flow is developed as follows:

```
%%powerflow
function [Tloss V
IL]=powerflow(newlinedata,busdata,nbus,Vbase);
linedata=newlinedata;
for k=1:nbus
    P(k)=busdata(k,2);
    Q(k)=busdata(k,3);
end
V=zeros(1,nbus);
V(1,:)=1;
delta=100;
A=linedata(:,2:3);
A(:,3)=A(:,2);
A(:,4)=A(:,1);
A(:,5)=linedata(:,4);
A(:,6)=linedata(:,5);
n=0;
m=A(1,1);
while (length(m)<nbus)
    z=m;
   for u=z
for k=1:length(A)
    if (A(k,1)==u && A(k,1)~=0)
        n=n+1;
        B(n,1:2)=A(k,1:2);
        B(n,3:4)=A(k,5:6);
        f=A(k,2);
        m=[m f];
        A(k,:)=0;
    end
    if (A(k,3)==u && A(k,1)~=0)
        n=n+1;
        B(n,1:2)=A(k,3:4);
        B(n,3:4)=A(k,5:6);
        f=A(k,4);
        m=[m f];
        A(k,:)=0;
    end
end
   end
end
for k=1:length(B)
    num(k,1)=k;
end
```

```
BB=[num B];
sizeLinedata=size(BB);
bran=sizeLinedata(1,1);
pathMAT=zeros(bran,nbus);
for k=1:bran
    pathMAT(:,BB(k,3))=pathMAT(:,BB(k,2));
    pathMAT(k,BB(k,3))=1;
end
linedata=BB;
iter=0;
dd=0.1;
while (delta>=dd)
    iter=iter+1;
    for k=1:nbus
        Ibus(k,1)=(P(k)-j*Q(k))./(conj(V(k)));
    end
    IL=pathMAT*Ibus;
V2(1)=1;
for k=1:length(linedata)
   V2(linedata(k,3))=V(linedata(k,2))-
(((linedata(k,4)+j*linedata(k,5))*(IL(k))));
end
    IL=abs(IL);
    for k=1:length(linedata)
        Loss(k)=((linedata(k,4))).*((IL(k)).^2);
    end
    Tloss=sum(Loss);
delta=max(abs(V2-V));
V=V2;
end

    Ibase=(100000)./((sqrt(3)).*Vbase);
IL=IL.*Ibase;
Tloss=Tloss.*100000;
end
```

4 Developed Source Code

In this section, the BGSO algorithm is comprehensively described based on MATLAB codes. The simulation comparisons are given by utilizing applying the BGSO algorithm to the 69-node and 119-node networks. The BGSO based methodology is produced by MATLAB 7.6 in 4 GHz, i7, pc. The outset amount of power flow division is 0.008. In this chapter, the highest amount of iterations is fixed to 200 and the community capacity is 20. The population size has direct bearings on the solution as well as computational burden of optimization process. If the population is assumed too large, slight better solution might be found while sacrificing computation efficiency of the optimization algorithm. On the contrary if the population is considered too small, a solution far from the optimal solution might be attained. Thus, a compromised decision should be made on the population size. In this chapter, the population size is attained based on a trial-and-error process in the simulation studies. The same discussion are valid regarding the value of maximum iterations number is identified based on a trial-and-error process in the simulation studies. Moreover, the upper and lower voltage ranges are assumed to be 1 p.u. and 0.9 p.u., respectively. In the following a source code which expresses step by step implementation of BGSO method to optimal network reconfiguration problem is provided.

```
clear
clc
linedata=xlsread('Blinedata.xlsx');
busdata=xlsread('Bbusdata.xlsx');
linedata(:,4)=linedata(:,4)./(1.602756);
linedata(:,5)=linedata(:,5)./(1.602756);
busdata(:,2)=busdata(:,2)./(100000);
busdata(:,3)=busdata(:,3)./(100000);
nbus=length(busdata);
Vbase=12.66; % In kv
Kleng=length(linedata);
secttionlizer=[1,1,1,1,1,1,1,1,1,1,1,1,1,1,1,1,1,1,1,1,
1,1,1,1,1,1,1,1,1,1,1,1;];
tie=[0 0 0 0 0];
baseTop=[sectionlizer,tie];
InMAT=baseTop;
[newlinedata]=NewDataMake(InMAT,linedata);
[radial]=radialChek(newlinedata,nbus);
[Tloss V IL]=powerflow(newlinedata,busdata,nbus,Vbase);
baseTLoss=Tloss;
Fin=[baseTLoss];
Open=length(tie);
brch=nbus-2;
%% Making Population
pup=0;
Osave=0;
popsize=15;
lim=round(popsize/Open);
for k=1:length(tie)
    npup=0;
```

```
    while (npup<lim)
        Sec=sectionlizer;
        Ti=tie;
        Ti(k)=1;
        Osec=ceil(brch.*rand+1);
        if (Osec==Osave)
            Osec=ceil(brch.*rand+1);
        end
        Osave=Osec;
        Sec(Osec)=0;
        InMAT=[Sec,Ti];
        [newlinedata]=NewDataMake(InMAT,linedata);
        [radial]=radialChek(newlinedata,nbus);
        if (radial==1)
            pup=pup+1;
            npup=npup+1;
            pupMAT(pup,:)=InMAT(1,:);
        end
    end
end
%% Power flow
for k=1:popsize
    InMAT=pupMAT(k,:);

    [newlinedata]=NewDataMake(InMAT,linedata);
    [Tloss V
IL]=powerflow(newlinedata,busdata,nbus,Vbase);
    LossMAT(k,1)=Tloss;
end
[LossMAT index]=sort(LossMAT);
pupMAT=pupMAT(index,:);
Fin=[Fin LossMAT(1,1)];
iter=0;
teta=length(Ti);
maxiter=50;
tic
while (iter<maxiter)

    iter=iter+1;
```

The source code associated with power flow is developed as the code in Sect. 3 and then, the source code associated with producer at binary search space is developed as the code presented in Sect. 2.2.1. The source code associated with scrounger at binary search space is developed as presented in Sect. 2.2.2. Finally, the source code associated with ranger at binary search space is developed as Sect. 2.2.3

```
%% Combine
for k=1:size(ScrPop,1)
    [newlinedata]=NewDataMake(ScrPop(k,:),linedata);
    [Tloss V
L]=powerflow(newlinedata,busdata,nbus,Vbas);
    SCLossMAT(k,1)=Tloss;
end
for k=1:size(Rangpop,1)
    [newlinedata]=NewDataMake(Rangpop(k,:),linedata);
    [Tloss V
IL]=powerflow(newlinedata,busdata,nbus,Vbase);
    RANGLossMAT(k,1)=Tloss;
end
pupMAT22=[pupMAT;ScrPop;Rangpop];
LossMAT22=[LossMAT;SCLossMAT;RANGLossMAT];
[LossMAT22 index]=sort(LossMAT22);
pupMAT22=pupMAT22(index,:);
pupMAT=pupMAT22(1:popsize,:);
LossMAT=LossMAT22(1:popsize,:);
Fin=[Fin LossMAT(1,1)];
plot(Fin,'r','Linewidth',2.5)
title(['Total Loss = ',num2str(min(Fin))]);
grid on
xlabel('Iteration')
ylabel('Loss [Kw]')
pause(0.0001)
end
%%NewDataMake
function [newlinedata]=NewDataMake(InMAT,linedata)
n=0;
for k=1:length(linedata)
    if (InMAT(k)==1)
        n=n+1;
        newlinedata(n,:)=linedata(k,:);
    end
end
for k=1:length(newlinedata)
nline(k,1)=k;
end
newlinedata(:,1)=nline;
end
```

The source code associated with radial check is developed as the source code in Sect. 3.

Table 1 69-node system reconfiguration results

State	Power loss (kW)	V_{min}	Tie switches
Before reconfiguration	221.98 kW	0.9082	42-11, 13-21
			15-46, 50-59
			27-65
Before reconfiguration	96.82 kW	0.9395	42-11, 13-21
			13-14, 55-56
			61-62

Table 2 Comparative result for 69-node distribution system

Method		Power loss (kW)	V_{min}	V_{max}
HPSO	[2]	99.67	0.9427	1
FEBE	[5]	101.01	0.927	1
TS	[6]	103.86	0.948	1
DP	[5]	99.06	0.93	1
SPSO	[7]	99.59	0.943	1
BGSO		97.998	0.9365	1

5 Test Results

5.1 69-Node System

This system involves 73 branches and 69 nodes. There are 68 sectionalizing switches and 5 tie switches [18]. Table 1 displays the simulation outcomes. According to Table 2, the power loss is decreased by 58% of its primary value. Moreover, comparison outcomes are listed in Table 2.

As can be seen from Table 2 the BGSO algorithm has a more reliable appearance in comparison with other introduced algorithms. Furthermore, the voltage profile is presented in Fig. 4 which is related to after and before reconfiguration. It is obvious that after reconfiguration the voltage profile is developed. The computational for this test system is 9.32 s.

5.2 119-Node System

Here, the 119-node test standard system is examined. This system is an 11 kV distribution system which includes 118 and 15 sectionalizing and tie switches, respectively. The complete data is accessible in [35]. The primary whole active power failure of the system is 1294.3 kW. Table 3 displays the simulation and comparison out-

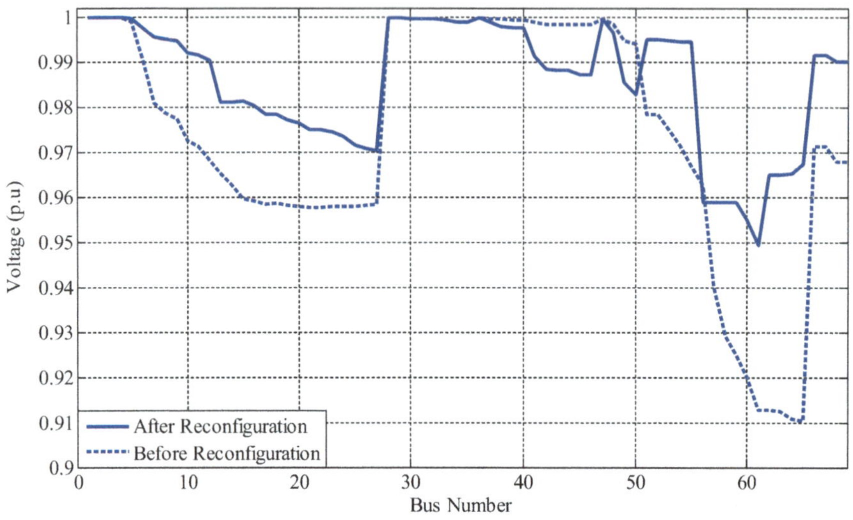

Fig. 4 Comparison of voltage profile for 69-node system before and after reconfiguration using BGSO method

comes which shows that after reconfiguration the active power failure is decreased by 37.71% of its primary value.

It can be understood from Table 3 that the BGSO algorithm has a more reliable appearance in comparison with other presented algorithms. Furthermore, in Fig. 5 the comparison of voltage profile, before and after reconfiguration, is displayed.

The minimum nodal voltage, before and after reconfiguration are 0.9825 and 0.9711, respectively. Additionally, the convergence characteristic of the BGSO method is depicted in Fig. 6. The computational for the 119-node test system is 18.34 s.

6 Conclusion

In this chapter, a BGSO method is introduced for determining the distribution network reconfiguration problem. The minimization of active power loss is the main purpose. A source code which expresses step by step implementation of BGSO method to optimal network reconfiguration problem is provided. The scientific outcomes confirm that the BGSO method is proficient of obtaining an optimal or near-optimal answer of two examined cases. Based on the whole simulation experience it could be assumed that the BGSO method looks to be a strong and secure binary optimization procedure.

Table 3 Comparative result for 119-node distribution system

	initial topology		Branch exchange [1]		TS [8]		ITS [21]		CLONR [22]		BGSO	
Tie lines	48	27	45	44	45	44	45	44	45	44	45	44
	17	27	17	27	27	26	27	26	25	26	25	26
	8	24	23	24	22	23	23	24	23	24	23	24
	56	45	53	52	54	53	54	53	52	53	52	53
	65	51	51	50	51	50	65	51	45	56	45	56
	38	65	64	65	64	65	61	62	61	62	61	62
	9	2	41	42	41	42	41	42	41	42	41	42
	61	100	61	100	61	100	95	100	95	100	94	100
	76	95	76	77	76	77	77	78	77	78	77	78
	91	78	74	75	74	75	74	75	74	75	74	75
	103	80	79	80	80	79	101	102	101	102	101	102
	113	86	85	86	85	86	86	113	86	86	86	113
	110	89	89	110	89	110	89	110	89	89	89	109
	115	123	114	115	114	115	114	115	114	111	111	112
	25	36	35	36	33	34	35	36	35	35	35	36
Power loss (kW)	1294.3		885.56		884.16		865.86		853.58		791.98	

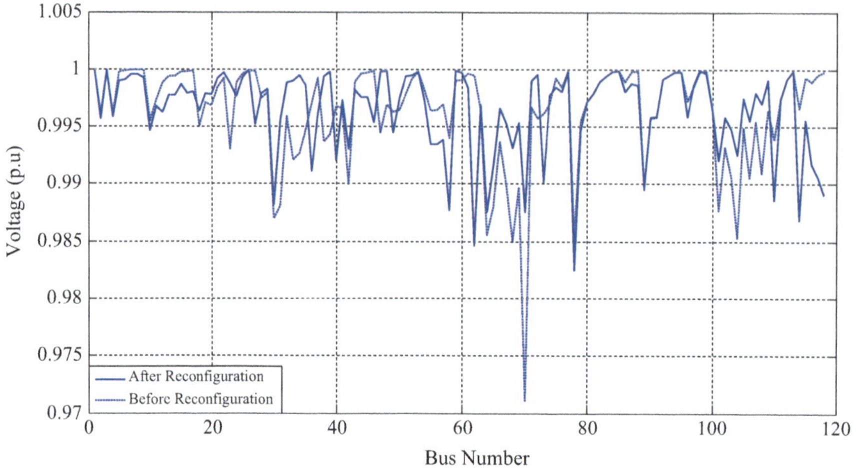

Fig. 5 Comparison of voltage profile for 119-node system before and after reconfiguration using BGSO method

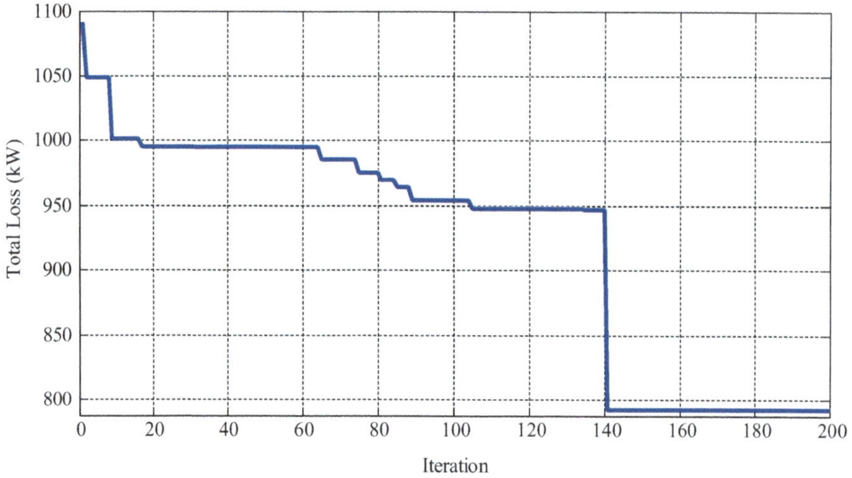

Fig. 6 The convergence curve of the BGSO method

MATLAB Code

MATLAB Codes are given within the chapter.

References

1. Baran, M.E., Wu, F.F.: Network reconfiguration in distribution systems for loss reduction and load balancing. IEEE Trans. Power Deliv. **4**, 1401–1407 (1989)
2. Li, Z., Chen, X., Yu, K., Sun, Y., Liu, H.: A hybrid particle swarm optimization approach for distribution network reconfiguration problem. In: Power and Energy Society General Meeting-Conversion and Delivery of Electrical Energy in the 21st Century, 2008 IEEE, 2008, pp. 1–7
3. Wagner, T., Chikhani, A., Hackam, R.: Feeder reconfiguration for loss reduction: an application of distribution automation. IEEE Trans. Power Deliv. **6**, 1922–1933 (1991)
4. Sarma, N., Prakasa Rao, K.: A new 0–1 integer programming method of feeder reconfiguration for loss minimization in distribution systems. Electr. Power Syst. Res. **33**, 125–131 (1995)
5. Liu, B.-S., Xie, K.-G., Zhou, J.-Q.: Electrical distribution networks reconfiguration using dynamic programming. Proc. Chin. Soc. Electr. Eng. **25**, 29 (2005)
6. Civanlar, S., Grainger, J., Yin, H., Lee, S.: Distribution feeder reconfiguration for loss reduction. IEEE Trans. Power Deliv. **3**, 1217–1223 (1988)
7. Das, D.: A fuzzy multiobjective approach for network reconfiguration of distribution systems. IEEE Trans. Power Deliv. **21**, 202–209 (2006)
8. Shirmohammadi, D., Hong, H.W.: Reconfiguration of electric distribution networks for resistive line losses reduction. IEEE Trans. Power Deliv. **4**, 1492–1498 (1989)
9. Ababei, C., Kavasseri, R.: Efficient network reconfiguration using minimum cost maximum flow-based branch exchanges and random walks-based loss estimations. IEEE Trans. Power Syst. **26**, 30–37 (2011)
10. Jeon, Y.-J., Kim, J.-C.: Application of simulated annealing and tabu search for loss minimization in distribution systems. Int. J. Electr. Power Energy Syst. **26**, 9–18 (2004)
11. Chiang, H.-D., Jean-Jumeau, R.: Optimal network reconfigurations in distribution systems. II. Solution algorithms and numerical results. IEEE Trans. Power Deliv. **5**, 1568–1574 (1990)
12. Jeon, Y.-J., Kim, J.-C., Kim, J.-O., Shin, J.-R., Lee, K.Y.: An efficient simulated annealing algorithm for network reconfiguration in large-scale distribution systems. IEEE Trans. Power Deliv. **17**, 1070–1078 (2002)
13. Nara, K., Shiose, A., Kitagawa, M., Ishihara, T.: Implementation of genetic algorithm for distribution systems loss minimum re-configuration. IEEE Trans. Power Syst. **7**, 1044–1051 (1992)
14. Su, C.-T., Chang, C.-F., Chiou, J.-P.: Distribution network reconfiguration for loss reduction by ant colony search algorithm. Electr. Power Syst. Res. **75**, 190–199 (2005)
15. Khalil, Gorpinich: Reconfiguration for loss reduction of distribution systems using selective particle swarm optimization. Int. J. Multidiscip. Sci. Eng. **3** (2012)
16. Wang, Zhang: Distribution network reconfiguration based on modified particle swarm optimization algorithm. In: International Conference on Machine Learning and Cybernetics, pp. 2076–2080 (2006)
17. Su, C.-T., Lee, C.-S.: Network reconfiguration of distribution systems using improved mixed-integer hybrid differential evolution. IEEE Trans. Power Deliv. **18**, 1022–1027 (2003)
18. Mustard, D.: Numerical integration over the. Math. Comput. **18**, 578–589 (1964)
19. He, S., Wu, Q.H., Saunders, J.: Group search optimizer: an optimization algorithm inspired by animal searching behavior. IEEE Trans. Evol. Comput. **13**, 973–990 (2009)
20. Mekhamer, S., Abdelaziz, A., Mohammed, F., Badr, M.: A new intelligent optimization technique for distribution systems reconfiguration. In: Power System Conference, 2008. MEPCON 2008. 12th International Middle-East, pp. 397–401 (2008)
21. Zhang, D., Fu, Z., Zhang, L.: An improved TS algorithm for loss-minimum reconfiguration in large-scale distribution systems. Electr. Power Syst. Res. **77**, 685–694 (2007)
22. Li, Z., Chen, X., Yu, K., Sun, Y., Liu, H.: A hybrid particle swarm optimization approach for distribution network reconfiguration problem. In: Power and Energy Society General Meeting-Conversion and Delivery of Electrical Energy in the 21st Century, 2008 IEEE, pp. 1–7 (2008)
23. Chiou, J.-P., Chang, C.-F., Su, C.-T.: Variable scaling hybrid differential evolution for solving network reconfiguration of distribution systems. IEEE Trans. Power Syst. **20**, 668–674 (2005)

24. Abdelaziz, A.Y., Mohamed, F., Mekhamer, S., Badr, M.: Distribution system reconfiguration using a modified Tabu Search algorithm. Electr. Power Syst. Res. **80**, 943–953 (2010)
25. Abdelaziz, A.Y., Mohammed, F., Mekhamer, S., Badr, M.: Distribution systems reconfiguration using a modified particle swarm optimization algorithm. Electr. Power Syst. Res. **79**, 1521–1530 (2009)
26. Kersting, Mendive, : An application of ladder theory to the solution of three-phase radial load-flow problem. IEEE Trans. Power Appar. Syst. **98**, 1060–1067 (1976)
27. Kersting, W.: A method to teach the design and operation of a distribution system. IEEE Trans. Power Appar.Us Syst., pp. 1945–1952 (1984)
28. Teng, J.-H.: A direct approach for distribution system load flow solutions. IEEE Trans. Power Deliv. **18**, 882–887 (2003)

Combined Heat and Power Economic Dispatch Using Particle Swarm Optimization

Farnaz Sohrabi, Farkhondeh Jabari, Pouya Pourghasem and Behnam Mohammadi-Ivatloo

Abstract Due to increased energy cost and limitations of fossil fuel energy sources, systems with higher efficiency such as combined heat and power (CHP) have become more popular. Optimal operation of the power system in the presence of CHP units which have non-linear and non-convex characteristics is getting more complicated. Difficulties of mentioned problem lead us to use heuristic and evolutionary methods. In this chapter, particle swarm optimization (PSO) is implemented in economic dispatch (ED) of CHP units. The main objective of ED problem is to obtain optimal output power and heat of each unit while the total generating cost is minimized and system operational constraints are satisfied. The results show the capability of this algorithm in solving CHP economic dispatch (CHPED) problem.

Keywords Combined heat and power · Economic dispatch · Particle swarm optimization

Nomenclature

$C_i(P_i^p)$	Operation cost of ith power-only unit for producing P_i^p MW
$C_j(P_j^c, H_j^c)$	Operation cost for jth co-generation unit for producing P_j^c MW electricity power and H_j^c MWth heat power

F. Sohrabi · F. Jabari · P. Pourghasem · B. Mohammadi-Ivatloo (✉)
Faculty of Electrical and Computer Engineering, University of Tabriz, Tabriz, Iran
e-mail: bmohammadi@tabrizu.ac.ir

F. Sohrabi
e-mail: farnaz_sohrabi95@ms.tabrizu.ac.ir

F. Jabari
e-mail: f.jabari@tabrizu.ac.ir

P. Pourghasem
e-mail: pouya.pourghasem@gmail.com

M. Pesaran Hajiabbas and B. Mohammadi-Ivatloo (eds.),
Optimization of Power System Problems, Studies in Systems, Decision and Control 262,
https://doi.org/10.1007/978-3-030-34050-6_6

$C_k(P_k^h)$	Operation cost of heat-only unit while producing H_k^h MWth heat power
N_p, N_c, N_h	Total number of power-only, CHP and heat-only units, respectively
i, j, k	Indices for power-only, CHP and heat-only units, respectively
$\alpha_i, \beta_i, \gamma_i$	Constant cost coefficients for ith power-only unit
$a_j, b_j, c_j, d_j, e_j, f_j$	Coefficients of cost function related to jth CHP unit
a_k, b_k, c_k	Coefficients for calculating the operation cost of heat-only units
P_d, H_d	Electrical and heat power demands
P_{loss}	Power system transmission loss
$P_i^{p\min}, P_i^{p\max}$	Lower and upper generation limits for power-only units, respectively
$P_j^{c\min}, H_j^{c\min}, P_j^{c\max}, H_j^{c\max}$	Minimum and maximum electric and heat powers outputs for CHP units, respectively
$H_k^{h\min}, H_k^{h\max}$	Limits for heat-only units
N	Total number of decision variables in the problem
ω	The inertia weight for PSO
r_1^n, r_2^n	Random numbers in the interval [0, 1]
$p_{best_{i,n}}^{iter-1}, g_{best,n}^{iter-1}$	Best position of ith particle in previous iteration and best position of entire swarm
C_1, C_2	Learning factors
$x_n^{\max}, x_n^{\min}$	Maximum and minimum limits of variables
r	Parameter to control the amount of change in velocity

1 Introduction

Lack of the conventional energy sources and high cost of energy production lead us to utilize power system in an optimized way. The optimal condition is achieved when the generation cost is minimized by considering system constraints. The purpose of economic dispatch (ED) is determining outputs of the units in the optimal condition. Hence, many authors investigated different methods to solve economic dispatch problem. One of the challenging optimization problems is the ED of CHP units due to non-linear and non-convex characteristics of these units and dual dependency between power and heat production.

2 Background

Recently, combined heat and power (CHP) units play an important role in producing energy because of their higher efficiency. The efficiency of CHP units is around 90% at best, while this amount is less than 60% for other combined cycle plants [1]. ED is applied to determine power and heat output of units while minimizing the operation cost. Heat-only, power-only and CHP units are three types of generating units in a cogeneration system. In combined heat and power economic dispatch (CHPED) problem, two types of demands are satisfied comprising power and heat demands. The generated power in CHP units depends on the produced heat and vice versa which makes the CHPED problem more complicated [2].

CHPED problem is solved using different mathematical and heuristic methods in the previous literature. In [3], CHPED problem is decomposed into two sub-problems that are connected using heat-power operation region constraints. Lagrangian method is used to solve optimization problem in this paper. Various heuristic methods are introduced and implemented into CHPED problem. Improved penalty function formulation for genetic algorithm (GA) is proposed in [4]. Firefly algorithm (FA) is applied in [5] which is inspired by behavior of fireflies in attracting each other using their luminosity. Group search optimization (GSO) as another heuristic optimization method that is based on searching behavior of animals, is implemented into CHPED [6].

In this chapter particle swarm optimization (PSO) is applied to solve CHPED problem. Results show that the implemented method is able to find optimum solution of the problem.

3 Problem Formulation

In CHPED problem, there are three types of units comprising power-only, CHP and heat-only units. The objective function of CHPED is to minimize the operation cost of the system while satisfying the constraints.

3.1 Objective Function

The formulation of CHPED problem is given in [7]. The objective function is defined as:

$$OF = \sum_{i=1}^{N_p} C_i(P_i^p) + \sum_{j=1}^{N_c} C_j(P_j^c, H_j^c) + \sum_{k=1}^{N_h} C_k(H_k^h) \quad (1)$$

where $C_i(P_i^p)$ is the operation cost of ith power-only unit for producing P_i^p MW. Operation cost for jth co-generation unit for producing P_j^c MW electricity power and H_j^c MWth heat power is denoted by $C_j(P_j^c, H_j^c)$. $C_k(P_k^h)$ is defined as the operation cost of heat-only unit while producing H_k^h MWth heat power. N_p, N_c and N_h are the total numbers of power-only, CHP and heat-only units, respectively. i, j and k are indices for above mentioned units, respectively. Cost functions of units are formulated as [8]:

$$C_i(P_i^p) = \alpha_i(P_i^p)^2 + \beta_i P_i^p + \gamma_i \quad (\$/h) \tag{2}$$

$$C_j(P_j^c, H_j^c) = a_j(P_j^c)^2 + b_j P_j^c + c_j + d_j(H_j^c)^2 + e_j H_j^c + f_j P_j^c H_j^c \quad (\$/h) \tag{3}$$

$$C_k(H_k^h) = a_k(H_k^h)^2 + b_k H_k^h + c_k \quad (\$/h) \tag{4}$$

In (2), α_i, β_i and γ_i are constant cost coefficients for ith power-only unit. In (3), a_j, b_j, c_j, d_j, e_j and f_j are coefficients of cost function related to jth CHP unit. a_k, b_k and c_k are coefficients for calculating the operation cost of kth heat-only units as mentioned in (4).

3.2 Constraints

The sum of produced power and heat should meet the power and heat demands, respectively [9]:

$$\sum_{i=1}^{N_p} P_i^p + \sum_{j=1}^{N_c} P_j^c = P_d + P_{loss} \tag{5}$$

$$\sum_{j=1}^{N_c} H_j^c + \sum_{k=1}^{N_h} H_k^h = H_d \tag{6}$$

where P_d and H_d are power and heat demands, respectively. P_{loss} is power system transmission loss. The produced electric and heat powers should be in the acceptable range for each unit:

$$P_i^{p\min} \le P_i^p \le P_i^{p\max} \quad i = 1, 2, 3, \ldots, N_p \tag{7}$$

$$P_j^{c\min}(H_j^c) \le P_j^c \le P_j^{c\max}(H_j^c) \quad j = 1, 2, 3, \ldots, N_c \tag{8}$$

$$H_j^{c\min}(P_j^c) \le H_j^c \le H_j^{c\max}(P_j^c) \quad j = 1, 2, 3, \ldots, N_c \tag{9}$$

$$H_k^{h\min} \le H_k^h \le H_k^{h\max} \quad k = 1, 2, 3, \ldots, N_h \tag{10}$$

where $P_i^{p\min}$ and $P_i^{p\max}$ are lower and upper generation limits for power-only units, respectively. $P_j^{c\min}$, $H_j^{c\min}$, $P_j^{c\max}$ and $H_j^{c\max}$ are minimum and maximum electric and heat powers outputs for CHP units, respectively. Also, the limits for heat-only units are denoted by $H_k^{h\min}$ and $H_k^{h\max}$.

3.3 Particle Swarm Optimization

PSO that has been introduced for the first time by Kennedy and Eberhart in 1995 [10], is an optimization algorithm based on swarm intelligence which is inspired by swarm behaviors of animals such as birds. PSO and PSO-based algorithms are widely used in the literature for solving different power systems problems [11–15]. The process of all evolutionary algorithms is based on producing random numbers and leading the numbers to those that have better fitness values according to goals of the problem. In PSO, positions of particles are considered as decision variables. Each particle i has a position vector X and a velocity vector V in each iteration number $iter$ that can be formulated as:

$$X_i^{iter} = [x_{i,1}^{iter}, x_{i,2}^{iter}, \ldots, x_{i,N}^{iter}] \tag{11}$$

$$V_i^{iter} = [v_{i,1}^{iter}, v_{i,2}^{iter}, \ldots, v_{i,N}^{iter}] \tag{12}$$

where N is the total number of decision variables in the problem. Each of particles in each iteration is willing to have better positions using its current velocity, its own experience of previous iterations and other particles' experiences. The mathematical formulations of this procedure are as follows:

$$v_{i,n}^{iter} = \omega \times v_{i,n}^{iter-1} + C_1 \times r_1^n \times (p_{best_{i,n}}^{iter-1} - x_{i,n}^{iter-1}) + C_2 \times r_2^n \times (g_{best,n}^{iter-1} - x_{i,n}^{iter-1}) \tag{13}$$

$$x_{i,n}^{iter} = x_{i,n}^{iter-1} + v_{i,n}^{iter} \tag{14}$$

where ω is the inertia weight, r_1^n and r_2^n are random numbers in the interval [0, 1]. $p_{best_{i,n}}^{iter-1}$ and $g_{best,n}^{iter-1}$ are the best positions of ith particle in previous iteration and the best position of entire swarm, respectively. C_1 and C_2 are learning factors that are usually equal. The updated velocities should be in a predefined range:

$$-v_n^{\max} \le v_{i,n} \le v_n^{\max} \tag{15}$$

$$v_n^{\max} = (x_n^{\max} - x_n^{\min})/r \tag{16}$$

where $x_n^{\max}$, $x_n^{\min}$ are the maximum and minimum limits of variables as mentioned in (7–10). r is a parameter to control the amount of change in velocity.

In this chapter, a simple and basic PSO algorithm is applied on CHPED problem in order to show the principles of evolutionary algorithms and investigate effectiveness of these methods in solving power system problems. The flowchart of implemented method is illustrated in Fig. 1.

4 Simulation, Results and Discussion

In this part of the chapter, a simple test system is taken from [2] and the discussed method is implemented into it. This test system consists of one power-only unit, two CHP units and one heat-only unit. The power system transmission losses are ignored in order to simplify the process. Power and heat demands are 200 MW and 115 MWth, respectively. The cost functions for power-only (17) and heat-only (18) units are considered to be linear:

$$C_1(P_1) = 50P_1\ ;\quad 0 \le P_1 \le 150 \tag{17}$$

$$C_4(H_4) = 23.4H_4\ ;\quad 0 \le H_4 \le 2695.2 \tag{18}$$

The cost coefficients for CHP units are provided in Table 1. Also, the heat-power feasible operation regions for co-generation units are depicted in Fig. 2 and Fig. 3.

In order to investigate the random nature of evolutionary algorithms, the proposed method is performed for 100 times and the variations of solutions are shown in Fig. 4. It should be noted that 40.41 s elapsed for 100 runs of the program. The best solution and variance of all solutions are reported in Table 2. Also, the convergence of method for the best solution is depicted in Fig. 5.

5 Conclusion

By extending the microgrid concept in power systems and increasing the number of distributed energy sources in power networks, the calculations are getting more complicated. Thus, evolutionary algorithms become more popular due to their higher speed and better results. In this chapter the combined heat and power economic dispatch problem, as a simple application of evolutionary algorithms in power systems, is solved using PSO in MATLAB software. Results show the performance and speed of this algorithm in solving non-linear CHPED problem. As future works, the dynamic and stochastic economic dispatch problems considering different objectives can be studied.

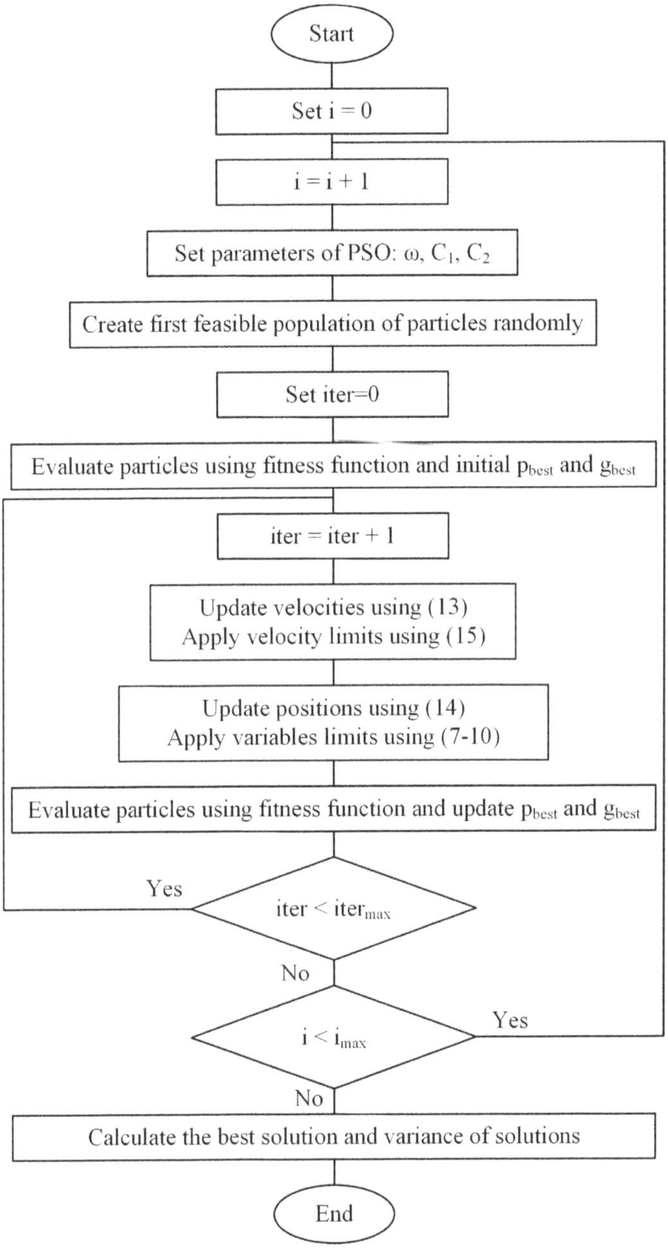

Fig. 1 Flowchart of implementing PSO into CHPED problem

Table 1 Cost coefficients for CHP units

Unit	*a*	*b*	*c*	*d*	*e*	*f*
2	0.0345	14.5	2650	0.03	4.2	0.031
3	0.0435	36	1250	0.027	0.6	0.011

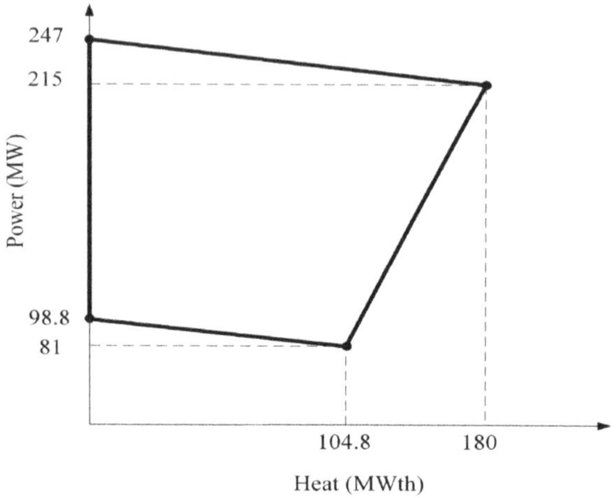

Fig. 2 Heat-power feasible operation region for unit 2

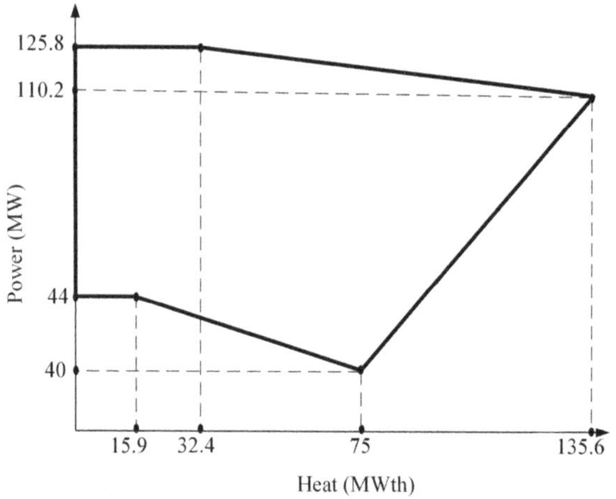

Fig. 3 Heat-power feasible operation region for unit 3

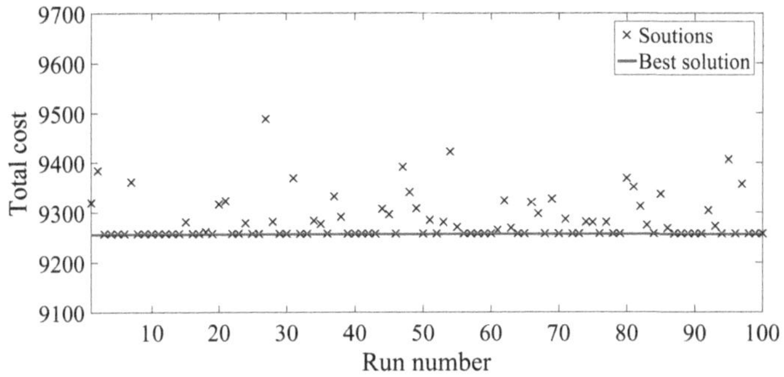

Fig. 4 Variations of solutions for different runs of program

Table 2 Best solution and variance of all solutions

Outputs	
P_1 (MW)	0
P_2 (MW)	160
P_3 (MW)	40
H_1 (MWth)	40
H_2 (MWth)	75
H_3 (MWth)	0
Total cost ($)	9257.07
Variance of solutions	1921.9

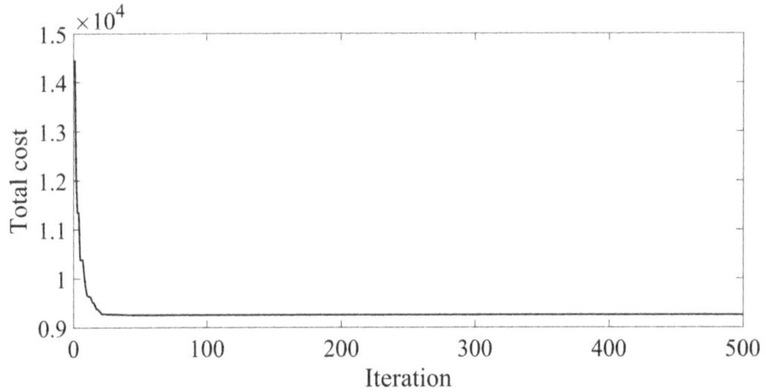

Fig. 5 Convergence of PSO for the best solution

MATLAB Codes

```
clc;
clear;
close all;

N=100;  %Number of iterations of the proposed method

%% Variables limits
bounds=[150     247     125.8   180     135.6   2695.2
        0       81      40      0       0       0];
%According to eq.(17), eq.(18), figure 2 and figure 3

%% PSO Parameters
particles=100;  %Number of particles
iterations=500; %Number of iterations for PSO
phi1=2.05;      %Adjustable parameter of PSO
phi2=2.05;      %Adjustable parameter of PSO
phi=phi1+phi2;  %Adjustable parameter of PSO
w=2/(phi-2+sqrt(phi^2-4*phi));  %Adjustable parameter
of PSO
c1=phi1*w;  %Adjustable parameter of PSO
c2=phi2*w;  %Adjustable parameter of PSO
r=10;       %Parameter for adjusting maximum variations
in positions

%% Main loop of the program
for z=1:N
    clc;
    display(z); %Display current iteration of the main
program

    %% PSO
    VarSize=size(bounds,2);                          %Ob-
taining number of variables
    x=zeros(particles,VarSize);
    v=zeros(particles,VarSize);
    VelocityMax=(bounds(1,:)-bounds(2,:))./r;    %Maxi-
mum variations in positions
    VelocityMin=-VelocityMax;                        %Mini-
mum variations in positions

    %% Initialize Positions
    for i = 1:particles
        for j = 1:VarSize
```

```
x(i,j)=unifrnd(bounds(2,j),bounds(1,j));     %Creating
initial positions for particles
        end
    end
    %% Evaluate First Particles
    for k=1:size(x,1)
        p1=x(k,1);  %Generated active power of the 1st
unit
        p2=x(k,2);  %Generated active power of the 2nd
unit
        p3=x(k,3);  %Generated active power of the 3rd
unit
        h2=x(k,4);  %Generated heat of the 2nd unit
        h3=x(k,5);  %Generated heat of the 3rd unit
        h4=x(k,6);  %Generated heat of the 4th unit
        F1=50*p1;   %Cost function of the 1st unit

F2=0.0345*(p2^2)+14.5*p2+2650+0.03*(h2^2)+4.2*h2+0.031*
p2*h2;   %Cost function of the 2nd unit

F3=0.0435*(p3^2)+36*p3+1250+0.027*(h3^2)+0.6*h3+0.011*p
3*h3;    %Cost function of the 3rd unit
        F4=23.4*h4;         %Cost function of the 4th
unit
        if (p1+p2+p3)<200   %According to eq.(5)
            pen1=4000;
        else
            pen1=0;
        end
        if (h2+h3+h4)<115   %According to eq.(6)
            pen2=4000;
        else
            pen2=0;
        end

        if (p2-247-((247-215)/(0-180))*(h2-0)>0)||(p2-
81-((215-81)/(180-104.8))*(h2-104.8)<0)||(p2-98.8-
((98.8-81)/(0-104.8))*(h2-0)<0) %According to figure 2
            pen3 = 4000;
        else
            pen3 = 0;
        end
        if (p3-110.2-((125.8-110.2)/(32.4-135.6))*(h3-
135.6)>0)||(p3-40-((40-110.2)/(75-135.6))*(h3-75)<0)
%According to figure 3
```

```
            pen4 = 4000;
        else
            pen4 = 0;
        end

        cost(k,1)=(F1+F2+F3+F4)+pen1+pen2+pen3+pen4;
%Fitness value of the particles
    end

    ParticlesBest.Cost=cost;              %Updating
ParticlesBest.Cost
    ParticlesBest.Position=x;             %Updating
ParticlesBest.Position
    [GlobalBest.Cost,index]=min(cost);    %Updating
GlobalBest.Cost by writing minimum cost of particles in
it
    GlobalBest.Position=x(index,:);       %Updating
GlobalBest.Position by writing positions with minimum
cost

    %% PSO Loop
    for iter = 1:iterations
        %% Update Velocity
        for i=1:particles

v(i,:)=w*v(i,:)+c1*rand(1,VarSize).*(ParticlesBest.Posi
tion(i,:)...
                                  -
x(i,:))+c2*rand(1,VarSize).*(GlobalBest.Position-
x(i,:));  %Calculating velocities

v(i,:)=min(max(v(i,:),VelocityMin(1,:)),VelocityMax(1,:
));      %Applying velocity constraints
        end
        x=x+v;  %Updating current positions

        %% Apply Constraints
        for i = 1:size(x, 1)

x(i,:)=min(max(x(i,:),bounds(2,:)),bounds(1,:));
%Applying variables limits
        end

        %% Evaluate Particles
        for k=1:size(x,1)
```

```
            p1=x(k,1);  %Generated active power of the
1st unit
            p2=x(k,2);  %Generated active power of the
2nd unit
            p3=x(k,3);  %Generated active power of the
3rd unit
            h2=x(k,4);  %Generated heat of the 2nd unit
            h3=x(k,5);  %Generated heat of the 3rd unit
            h4=x(k,6);  %Generated heat of the 4th unit
            F1=50*p1;   %Cost function of the 1st unit

F2=0.0345*(p2^2)+14.5*p2+2650+0.03*(h2^2)+4.2*h2+0.031*
p2*h2;    %Cost function of the 2nd unit

F3=0.0435*(p3^2)+36*p3+1250+0.027*(h3^2)+0.6*h3+0.011*p
3*h3;     %Cost function of the 3rd unit
            F4=23.4*h4;         %Cost function of the
4th unit
            if (p1+p2+p3)<200   %According to eq.(5)
                pen1=4000;
            else
                pen1=0;
            end
            if (h2+h3+h4)<115   %According to eq.(6)
                pen2=4000;
            else
                pen2=0;
            end

            if (p2-247-((247-215)/(0-180))*(h2-
0)>0)||(p2-81-((215-81)/(180-104.8))*(h2-
104.8)<0)||(p2-98.8-((98.8-81)/(0-104.8))*(h2-0)<0)
%According to figure 2
                pen3 = 4000;
            else
                pen3 = 0;
            end
            if (p3-110.2-((125.8-110.2)/(32.4-
135.6))*(h3-135.6)>0)||(p3-40-((40-110.2)/(75-
135.6))*(h3-75)<0)  %According to figure 3
                pen4 = 4000;
            else
                pen4 = 0;
            end
```

```
cost(k,1)=(F1+F2+F3+F4)+pen1+pen2+pen3+pen4; %Fitness
value of the particles
        end

        for i = 1:length(cost)
                if cost(i)<ParticlesBest.Cost(i)
%Comparing current fitness values with previous values
                ParticlesBest.Cost(i)=cost(i);
%updating ParticlesBest.Cost
                ParticlesBest.Position(i,:)=x(i, :);
%updating ParticlesBest.Position
                end
        end
        if min(cost)<GlobalBest.Cost
                [GlobalBest.Cost,index]=min(cost);
%updating GlobalBest.Cost
                GlobalBest.Position=x(index,:);
%Updating GlobalBest.Position
        end
        BEST_COST_iterations(z,iter) = GlobalBest.Cost;
%Updating BEST_COST_iterations
    end

    %% Results
    BestCost(z,1) = GlobalBest.Cost;            %Best
costs in each iteration of the main program
    BestParticles(z,:) = GlobalBest.Position;   %Best
Particles in each iteration of the main program
end

[optimal_cost index]=min(BestCost); %Searching for the
best solution

figure(1)   %Plotting figure 5
plot(BEST_COST_iterations(index,:));

figure(2)   %Plotting figure 4
plot(BestCost,'x');
hold on
plot(optimal_cost*ones(1,N));

%% variance
var(BestCost)   %Calculating variance using var func-
tion of the MATLAB
```

References

1. Shi, B., Yan, L.-X., Wu, W.: Multi-objective optimization for combined heat and power economic dispatch with power transmission loss and emission reduction. Energy **56**, 135–143 (2013)
2. Mohammadi-Ivatloo, B., Moradi-Dalvand, M., Rabiee, A.: Combined heat and power economic dispatch problem solution using particle swarm optimization with time varying acceleration coefficients. Electr. Power Syst. Res. **95**, 9–18 (2013)
3. Guo, T., Henwood, M.I., Van Ooijen, M.: An algorithm for combined heat and power economic dispatch. IEEE Trans. Power Syst. **11**, 1778–1784 (1996)
4. Song, Y., Xuan, Q.: Combined heat and power economic dispatch using genetic algorithm based penalty function method. Electr. Mach. Power Syst. **26**, 363–372 (1998)
5. Yazdani, A., Jayabarathi, T., Ramesh, V., Raghunathan, T.: Combined heat and power economic dispatch problem using firefly algorithm. Front. Energy **7**, 133 (2013)
6. Basu, M.: Group search optimization for combined heat and power economic dispatch. Int. J. Electr. Power Energy Syst. **78**, 138–147 (2016)
7. Narang, N., Sharma, E., Dhillon, J.: Combined heat and power economic dispatch using integrated civilized swarm optimization and Powell's pattern search method. Appl. Soft Comput. **52**, 190–202 (2017)
8. Davoodi, E., Zare, K., Babaei, E.: A GSO-based algorithm for combined heat and power dispatch problem with modified scrounger and ranger operators. Appl. Therm. Eng. **120**, 36–48 (2017)
9. Basu, M.: Combined heat and power economic dispatch using opposition-based group search optimization. Int. J. Electr. Power Energy Syst. **73**, 819–829 (2015)
10. Kennedy, J., Eberhart, R.: Particle swarm optimization. In: Proceedings of IEEE International Conference on Neural Networks IV (1995)
11. Gaing, Z.-L.: Particle swarm optimization to solving the economic dispatch considering the generator constraints. IEEE Trans. Power Syst. **18**, 1187–1195 (2003)
12. Park, J.-B., Lee, K.-S., Shin, J.-R., Lee, K.Y.: A particle swarm optimization for economic dispatch with nonsmooth cost functions. IEEE Trans. Power Syst. **20**, 34–42 (2005)
13. Abido, M.: Multiobjective particle swarm optimization for environmental/economic dispatch problem. Electr. Power Syst. Res. **79**, 1105–1113 (2009)
14. Yoshida, H., Kawata, K., Fukuyama, Y., Takayama, S., Nakanishi, Y.: A particle swarm optimization for reactive power and voltage control considering voltage security assessment. IEEE Trans. Power Syst. **15**, 1232–1239 (2000)
15. Abido, M.: Optimal power flow using particle swarm optimization. Int. J. Electr. Power Energy Syst. **24**, 563–571 (2002)

Combined Heat and Power Stochastic Dynamic Economic Dispatch Using Particle Swarm Optimization Considering Load and Wind Power Uncertainties

Pouya Pourghasem, Farnaz Sohrabi, Farkhondeh Jabari, Behnam Mohammadi-Ivatloo and Somayeh Asadi

Abstract Due to the increased cost of energy sources and related environmental problems, systems with higher efficiency such as combined heat and power (CHP) units are getting more popular. Renewable energy sources can be another alternative solution for the above mentioned problems. Scheduling of renewable-based systems are getting more complicated due to the intermittent behavior of these sources. In this chapter, a stochastic programming framework is utilized to model uncertainties in dynamic economic dispatch (DED) problem of CHP based systems integrating wind energy. Forecast errors of electrical load and wind power are assumed as the two sources of uncertainty. A heuristic method called particle swarm optimization (PSO) is used to attain optimal solution of the problem due to non-linearity, non-convexity, and complexity of the problem. The stochastic programming provides more comprehensive and realistic viewpoint about dispatch problem by considering a variety of most probable scenarios compared to a single scenario.

P. Pourghasem · F. Sohrabi · F. Jabari · B. Mohammadi-Ivatloo
Faculty of Electrical and Computer Engineering, University of Tabriz, Tabriz, Iran
e-mail: pouya.pourghasem@gmail.com

F. Sohrabi
e-mail: sohrabi.farnaz@gmail.com

F. Jabari
e-mail: f.jabari@tabrizu.ac.ir

B. Mohammadi-Ivatloo
e-mail: bmohammadi@tabrizu.ac.ir

S. Asadi (✉)
Department for Management of Science and Technology Development, Ton Duc Thang University, Ho Chi Minh City, Vietnam
e-mail: somayehasadi@tdtu.edu.vn

Faculty of Applied Sciences, Ton Duc Thang University, Ho Chi Minh City, Vietnam

M. Pesaran Hajiabbas and B. Mohammadi-Ivatloo (eds.),
Optimization of Power System Problems, Studies in Systems, Decision and Control 262,
https://doi.org/10.1007/978-3-030-34050-6_7

Nomenclature

$P_{d,t,s}$, $P_{w,t,s}$	Electrical load demand and output power of *w*th wind unit at time *t* in scenario *s*, respectively
$P_{d,t}^{forecasted}$, $P_{w,t}^{forecasted}$	Forecasted values for electrical load demand and output wind units at time *t*, respectively
$\Delta P_{d,t,s}$, $\Delta P_{w,t,s}$	Forecast errors related to electrical load demand and output power of wind unit *w* at time *t* in scenario *s*, respectively
N_s	Total number of scenarios
$B_{(\text{interval},t,s)}^{L}$, $B_{(\text{interval},t,s)}^{W}$	Binary parameters of intervals at time *t* in scenario *s* for electrical load demand and wind power, respectively
π_s	Probability of scenario *s*
$\alpha_{i,t}$, $\beta_{j,t}$	Probabilities of electrical load demand and wind power for intervals *i* and *j* at time *t*, respectively
$C_i(P_i^p)$	Operation cost of *i*th power-only unit for producing P_i^p MW
$C_j(P_j^c, H_j^c)$	Operation cost for *j*th co-generation unit for producing P_j^c MW electricity power and H_j^c MWth heat power
$C_k(P_k^h)$	Operation cost of heat-only unit while producing H_k^h MWth heat power
N_p, N_c, N_h	Total number of power-only, CHP and heat-only units, respectively
i, j, k	Indices for power-only, CHP and heat-only units, respectively
α_i, β_i, γ_i, λ_i, ρ_i	Constant cost coefficients for *i*th power-only unit
a_j, b_j, c_j, d_j, e_j, f_j	Coefficients of cost function related to *j*th CHP unit
a_k, b_k, c_k	Coefficients for calculating the operation cost of heat-only units
$P_{d,t,s}$	Electrical power demand at time *t* in scenario *s*
$H_{d,t}$	Heat power demand at time *t*
$P_i^{p,\min}$, $P_i^{p,\max}$	Lower and upper generation limits for power-only units, respectively
$P_j^{c,\min}$,$H_j^{c,\min}$, $P_j^{c,\max}$, $H_j^{c,\max}$	Minimum and maximum electric and heat powers outputs for CHP units, respectively
$H_k^{h,\min}$, $H_k^{h,\max}$	Limits for heat-only units
V^{CO}, V^{CI}, V^R	Cut-off, cut-in and rated speed of wind turbine
$P^{\max}$	Rated power of the wind turbine
V_t	Forecasted wind speed at time *t*
N	Total number of decision variables in the problem
ω	The inertia weight for PSO
r_1^n, r_2^n	Random numbers in the interval [0, 1]

$p_{best_{i,n}}^{iter-1}, g_{best,n}^{iter-1}$	Best position of ith particle in previous iteration and best position of entire swarm
C_1, C_2	Learning factors of PSO
$x_n^{\max}, x_n^{\min}$	Maximum and minimum limits of variables
r	Parameter to control the amount of change in velocity in PSO

1 Introduction

The energy crisis is getting more important due to the limited sources of fossil fuels and environmental problems. As a result, renewable energy sources and high-efficiency power production units, such as combined heat and power (CHP) units, are increasing significantly. The intermittent nature of the systems with renewable energy sources makes the scheduling problem more complicated. Also, the methods used for load forecasting are not able to predict the exact amount of load demand. Thus, the stochastic programming is applied to dynamic economic dispatch (DED) problem in order to deal with uncertainties related to electrical load demand and wind power generation. The application of the stochastic framework in economic dispatch (ED) problem provides more actual solutions for different possible scenarios.

2 Background

Combined heat and power (CHP) dispatch problem has been investigated in many previous studies. Different methods have been used to solve combined heat and power economic dispatch (CHPED). Due to non-linearity, non-convexity and complexity of CHPED problem, evolutionary and heuristic algorithms have been widely used to solve this problem in the literature. In [1], an optimization algorithm that is based on particle swarm optimization (PSO) and society civilization algorithm (SCA) is used to solve CHPED problem. The proposed method in this reference uses civilized swarm optimization (CSO) as a global search technique and Powell's pattern search (PPS) as a local search technique. Cuckoo optimization algorithm which is based on egg laying and lifestyle of a bird family is implemented into the CHPED problem considering valve-point effects in [2]. A modified version of the conventional cuckoo search algorithm (CSA) namely effective cuckoo search algorithm (ECSA) is proposed in [3] to find optimum outputs of the units in a CHP system. Modified group search optimizer (MGSO) algorithm that is another population-based method is proposed in [4] which is inspired by animal behavior.

Stochastic programming is implemented into different power system's issues. In [5], stochastic scheduling of a micro-grid comprised of CHP units, wind turbine, fuel cell, and energy storage devices in the presence of demand response programs (DRPs) in proposed. A multi-objective economic and environmental dynamic dispatch problem by consideration of forecasted load demand and wind power uncertainties is solved using a scenario-based method in [6]. An enhanced PSO method is used in [7] to solve multi-objective and stochastic dispatch problem and deal with load and wind uncertainties. The roulette wheel mechanism is applied in this reference for the scenario generation process.

In this chapter, the PSO algorithm is used to solve stochastic dynamic economic dispatch (SDED) problem. Valve point effects and non-convexity are taken into account to model the economic dispatch problem more precisely and PSO is used as an optimization algorithm. A scenario-based approach is selected to model uncertainties related to electrical load demand and wind power. Roulette wheel mechanism is applied in the scenario generation process. In order to reduce the number of generated scenarios, the backward scenario reduction method is utilized.

3 Uncertainty Modeling

In order to model uncertainties in the problem, scenario-based model is selected. Scenario generation and scenario reduction processes are discussed in the following.

3.1 Scenario Generation

The electrical load demand and wind power are assumed to be uncertain in this problem. In order to model uncertainties, the forecast errors related to load demand and wind power are considered as random variables with known probability density function (PDF). Roulette wheel mechanism is applied to generate scenarios [8]. The electrical load demand and wind power for each scenario s are obtained as follows:

$$P_{d,t,s} = P_{d,t}^{forecasted} + \Delta P_{d,t,s} \quad t = 1, \ldots, 24; s = 1, \ldots, N_s \tag{1}$$

$$P_{w,t,s} = P_{w,t}^{forecasted} + \Delta P_{w,t,s} \quad t = 1, \ldots, 24; s = 1, \ldots, N_s \tag{2}$$

where $P_{d,t,s}$ and $P_{w,t,s}$ are electrical load demand and output power of wind units at time t in scenario s, respectively. $P_{d,t}^{forecasted}$ and $P_{w,t}^{forecasted}$ are forecasted values for electrical load demand and output of wind units at time t, respectively. $\Delta P_{d,t,s}$ and $\Delta P_{w,t,s}$ are forecast errors related to electrical load demand and output power of wind units at time t in scenario s, respectively. N_s is a total number of scenarios.

To generate scenarios for load and wind power, the PDF of each random variable must be discretized as depicted in Fig. 1. As shown in this figure, seven intervals are considered that are centered on zero mean and width of each interval is equal

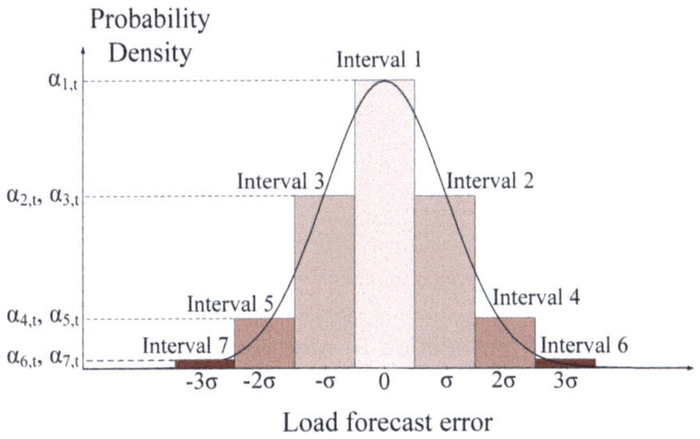

Fig. 1 Discretized probability density function of load forecast error

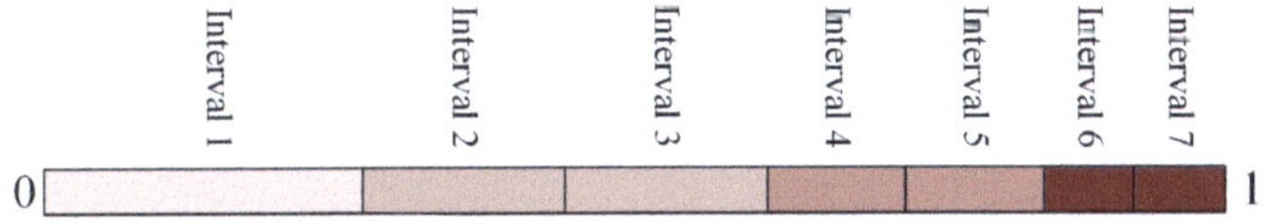

Fig. 2 Accumulated and normalized probabilities of intervals

to standard deviation of forecast error (σ) which is equal to 10% of the forecasted value. The probability of each interval at time t is denoted by $\alpha_{\text{interval},t}$ in this figure.

Afterward, the probabilities of intervals are normalized in such a way that the sum of probabilities becomes unity. In the next step, as shown in Fig. 2, the accumulated probabilities of intervals are calculated.

To create a scenario, a random number in the interval [0, 1] is generated and compared with accumulated probabilities of intervals beginning from the last interval. The first interval with accumulated probability equal or less than the random number is selected and the binary parameter related to this interval becomes equal to 1. So there is a binary vector for each scenario that shows the binary parameters of electrical load demand and wind power intervals as shown in (3).

$$\text{scenario (s)} = [B^L_{(1,t,s)}, \ldots, B^L_{(7,t,s)}, B^W_{(1,t,s)}, \ldots, B^W_{(7,t,s)}]_{t=1,\ldots,24} \tag{3}$$

where $B^L_{(\text{interval},t,s)}$ and $B^W_{(\text{interval},t,s)}$ are binary parameters of intervals at time t in scenario s for electrical load demand and wind power, respectively. The probability of each scenario is calculated using:

$$\pi_s = \frac{\prod_{t=1}^{24}\left(\sum_{i=1}^{7}(B^L_{i,t,s} \times \alpha_{i,t}) \times \sum_{j=1}^{7}(B^W_{j,t,s} \times \beta_{j,t})\right)}{\sum_{s=1}^{N_s}\left(\prod_{t=1}^{24}\left(\sum_{i=1}^{7}(B^L_{i,t,s} \times \alpha_{i,t}) \times \sum_{j=1}^{7}(B^W_{j,t,s} \times \beta_{j,t})\right)\right)} \tag{4}$$

where π_s is the probability of scenario s. $\alpha_{i,t}$ and $\beta_{j,t}$ are probabilities of electrical load demand and wind power for intervals i and j at time t, respectively that are extracted from Fig. 1.

3.2 Scenario Reduction

A large number of scenarios are needed to model the uncertainties more precisely. Thus, the computational time and needed memory will be increased significantly. In order to overcome these problems, the number of scenarios should be reduced after the scenario generation process using scenario reduction techniques. Different algorithms are proposed for scenario reduction purpose in the literature [6–11]. In this chapter, the fast forward algorithm is applied to reduce the number of generated scenarios [11]. Consider the original scenario set Ω with N_s scenarios and selected scenario set Ω_S and final selected scenario set Ω_S^* with N_s^* scenarios. The steps of the algorithm are as follows:

Step 1: Calculate the distance between each pair of scenarios using:

$$v(\omega, \omega') = \sum_{\text{load\&wind}} |P_\omega - P_{\omega'}| \tag{5}$$

Step 2: Calculate the average distance between each scenario and the other scenarios. The scenario with minimum distance is selected in this step as the first selected scenario and the selected scenario set Ω_S and non-selected scenario set Ω_J will be updated:

$$\omega_1 = \arg\{\min_{\omega' \in \Omega} \sum_{\omega \in \Omega} \pi_\omega v(\omega, \omega')\} \tag{6}$$

$$\Omega_S = \Omega_S \cup \{\omega_1\} \tag{7}$$

$$\Omega_J = \Omega \backslash \{\omega_1\} \tag{8}$$

Step i: Calculate the distance between each non-selected scenario and selected scenarios. The scenario that minimizes this distance is the next selected scenario and the selected and non-selected scenario sets will be updated:

$$\omega_i = \arg\{\min_{\omega' \in \Omega_J} \sum_{\omega \in \Omega_J \backslash \{\omega'\}} \pi_\omega \min_{\omega'' \in \Omega_S \cup \{\omega\}} v(\omega, \omega'')\} \tag{9}$$

$$\Omega_S = \Omega_S \cup \{\omega_i\} \tag{10}$$

$$\Omega_J = \Omega_J \backslash \{\omega_i\} \tag{11}$$

Last step: In the last step, the probability of each non-selected scenario is added to its closest selected scenario probability:

$$\pi_{\omega}^{*} = \pi_{\omega} + \sum_{\omega' \in J(\omega)} \pi_{\omega'} \tag{12}$$

where $J(\omega)$ is the closest non-selected scenario set to the selected scenario ω:

$$\omega = \arg\{\min_{\omega'' \in \Omega_S} v(\omega'', \omega')\} \ ; \omega' \in \Omega_J \tag{13}$$

4 Stochastic Dynamic Economic Dispatch

The main goal of the SDED problem is to find the optimal output of each generating unit at each hour and in each scenario while electrical and heat load demands are satisfied and other operational constraints are met. The studied system consists of power-only, CHP, heat-only units and wind turbines. The mathematical formulation of the SDED problem will be presented in this section.

4.1 Objective Function

The objective function of the SDED problem is defined as

$$OF = \sum_{s=1}^{N_s} \pi_s \sum_{t=1}^{24} \left(\sum_{i=1}^{N_p} C_i(P_{i,t,s}^{p}) + \sum_{j=1}^{N_c} C_j(P_{j,t,s}^{c}, H_{j,t,s}^{c}) + \sum_{k=1}^{N_h} C_k(H_{k,t,s}^{h}) \right) \tag{14}$$

where $C_i(P_i^p)$ is the generation cost of the ith power-only unit while producing P_i^p MW electric power. Generation cost for jth co-generation unit for producing P_j^c MW electrical power and H_j^c MWth heat power is shown by $C_j(P_j^c, H_j^c)$. $C_k(P_k^h)$ is the generation cost of the heat-only unit while producing H_k^h MWth heat power. N_p, N_c and N_h are the total numbers of power-only, CHP and heat-only units, respectively. i, j and k are indices for above-mentioned units, respectively. N_s is the total number of scenarios and s is the scenario number index. Cost functions of mentioned units are as follows [4]:

$$C_i(P_{i,t,s}^{p}) = \alpha_i (P_{i,t,s}^{p})^2 + \beta_i P_{i,t,s}^{p} + \gamma_i + \left| \lambda_i \sin(\rho_i (P_i^{p,\min} - P_{i,t,s}^{p})) \right| \ (\$/h) \tag{15}$$

$$C_j(P_{j,t,s}^{c}, H_{j,t,s}^{c}) = a_j (P_{j,t,s}^{c})^2 + b_j P_{j,t,s}^{c} + c_j + d_j (H_{j,t,s}^{c})^2$$

$$+ e_j H^c_{j,t,s} + f_j P^c_{j,t,s} H^c_{j,t,s} \quad (\$/h) \tag{16}$$

$$C_k(H^h_{k,t,s}) = a_k (H^h_{k,t,s})^2 + b_k H^h_{k,t,s} + c_k \quad (\$/h) \tag{17}$$

where α_i, β_i, γ_i, λ_i and ρ_i are cost coefficients of power-only units. a_j, b_j, c_j, d_j, e_j and f_j are cost coefficients for CHP units. a_k, b_k and c_k are coefficients of cost function related to heat-only units.

4.2 Constraints

The electrical and heat demands balance equations are the only equality constraints in the SDED problem [12]:

$$\sum_{i=1}^{N_p} P^p_{i,t,s} + \sum_{j=1}^{N_c} P^c_{j,t,s} = P_{d,t,s} \ \forall t, \forall s \tag{18}$$

$$\sum_{j=1}^{N_c} H^c_{j,t,s} + \sum_{k=1}^{N_h} H^h_{k,t,s} = H_{d,t} \quad \forall t, \forall s \tag{19}$$

where $P_{d,t,s}$ and $H_{d,t}$ are electrical and heat power demands at time t in scenario s, respectively. The production of each unit must be in the acceptable range as follows:

$$P_i^{p,\min} \leq P^p_{i,t,s} \leq P_i^{p,\max} \quad \forall i, \ \forall t, \forall s \tag{20}$$

$$P_j^{c,\min}(H^c_j) \leq P^c_{j,t,s} \leq P_j^{c,\max}(H^c_j) \quad \forall j, \ \forall t, \forall s \tag{21}$$

$$H_j^{c,\min}(P^c_j) \leq H^c_{j,t,s} \leq H_j^{c,\max}(P^c_j) \quad \forall j, \ \forall t, \forall s \tag{22}$$

$$H_k^{h,\min} \leq H^h_{k,t,s} \leq H_k^{h,\max} \quad \forall k, \ \forall t, \forall s \tag{23}$$

where $P_i^{p,\min}$ and $P_i^{p,\max}$ are minimum and maximum limits for output of power-only units, respectively. $P_j^{c,\min}$ and $P_j^{c,\max}$ are minimum and maximum electric power output for CHP units. $H_j^{c,\min}$ and $H_j^{c,\max}$ are the limits for the heat output of CHP units. The upper and lower limits for heat-only units are denoted by $H_k^{h,\min}$ and $H_k^{h,\max}$, respectively.

4.3 Wind Turbine Formulation

The generated power of a wind turbine is a function of current wind speed and characteristics of the wind turbine which can be formulated as follows:

$$P_{w,t}^{forecasted} = \begin{Bmatrix} 0 & V_t > V^{CO}, V_t < V^{CI} \\ P^{\max} \times (\frac{V_t - V^{CI}}{V^R - V^{CI}}) & V^{CI} \leq V_t < V^R \\ P^{\max} & V^R \leq V_t \leq V^{CO} \end{Bmatrix} \quad (24)$$

where $P_{w,t}^{forecasted}$ is the forecasted output of the wind turbine at time t. V^{CO}, V^{CI} and V^R are the cut-off, cut-in and rated speed of wind turbine. Also, the rated power of the wind turbine is denoted by $P^{\max}$ and V_t is the forecasted wind speed at time t.

4.4 Particle Swarm Optimization

PSO is a population-based optimization algorithm that is inspired by the social behavior of animals, such as birds and fishes. In this algorithm, each population member, called particle, changes its position with respect to its experience and the best particle's experience in a multi-dimensional space [13]. The values of random variables are assumed as positions of the particles. Hence, the positions of an ith particle in each iteration is represented as $X_i^{iter} = (x_{i,1}^{iter}, x_{i,2}^{iter}, \ldots, x_{i,N}^{iter})$ and the speed of particles are stated as $V_i^{iter} = (v_{i,1}^{iter}, v_{i,2}^{iter}, \ldots, v_{i,N}^{iter})$. It should be noted that the total number of decision variables in the optimization problem is denoted by N. The changing in velocity and position of each particle can be formulated as:

$$v_{i,n}^{iter} = \omega \times v_{i,n}^{iter-1} + C_1 \times r_1^n \times (p_{best_{i,n}}^{iter-1} - x_{i,n}^{iter-1}) + C_2 \times r_2^n \times (g_{best,n}^{iter-1} - x_{i,n}^{iter-1}) \quad (25)$$

$$x_{i,n}^{iter} = x_{i,n}^{iter-1} + v_{i,n}^{iter} \quad (26)$$

where ω is the inertia weight, r_1^n and r_2^n are random numbers between 0 and 1. $p_{best,i,n}^{iter-1}$ and $g_{best,n}^{iter-1}$ are the best positions of particle i in the previous iteration and the best position of all particles in all iterations, respectively. C_1 and C_2 are learning factors that are usually equal. The updated velocity must be in the acceptable range as in (27).

$$-v_n^{\max} \leq v_{i,n} \leq v_n^{\max} \quad (27)$$

$$v_n^{\max} = (x_n^{\max} - x_n^{\min})/r \quad (28)$$

where $x_n^{\max}$, $x_n^{\min}$ are the maximum and minimum limits of variables. r is a parameter to control the amount of change in velocity.

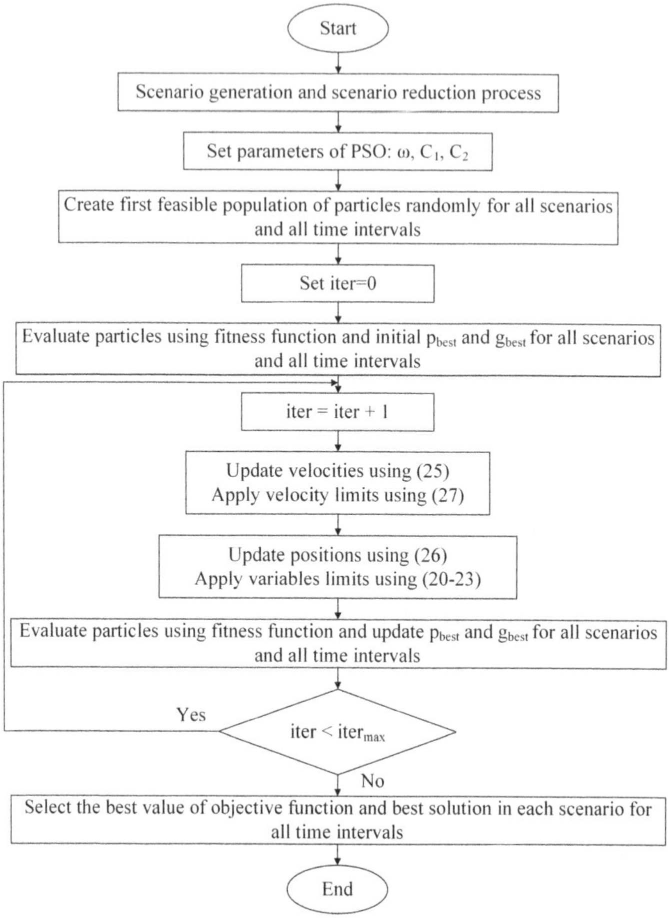

Fig. 3 Flowchart of implementing PSO into SDED problem

In this chapter, the SDED problem is solved using a simple PSO algorithm to show the principles of the method. The flowchart of the implemented method is illustrated in Fig. 3.

5 Simulation, Results, and Discussion

In this part of the chapter, the proposed method is implemented into a test system that is taken from [14] and modified. This test system consists of four power-only units, two CHP units, one heat-only unit, and 55×2-MW wind turbines. Valve point effects are considered to show the capability of the proposed method. Electrical power and heat demand profiles as forecasted values are taken from [15] and scaled to 600 MW and 150 MWth, respectively. Wind speed profile for calculating the forecasted wind power can be found in [16]. The coefficients of the cost function for power-only units are provided in Table 1. The cost coefficients of CHP units are tabulated in Table 2 and coefficients of the heat-only unit is presented in Table 3. Also, the heat-power feasible operation regions for co-generation units are depicted in Figs. 4 and 5. The parameters related to wind turbines are provided in Table 4.

First, the deterministic dynamic economic dispatch problem is solved for ten times and results are reported in Table 5. Next, the stochastic problem is solved. In order to solve the SDED problem, 1000 scenarios are generated and reduced to 10 scenarios using fast forward algorithm. The generated scenarios for electrical load demand and wind power are depicted in Figs. 6 and 7, respectively. The best case, mean and worst case results for different scenarios of SDED problem are presented in Table 6. It can be seen that the total cost for stochastic problem increases in all cases by taking into account the system uncertainties due to consideration of different scenarios that can happen in the actual situations. For example, the difference between the worst value of the deterministic problem and the objective function of the stochastic problem is $201,732.8 that increased by 77.98%. The contribution of different scenarios with uncertainties in the value of the objective function in the stochastic problem causes this increase in comparison with a deterministic problem that relies on a single scenario. Thus, the results of the stochastic problem are more realistic and more reliable than the deterministic problem.

Table 1 Cost coefficients for power-only units

Unit	α	β	γ	λ	ρ	$P^{p,\min}$	$P^{p,\max}$
1	0.008	2	25	100	0.042	10	75
2	0.003	1.8	60	140	0.04	20	125
3	0.0012	2.1	100	160	0.038	30	175
4	0.001	2	120	180	0.037	40	250

Table 2 Cost coefficients for CHP units

Unit	a	b	c	d	e	f
5	0.0345	14.5	2650	0.03	4.2	0.031
6	0.0435	36	1250	0.027	0.6	0.011

Table 3 Cost coefficients for heat-only unit

Unit	a	b	c	$H^{\min}$	$H^{\max}$
7	0.038	2.0109	950	0	2695.2

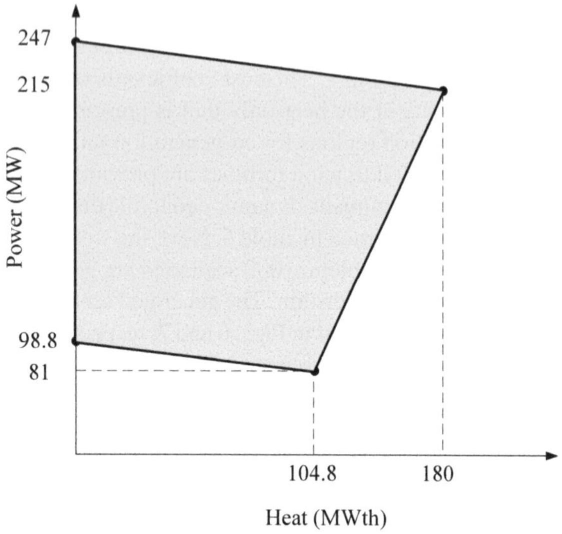

Fig. 4 Heat-power feasible operation region for unit 5

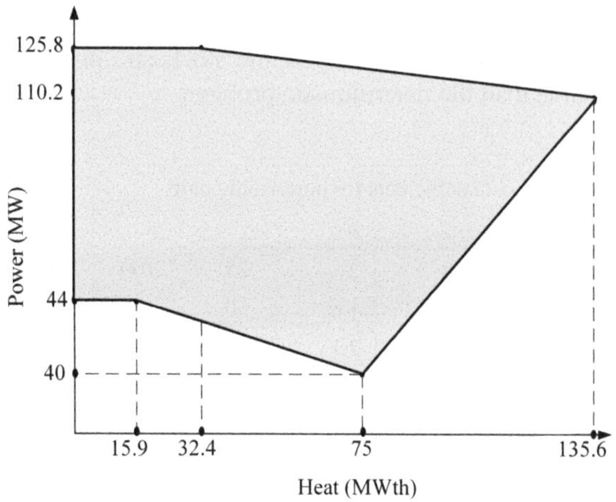

Fig. 5 Heat-power feasible operation region for unit 6

Table 4 Wind turbines parameters

Rated power (MW)	Cut-in speed (m/s)	Rated speed (m/s)	Cut-off speed (m/s)
2	3	13	25

Table 5 Results for deterministic dynamic economic dispatch problem

Total cost ($)		
Best value	Mean value	Worst value
241,992.4	249,720.6	258,711.7

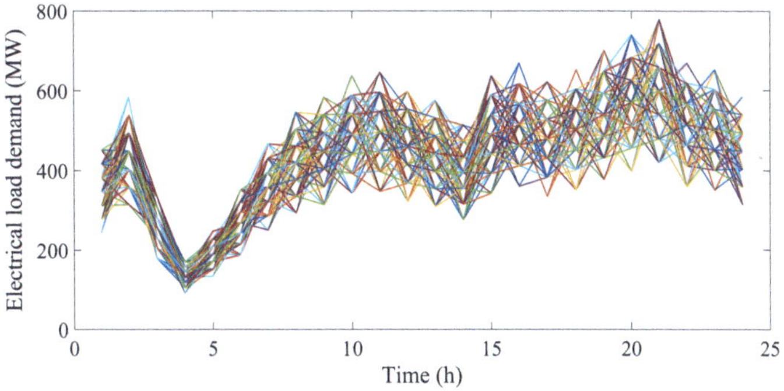

Fig. 6 Generated scenarios for electrical load demand

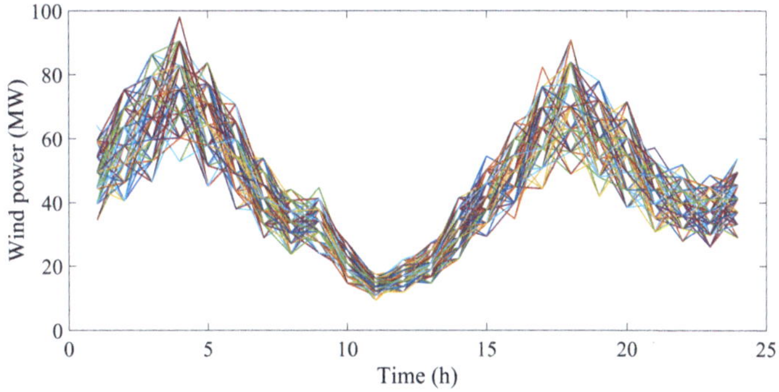

Fig. 7 Generated scenarios for wind power

Table 6 Results for stochastic dynamic economic dispatch problem

Total cost ($)			
Objective function value	Best case	Mean	Worst case
460,444.5	385,919.2	467,275.3	614,446.5

6 Conclusions

This chapter presents a stochastic framework for solving economic dispatch of CHP systems in the presence of wind power. The uncertainties related to electrical load demand and generation of wind turbines are included in this framework. Non-convexity and valve-point effects are considered that increase the complexity of the problem. To overcome the complexities of this problem, the PSO algorithm is used as the heuristic optimization method. By applying the scenario-based method to model the uncertainties related to electrical load demand and wind power forecasts errors, more realistic and reliable results are provided for SDEED problem. The stochastic framework provides better information for the dispatcher to use energy resources in an optimal way. As future work, the other objectives such as emission reduction can be added to SDED problem and the multi-objective stochastic problems can be solved.

MATLAB Codes

The main m-file of the program is as follows:

```
clc;
clear;
close all;
global demand_e demand_h power_wind Ns Nt particles
scenario_prob;
particles=150;
iterations=2000;
[~,demand_h]=demand( );
Ns=10;
Nt=24;
load('Reduced Scenarios (fast)');
demand_e=Reduced_Scenarios(:,1);
power_wind=Reduced_Scenarios(:,2);
scenario_prob=Probability;
[BestCost,BestParticles,BEST_COST]=PSO(particles, it-
erations);
```

The function of PSO algorithm that is used for optimization procedure is provided in the following. The number of particles and total number of iterations are inputs of this function, while the optimal cost, best solutions and best cost of each iteration are outputs of this function:

```
function [Best-
Cost,BestParticles,BEST_COST]=PSO(particles, itera-
tions)
global Ns Nt VarSize;
%% PSO Parameters
phi1=2.05;
phi2=2.05;
phi=phi1+phi2;
w=2/(phi-2+sqrt(phi^2-4*phi));
c1=phi1*w;
c2=phi2*w;
r=10;
%% Problem Definition
[bounds]=unit_data();
VarSize=size(bounds,2);
x=zeros(particles,VarSize);
x=repmat(x,1,Nt);
x=repmat(x,Ns,1);
v=zeros(particles,VarSize);
v=repmat(v,1,Nt);
v=repmat(v,Ns,1);
VelocityMax=(bounds(1,:)-bounds(2,:))./r;
VelocityMin=-VelocityMax;
%% Initialize Positions
for s=1:Ns
for t=1:Nt
x_temp=x(particles*(s-1)+1:particles*s,VarSize*(t-
1)+1:VarSize*t);
for i=1:particles
for j=1:VarSize
x_temp(i,j)=unifrnd(bounds(2,j),bounds(1,j));
end
end
x(particles*(s-1)+1:particles*s,VarSize*(t-
1)+1:VarSize*t)=x_temp;
end
end
%% Evaluate First Particles
[cost]=Fitness(x);
ParticlesBest.Cost=cost;
ParticlesBest.Position=x;
[GlobalBest.Cost,index]=min(cost);
for s=1:Ns
x_temp=x(particles*(s-1)+1:particles*s,:);
GlobalBest.Position(s,:)=x_temp(index,:);
end
%% PSO Loop
```

```
for iter=1:iterations
%% Update Velocity
for s=1:Ns
for t=1:Nt
v_temp=v(particles*(s-1)+1:particles*s,VarSize*(t-
1)+1:VarSize*t);
ParticlesBest_temp=ParticlesBest.Position(particles*(s-
1)+1:particles*s,VarSize*(t-1)+1:VarSize*t);
GlobalBest_temp=GlobalBest.Position(s,VarSize*(t-
1)+1:VarSize*t);
x_temp=x(particles*(s-1)+1:particles*s,VarSize*(t-
1)+1:VarSize*t);
for i=1:particles
v_temp(i,:)=w*v_temp(i,:)+c1*rand(1,VarSize).*(Particle
sBest_temp(i,:)x_temp(i,:))+c2*rand(1,VarSize).*(Global
Best_temp-x_temp(i,:));
v_temp(i,:)=min(max(v_temp(i,:),VelocityMin(1,:)),Veloc
ityMax(1,:));
end
x_temp=x_temp+v_temp;
x(particles*(s-1)+1:particles*s,VarSize*(t-
1)+1:VarSize*t)=x_temp;
v(particles*(s-1)+1:particles*s,VarSize*(t-
1)+1:VarSize*t)=v_temp;
end
end
%% Apply Constraints
for s=1:Ns
for t=1:Nt
x_temp=x(particles*(s-1)+1:particles*s,VarSize*(t-
1)+1:VarSize*t);
for i=1:size(x_temp, 1)
x_temp(i,:)=min(max(x_temp(i,:),bounds(2,:)),bounds(1,:
));
end
x(particles*(s-1)+1:particles*s,VarSize*(t-
1)+1:VarSize*t)=x_temp;
end
end
%% Evaluate Particles
cost=Fitness(x);
for i=1:length(cost)
if cost(i)<ParticlesBest.Cost(i)
ParticlesBest.Cost(i)=cost(i);
for s=1:Ns
x_temp=x(particles*(s-1)+1:particles*s,:);
```

```
ParticlesBest_temp=ParticlesBest.Position(particles*(s-
1)+1:particles*s,:);
ParticlesBest_temp(i,:)=x_temp(i,:);
ParticlesBest.Position(particles*(s1)+1:particles*s,:)=
ParticlesBest_temp;
end
end
end
if min(cost)<GlobalBest.Cost
[GlobalBest.Cost,index]=min(cost);
for s=1:Ns
x_temp=x(particles*(s-1)+1:particles*s,:);
GlobalBest.Position(s,:)=x_temp(index,:);
end
end
BEST_COST(iter)=GlobalBest.Cost;
end
%% Results
BestCost=GlobalBest.Cost;
BestParticles=GlobalBest.Position;
plot(BEST_COST);
end
```

Fitness function for calculating the objective function that receives particles as input and returns total cost as output is as follows:

```
function [cost]=Fitness(x)
global demand_e demand_h Ns Nt particles power_wind
scenario_prob VarSize;
for i=1:particles
for s=1:Ns
for t=1:Nt
x_temp=x(particles*(s-1)+1:particles*s,VarSize*(t-
1)+1:VarSize*t);
p1=x_temp(i,1);
p2=x_temp(i,2);
p3=x_temp(i,3);
p4=x_temp(i,4);
p5=x_temp(i,5);
p6=x_temp(i,6);
h5=x_temp(i,7);
h6=x_temp(i,8);
h7=x_temp(i,9);
F1=25+2*p1+0.008*p1^2+abs(100*sin(0.042*(p1-10)));
F2=60+1.8*p2+0.003*p2^2+abs(140*sin(0.04*(p2-20)));
```

```
F3=100+2.1*p3+0.0012*p3^2+abs(160*sin(0.038*(p3-30)));
F4=120+2*p4+0.001*p4^2+abs(180*sin(0.037*(p4-40)));
F5=2650+14.5*p5+0.0345*p5^2+4.2*h5+0.03*h5^2+0.031*p5*h
5;
F6=1250+36*p6+0.0435*p6^2+0.6*h6+0.027*h6^2+0.011*p6*h6
;
F7=950+2.0109*h7+0.038*h7^2;
pen=0;
if (p1+p2+p3+p4+p5+p6)<(demand_e(Nt*(s-1)+t,1)-
power_wind(Nt*(s-1)+t,1))
pen(1)=1e+15;
else
pen(1)=0;
end
if (h5+h6+h7)<demand_h(t,1)
pen(2)=1e+15;
else
pen(2)=0;
end
if (p5-247-((247-215)/(0-180))*(h5-0)>0)||(p5-81-((215-
81)/(180-104.8))*(h5-104.8)<0)||(p5-98.8-((98.8-81)/(0-
104.8))*(h5-0)<0)
pen(3)=1e+15;
else
pen(3)=0;
end
if (p6-110.2-((125.8-110.2)/(32.4-135.6))*(h6-
135.6)>0)||(p6-40-((40-110.2)/(75-135.6))*(h6-75)<0)
pen(4)=1e+15;
else
pen(4)=0;
end
cost_temp(s,t)=F1+F2+F3+F4+F5+F6+F7+sum(pen);
end
end
cost_scen=sum(cost_temp,2);
cost_weighted=cost_scen.*scenario_prob;
cost(i,1)=sum(cost_weighted);
end
end
```

The MATLAB code for scenario generation process is as follows:

```
clear;
clc;
close all;
Ns=1000;
Nvariables=2;
Nt=24;
Nsigma=7;
[scenario,prob_scen]=BinaryScenarios(
Ns,Nvariables,Nt,Nsigma );
[demand_e,demand_h,power_wind,Scenario_All,SA]
=scenario_analyze(scenario,Nsigma,Nvariables,Nt);
save('Non-Reduced Scenarios');
```

The function that is used for producing binary scenarios is provided in the following:

```
function [sce-
nario,prob_scen]=BinaryScenarios(Ns,Nvariables,
Nt,Nsigma)
down=0;
[demand_e,demand_h]=demand();
[power_wind]=wind_data();
sigma_L=0.1.*demand_e;
sigma_W=0.1.*power_wind;
for t=1:Nt
alpha(1,t)=(1/sigma_L(t,1))*(2*pi)^(-0.5)*exp(0);
alpha(2,t)=(1/sigma_L(t,1))*(2*pi)^(-0.5)*exp(-0.5);
alpha(3,t)=(1/sigma_L(t,1))*(2*pi)^(-0.5)*exp(-0.5);
alpha(4,t)=(1/sigma_L(t,1))*(2*pi)^(-0.5)*exp(-0.5*4);
alpha(5,t)=(1/sigma_L(t,1))*(2*pi)^(-0.5)*exp(-0.5*4);
alpha(6,t)=(1/sigma_L(t,1))*(2*pi)^(-0.5)*exp(-0.5*9);
alpha(7,t)=(1/sigma_L(t,1))*(2*pi)^(-0.5)*exp(-0.5*9);
beta(1,t)=(1/sigma_W(t,1))*(2*pi)^(-0.5)*exp(0);
beta(2,t)=(1/sigma_W(t,1))*(2*pi)^(-0.5)*exp(-0.5);
beta(3,t)=(1/sigma_W(t,1))*(2*pi)^(-0.5)*exp(-0.5);
beta(4,t)=(1/sigma_W(t,1))*(2*pi)^(-0.5)*exp(-0.5*4);
beta(5,t)=(1/sigma_W(t,1))*(2*pi)^(-0.5)*exp(-0.5*4);
beta(6,t)=(1/sigma_W(t,1))*(2*pi)^(-0.5)*exp(-0.5*9);
beta(7,t)=(1/sigma_W(t,1))*(2*pi)^(-0.5)*exp(-0.5*9);
alpha(:,t)=alpha(:,t)/sum(alpha(:,t)); % Normalize to
one
beta(:,t)=beta(:,t)/sum(beta(:,t));
Intervals_L(1,t)=0;
Intervals_W(1,t)=0;
for k=1:Nsigma
```

```
Intervals_L(k+1,t)=alpha(k,t)+Intervals_L(k,t);
end
for k=1:Nsigma
Intervals_W(k+1,t)=beta(k,t)+Intervals_W(k,t);
end
end
for s=1:Ns
index=0;
w=zeros(Nvariables*Nsigma,Nt);
w_L=zeros(Nsigma,Nt);
w_W=zeros(Nsigma,Nt);
top=1;
for t=1:Nt
sumAlphaW=0;
sumBetaW=0;
r1=rand;
r2=rand;
for j=1:Nsigma
if r1>Intervals_L(j,t) && r1<Intervals_L(j+1,t)
w_L(j,t)=1;
end
end
for j=1:Nsigma
if r2>Intervals_W(j,t)&&r2<Intervals_W(j+1,t)
w_W(j,t)=1;
end
end
w=[w_L;w_W];
scenario(s,index+1:index+Nvariables*Nsigma)=w(:,t)';
index=index+Nvariables*Nsigma;
AlphaW=alpha(:,t).*w_L(:,t);
sumAlphaW=sum(AlphaW);
BetaW=beta(:,t).*w_W(:,t);
sumBetaW=sum(BetaW);
multiply=sumAlphaW*sumBetaW;
top=top*multiply;
end
prob_scen(s,1)=top;
down=down+top;
end
prob_scen=prob_scen./down;
end
```

The function for converting binary scenarios to scenarios with real values is as follows:

```
function [demand_e,demand_h,power_wind,Scenario_All,SA]
=scenario_analyze( scenario,Nsigma,Nvariables,Nt )
kk=1;
[demand_e_all,demand_h_all]=demand();
[power_wind_all]=wind_data();
demand_h=demand_h_all';
for s=1:size(scenario,1)
for t=1:Nt
w_s_t_load=scenario(s,(t-1)*Nsigma+1:(t-
1)*Nsigma+Nsigma);
w_s_t_wind=scenario(s,(t-1)*Nsigma+Nsigma+1:(t-1)*
Nsigma+Nvariables*Nsigma);
for j=1:Nsigma
if w_s_t_load(1,j)==1
break;
end
end
if j==1
demand_e(s,t)=demand_e_all(t,1);
elseif j==2
demand_e(s,t)=demand_e_all(t,1)+0.1*demand_e_all(t,1);
elseif j==3
demand_e(s,t)=demand_e_all(t,1)-0.1*demand_e_all(t,1);
elseif j==4
demand_e(s,t)=demand_e_all(t,1)+0.2*demand_e_all(t,1);
elseif j==5
demand_e(s,t)=demand_e_all(t,1)-0.2*demand_e_all(t,1);
elseif j==6
demand_e(s,t)=demand_e_all(t,1)+0.3*demand_e_all(t,1);
elseif j==7
demand_e(s,t)=demand_e_all(t,1)-0.3*demand_e_all(t,1);
end
for i=1:Nsigma
if w_s_t_wind(1,i)==1
break;
end
end
if i==1
power_wind(s,t)=power_wind_all(t,1);
elseif i==2
power_wind(s,t)=power_wind_all(t,1)+0.1*power_wind_all(
t,1);
elseif i==3
```

```
power_wind(s,t)=power_wind_all(t,1)-
0.1*power_wind_all(t,1);
elseif i==4
power_wind(s,t)=power_wind_all(t,1)+0.2*power_wind_all(
t,1);
elseif i==5
power_wind(s,t)=power_wind_all(t,1)-
0.2*power_wind_all(t,1);
elseif i==6
power_wind(s,t)=power_wind_all(t,1)+0.3*power_wind_all(
t,1);
elseif i==7
power_wind(s,t)=power_wind_all(t,1)-
0.3*power_wind_all(t,1);
end
Scenario_All(s,(2*t)-1)=demand_e(s,t);
Scenario_All(s,2*t)=power_wind(s,t);
SA(kk,:)=[demand_e(s,t),power_wind(s,t)];
kk=kk+1;
end
end
end
```

The fast forward scenario reduction MATLAB code is as follows:

```
clear;
clc;
load('Non-Reduced Scenarios');
Ns=1000;
Nt=24;
N_desired=10;
for i=1:Ns
x=sum(SA(Nt*i-(Nt-1):Nt*i,:));
w_j(i,:)=sum(x);
end
for i=1:Ns
for j=1:Ns
v_main(i,j)=abs(w_j(i,1)-w_j(j,1));
end
end
for i=1:Ns
d(i,1)=sum(prob_scen.*v_main(:,i));
end
```

```
[min_d,index]=min(d);
w_s(1,1)=index;
w_j_main=w_j;
w_j(index,:)=[];
prob_scen_main=prob_scen;
prob_scen(index,:)=[];
v=v_main;
for i=2:N_desired
d=0;
x=0;
v_new=0;
for m=1:size(w_j,1)+1
if ~sum(m==w_s(i-1,1))
x=x+1;
y=0;
for n=1:size(w_j,1)+1
if ~sum(n==w_s(i-1,1))
y=y+1;
v_new(x,y)=min(v(m,n),v(m,index));
end
end
end
end
v=v_new;
for m=1:size(v_new,2)
d(m,1)=sum(v_new(:,m).*prob_scen);
end
[min_d,index]=min(d);
x=find(w_j_main==w_j(index));
w_s(i,1)=x;
w_j(index,:)=[];
prob_scen(index,:)=[];
end
for i=1:N_desired
Reduced_Scenarios(Nt*i-(Nt-1):Nt*i,:)=SA(Nt*w_s(i,1)-
(Nt-1):Nt*w_s(i,1),:);
end
for i=1:size(w_j,1)
x=find(w_j_main==w_j(i));
for j=1:size(w_s,1)
v_prob(i,j)=v_main(x,w_s(j,1));
end
end
for i=1:size(w_s,1)
Probability(i,1)=prob_scen_main(w_s(i,1));
end
for i=1:size(w_j,1)
```

```
[~,index]=min(v_prob(i,:));
x=find(w_j_main==w_j(i));
Probabil-
ity(index,1)=Probability(index,1)+prob_scen_main(x,1);
end
save('Reduced Scenarios
(fast)','Reduced_Scenarios','Probability')
```

The function for creating electrical and heat demands as forecasted values is coded as:

```
function [demand_e,demand_h]=demand()
scale_e = 600;
scale_h = 150;
%% electrical demand
demand_e=[
0.579741
0.75
0.428879
0.219828
0.318966
0.450431
0.599138
0.700431
0.75
0.818966
0.829741
0.769397
0.739224
0.659483
0.818966
0.859914
0.799569
0.838362
0.900862
0.950431
1
0.859914
0.838362
0.75];
%% heat demand
demand_h=[
0.509091
0.539394
0.421212
```

```
0.230303
0.330303
0.639394
0.690909
0.760606
0.660606
0.678788
0.830303
0.639394
0.790909
0.6
0.878788
0.939394
0.7
0.890909
0.778788
0.9
1
1
0.821212
0.639394];
%% demand scaling
demand_e=scale_e .*demand_e;
demand_h=scale_h .*demand_h;
end
```

The function to model the wind turbine and forecasted wind speed can be coded as:

```
function [power_wind]=wind_data()
v_cut_in=3;       %Cut-in speed of the wind turbine (m/s)
v_cut_out=25;     %Cut-out speed of the wind turbine
(m/s)
v_rated=13;       %Rated wind speed of the wind turbine
(m/s)
p_rated=2;        %Rated power of the wind turbine (MW)
wind_speed=[
1    7.95
2    8.8
3    9.65
4    10.55
5    9.45
6    8.45
7    7.15
8    6.4
```

```
9   6.45
10  5.1
11  4.35
12  4.7
13  5.1
14  6.2
15  7.2
16  8
17  9.35
18  10
19  9
20  8.5
21  7.4
22  7
23  6.75
24  7.15
];
for time=1:24
if
(wind_speed(time,2)>v_cut_out)&&(wind_speed(time,2)<v_c
ut_in)
power_wind(time,1)=0;
elseif
(v_cut_in<wind_speed(time,2))&&(wind_speed(time,2)<v_ra
ted)
power_wind(time,1)=p_rated*((wind_speed(time,2)-
v_cut_in)/(v_rated-v_cut_in));
elseif
(v_rated<wind_speed(time,2))&&(wind_speed(time,2)<v_cut
_out)
power_wind(time,1)=p_rated;
end
end
power_wind=power_wind.*55;
end
```

References

1. Narang, N., Sharma, E., Dhillon, J.: Combined heat and power economic dispatch using integrated civilized swarm optimization and Powell's pattern search method. Appl. Soft Comput. **52**, 190–202 (2017)
2. Mehdinejad, M., Mohammadi-ivatloo, B., Dadashzadeh-Bonab, R.: Energy production cost minimization in a combined heat and power generation systems using cuckoo optimization algorithm. Energ. Effi. **10**, 81–96 (2017)

3. Nguyen, T.T., Nguyen, T.T., Vo, D.N.: An effective cuckoo search algorithm for large-scale combined heat and power economic dispatch problem. Neural Comput. Appl. 1–20 (2017)
4. Davoodi, E., Zare, K., Babaei, E.: A GSO-based algorithm for combined heat and power dispatch problem with modified scrounger and ranger operators. Appl. Therm. Eng. **120**, 36–48 (2017)
5. Alipour, M., Mohammadi-ivatloo, B., Zare, K.: Stochastic scheduling of renewable and CHP-based microgrids. IEEE Trans. Industr. Inf. **11**, 1049–1058 (2015)
6. Bahmani-Firouzi, B., Farjah, E., Azizipanah-Abarghooee, R.: An efficient scenario-based and fuzzy self-adaptive learning particle swarm optimization approach for dynamic economic emission dispatch considering load and wind power uncertainties. Energy **50**, 232–244 (2013)
7. Aghaei, J., Niknam, T., Azizipanah-Abarghooee, R., Arroyo, J.M.: Scenario-based dynamic economic emission dispatch considering load and wind power uncertainties. Int. J. Electr. Power Energy Syst. **47**, 351–367 (2013)
8. Mohammadi, S., Soleymani, S., Mozafari, B.: Scenario-based stochastic operation management of microgrid including wind, photovoltaic, micro-turbine, fuel cell and energy storage devices. Int. J. Electr. Power Energy Syst. **54**, 525–535 (2014)
9. Niknam, T., Azizipanah-Abarghooee, R., Narimani, M.R.: An efficient scenario-based stochastic programming framework for multi-objective optimal micro-grid operation. Appl. Energy **99**, 455–470 (2012)
10. Aalami, H.A., Nojavan, S.: Energy storage system and demand response program effects on stochastic energy procurement of large consumers considering renewable generation. IET Gener. Transm. Distrib. **10**, 107–114 (2016)
11. Pineda, S., Conejo, A.: Scenario reduction for risk-averse electricity trading. IET Gener. Transm. Distrib. **4**, 694–705 (2010)
12. Basu, M.: Combined heat and power economic dispatch using opposition-based group search optimization. Int. J. Electr. Power Energy Syst. **73**, 819–829 (2015)
13. Gaing, Z.-L.: Particle swarm optimization to solving the economic dispatch considering the generator constraints. IEEE Trans. Power Syst. **18**, 1187–1195 (2003)
14. Mohammadi-ivatloo, B., Moradi-Dalvand, M., Rabiee, A.: Combined heat and power economic dispatch problem solution using particle swarm optimization with time varying acceleration coefficients. Electr. Power Syst. Res. **95**, 9–18 (2013)
15. Nazari-Heris, M., Abapour, S., Mohammadi-ivatloo, B.: Optimal economic dispatch of FC-CHP based heat and power micro-grids. Appl. Therm. Eng. **114**, 756–769 (2017)
16. Mazidi, M., Monsef, H., Siano, P.: Robust day-ahead scheduling of smart distribution networks considering demand response programs. Appl. Energy **178**, 929–942 (2016)

Economic Dispatch of Multiple-Chiller Plants Using Wild Goats Algorithm

Farkhondeh Jabari, Alireza Akbari Dibavar and Behnam Mohammadi-Ivatloo

Abstract Use of multiple-chiller systems in air-conditioning applications is known as a major factor in increasing electricity consumption. To obtain a significant energy saving in building space cooling, optimal operation of chillers as an energy-efficient manner is necessary. Therefore, this chapter aims to obtain the best performance of the multi-chiller systems, which can be attained by minimizing the total power consumption of chillers considering their partial load ratios (PLRs) as decision variables. In this chapter, optimal chiller loading (OCL) problem is implemented on three different case studies using a novel evolutionary algorithm, called wild goats algorithm (WGA). This algorithm is inspired from wild goats' climbing, living in groups and based on cooperation between members of groups. Numerical results show the effectiveness of the WGA to solve the OCL problem.

Keywords Economic dispatch · Optimal chiller loading (OCL) · Non-linear optimization problem · Electricity consumption

Indices

t	Time (h)
i	Chiller number

F. Jabari (✉) · A. Akbari Dibavar · B. Mohammadi-Ivatloo
Faculty of Electrical and Computer Engineering, University of Tabriz, Tabriz, Iran
e-mail: f.jabari@tabrizu.ac.ir

A. Akbari Dibavar
e-mail: a.dibavar96@ms.tabrizu.ac.ir

B. Mohammadi-Ivatloo
e-mail: bmohammadi@tabrizu.ac.ir

M. Pesaran Hajiabbas and B. Mohammadi-Ivatloo (eds.),
Optimization of Power System Problems, Studies in Systems, Decision and Control 262,
https://doi.org/10.1007/978-3-030-34050-6_8

Symbols

P_i^t	Power consumption of chiller i (kW)
PLR_i^t	Partial load ratio of unit i
u_i^t	Binary variable that is 1 if unit i is on, else it is 0
RT_i	Capacity of chiller i (Refrigeration ton)
CL_t	Cooling demand (kW)
f	Fitness function
N_{wg}	Number of wild goats
N_{var}	Dimension of optimization problem
W	Weight of each wg
V_i	Movement vector of wild goats

1 Motivation and Literature Review

Under warm climate, multiple-chiller systems usually consume high electricity to generate cool. Therefore, 30% of summer peak-electrical demand is associated with these systems. If a mixed-integer non-linear optimization program is developed for multiple-chiller plants, on or off status and refrigeration outputs of chillers will be found in a way that their energy consumption is minimized as low as possible [1]. In [2], the branch and bound method based Lagrangian algorithm is introduced, that is able to determine the partial load ratios (PLRs) of the chillers. The objective function of the optimization problem is the hourly electrical power consumption of the chillers. Reference [3] presented the gradient method (GM) with less calculation time and better objective functions than those of found by Lagrangian approach. Simulated annealing (SA) is capable to solve find more accurate PLRs than Lagrangian problem [4]. Gaussian distribution based firefly algorithm [5], differential cuckoo search algorithm (DCSA) [6], evolution strategy (ES) [7], differential evolution algorithm [8], BONMIN optimization process [9], teaching and learning optimization approach [10], genetic algorithm (GA) [11, 12], ripple bee swarm optimization (RBSO) [13], invasive weed optimization (IWO) [14] are some search strategies, which have been proposed by researches to solve the optimal chiller loading (OCL) problem. In these works, the refrigeration production of the chillers is equal to the cooling demand. In addition, the refrigeration capability of each chiller is less than or equal to its cooling capacity. A binary decision variable is used for modeling the on/off status of each chiller. Then, the refrigeration produced by chiller is computed based on its binary variable. If some of refrigeration productions of chillers is equal to cooling load as well as the refrigeration capacity constraint is satisfied for all units, the hourly energy consumption of the plant will be calculated until its minimum value is found. This chapter applies wild goats algorithm (WGA) on OCL problem to find good operating scenarios for 3 standard systems, while satisfying balance and capacity criteria.

The remainder of the present chapter are as follows: WGA based OCL problem is described in Sect. 2. Case study and analysis are provided in Sect. 3. Conclusion is presented in Sect. 4.

2 Problem Formulation

2.1 Economic Dispatch of Multiple-Chiller Systems

In OCL problem, sum of electrical power consumption of N chillers is minimized as objective function (1). where, P_i^t is the electrical power consumed by ith unit at time t and is obtained from (2)–(4). In (2), u_i^t is a binary variable that is 1 if unit i is on, else it is 0. The refrigeration capacity constraint is modeled as (3). It is obvious that a random variable, which belongs to [0,1], is generated as PLR of chiller i. Then, the cooling load of this unit is calculated based on its random number and refrigeration capacity. Obviously, the cooling capability of the chiller i is less than (if random number is less than 1) or equal to (if rand = 1) its refrigeration capacity. Relation (4) indicates that if chiller i is on at time horizon t, its power consumption depends on coefficients a_i, b_i, c_i and d_i and cooling demand satisfied by this chiller, else, it will be 0. Load-generation balance constraint is investigated by Eq. (5). The parameters CL_t and RT_i are the cooling load at time t and the refrigeration capacity of the chiller i, respectively.

$$\text{Objective function} = \sum_{i=1}^{N} P_i^t \tag{1}$$

$$u_i^t = \begin{cases} 0 \text{ if chiller } i \text{ is off} \\ 1 \text{ if chiller } i \text{ is on} \end{cases} \tag{2}$$

$$PLR_i^t = \begin{cases} 0 & \text{if unit } i \text{ is off} \\ \frac{\text{cooling load of unit } i \text{ at time } t}{\text{Refrigeration capacity of unit } i} = \text{rand}; & \text{if unit } i \text{ is on} \end{cases} \tag{3}$$

$$P_i^t = \begin{Bmatrix} 0; & \text{if } u_i^t = 0 \\ a_i + b_i PLR_i^t + c_i \left(PLR_i^t\right)^2 + d_i \left(PLR_i^t\right)^3; & \text{if } u_i^t = 1 \end{Bmatrix} \tag{4}$$

$$CL_t = \sum_{i=1}^{N} PLR_i^t \times RT_i \tag{5}$$

2.2 Proposed Optimization Algorithm

In this research, wild goats algorithm (WGA) is applied to solve the OCL problem. The WGA is recently introduced and is founded based on group life of wild goats which are widespread species of goats and ancestor of domestic goats. Wild goats often have group life and the herds live in the mountains. Each herd usually has an older and strong leader. The wild goats reside mountainous areas and usually feed on mountain plants and shrubs. Members of each group follow their leader, these members called as "Followers". In first step, movement of leader is in direction of its personal experiences but in later steps, in addition to personal experiences, it will track the movement of other successful leaders. The first step of the iterative process of a population-based algorithm is initialization of all population members.

$$wg_i = \left[x_{i,1}, \ldots, x_{i,N_{var}}\right], \quad i = 1, \ldots, N_{wg} \tag{6}$$

In this algorithm, after initialization, fitness values of each "wg_i", which is an optimal solution candidate, must be evaluated respect to objective function.

$$f(wg_i) = f\left(\left[x_{i,1}, \ldots, x_{i,N_{var}}\right]\right), \quad i = 1, \ldots, N_{wg} \tag{7}$$

In order to have a better comparison between wgs, a weight is defined for each wg as (8), and the solutions with highest weight are selected as leaders of groups. The high weighted leaders absorb more followers. The number of groups determination and groups' formation process is described more in [15].

$$W_i = \exp\left(-N_{var} \frac{f(wg_i) - \min_j \{f(wg_i)\}}{\sum_{j=1}^{N_{wg}} \left(f\left(wg_j\right) - \min_j \{f(wg_i)\}\right)}\right), \quad i = 1, \ldots, N_{wg} \tag{8}$$

The leader and followers of each group travel towards the best point of the search space (feasible region in optimization problems). This adapt with the attempt of wild goats to climbing the mountain. The initial movement vectors of all population is zero at the first time. The best leader which has the highest weight, proceeds only in the direction of its movement vector. The other leaders in addition to pursuing of the best leader's direction, proceed in the direction of the leaders who have a higher weight than them. The other goats move like the leaders. Here, followers in addition to moving towards their movement vector and their best attempt, pursue group's leader and all other goats in their groups which have a higher weight than those. After these, the weight of each *wg* evaluated compared to the other members of the group; if its weight reduced, then it follows other followers with high weight, vice versa if its weight gets rise, other members follow it. Like PSO algorithm, the position of each wg_i, is determined by its previous position and the current iteration's movement vector, as (9).

$$wg_i(t+1) = wg_i(t) + v_i(t+1), \quad i = 1, \ldots, N_{wg} \tag{9}$$

Like the other heuristic algorithms, WGA also has mutation and cooperation concepts. The wild goats' groups reach together in the slopes of mountain while they go to summit or come down to the foothills and exchange their experiences with together and the leaders of different groups will confer together about the direction and quality of routing, which creates the cooperation concept. The groups which have the better experience about the path, attract followers of poor groups who have not traveled the path properly. Finally, the leader of the weaker group also joins into a stronger herd. New groups will be created if young goats exhibit their ability in climbing, which creates mutation in algorithm, otherwise they turn back to one of the groups. It is notable that the number of mutated wild goat(s) should be less than the number of groups for each iteration. At the end of the algorithm, only one group is endured, and the leader of that group, will be reach to the best point and in fact, is the optimum solution. More information and explanation about this algorithm can be found in [15]. The developed MATLAB codes for WGA based optimal chiller loading problem is presented in the final section.

3 Case Studies and Discussions

In order to demonstrate the speed and accuracy of the proposed strategy in solving OCL problem and compare the obtained numerical results with the best solutions of other approaches, three test systems with six, four and three compression chillers have been studied. In the next subsection, the specifications of the benchmark chiller plants and the simulation results achieved from WGA and recently published works have been presented.

Table 1 Constant factors of chillers in case 1

Chiller	a_i	b_i	c_i	RT_i
1	399.345	−122.12	770.46	1280
2	287.116	80.04	700.48	1280
3	−120.505	1525.99	−502.14	1280
4	−19.121	898.76	−98.15	1280
5	−95.029	1202.39	−352.16	1280
6	191.750	224.86	524.04	1280

Case 1

In the first case study, six 1280 RT electric chillers are considered for supplying the cooling demand. Table 1 represents the constant power consumption factors of the chillers [14]. The optimization problem (1)–(5) is solved using the WGA under MATLAB software to minimize the hourly electricity consumption of the air conditioner and determine the optimum values of the partial load ratios, when the cooling load-generation balance constraint is satisfied. In Table 2, the partial load ratio of each chiller and the objective function in different cooling load scenarios found by WGA algorithm is compared with SA [4], PSO [16], ES [7], DCSA [6], and IWO [14]. It is obvious that total electrical power consumed by chillers in all hours of this test system is lower than that of mentioned optimization algorithms.

Case 2

A four-unit plant [14] with power consumption coefficients and cooling capacity reported in Table 3 is studied in this case. The optimum operating point of the system under variable cooling load are reported in Table 4. Obviously, WGA reaches a good solution vector with total power consumption lower than or equal to that of above mentioned procedures.

Case 3

According to Table 5, three 800 RT chillers [14] is considered in case study 3. The optimization problem is conducted on test system for generating 2160, 1920, 1680, 1440, 1200 and 960 RT cool. Economic solutions of GA [11, 12], DE [8], PSO [16], DCSA [6], IWO [14], and WGA are provided in Table 6. It is shown that WGA is able to find solutions with objective functions equal to or less than that of other algorithms.

Table 2 (a) PLRs and objective functions found by SA, PSO and ES in case 1. (b) PLRs and objective functions found by DCSA, IWO and WGA in case 1

(a)

CL_t	Chiller	SA [4]		PSO [16]		ES [7]	
	i	PLR_i^t	P_{total}^t	PLR_i^t	P_{total}^t	PLR_i^t	P_{total}^t
6858 (90%)	1	0.7789	4777.03	0.8026	4739.53	0.82	4738.76
	2	0.7587		0.7799		0.75	
	3	0.9791		0.9996		1	
	4	0.9781		0.9998		1	
	5	0.9820		0.9999		1	
	6	0.9265		0.8183		0.83	
6477 (85%)	1	0.8051	4453.67	0.7606	4423.04	0.74	4422.06
	2	0.6056		0.6555		0.64	
	3	0.9689		1		1	
	4	0.9941		1		1	
	5	0.9866		1		1	
	6	0.7432		0.6835		0.72	
6096 (80%)	1	0.5635	4178.73	0.6591	4147.69	0.64	4144.12
	2	0.5743		0.5798		0.55	
	3	0.9675		0.9991		1	
	4	0.9798		0.9979		0.998	
	5	0.9845		0.9921		1	
	6	0.7338		0.5710		0.61	
5717 (75%)	1	0.6140	3925.51	0.7713	3921.07	0.57	3906.19
	2	0.4429		0.7177		0.46	
	3	0.9891		0.3		1	
	4	0.8867		0.9991		1	
	5	0.9841		1		1	
	6	0.5878		0.7187		0.47	
5334 (70%)	1	0.6265	3675.34	0.6418	3642.55	0.63	3627.46
	2	0.7403		0.6621		0.6	
	3	0.3093		0.3301		0.3	
	4	0.9546		0.9906		1	
	5	0.9511		0.999		1	
	6	0.6250		0.5806		0.67	

(continued)

Table 2 (continued)

(**b**)

CL_t	Chiller	DCSA [6]		IWO [14]		WGA	
	i	PLR_i^t	P_{total}^t	PLR_i^t	P_{total}^t	PLR_i^t	P_{total}^t
6858 (90%)	1	0.812726	4738.575	0.8127	4738.575	0.7936	4690.546
	2	0.749619		0.7492		0.7284	
	3	1		1		1	
	4	1		1		0.9999	
	5	1		1		1	
	6	0.838559		0.8390		0.8357	
6477 (85%)	1	0.727731	4421.649	0.7275	4421.649	0.7101	4382.03
	2	0.656132		0.6563		0.6368	
	3	1		1		1	
	4	1		1		0.9999	
	5	1		1		1	
	6	0.716524		0.7166		0.7131	
6096 (80%)	1	0.642735	4143.706	0.6427	4143.706	0.6268	4111.756
	2	0.562645		0.5628		0.5450	
	3	1		1		1	
	4	1		1		1	
	5	1		1		1	
	6	0.59449		0.5944		0.5906	
5717 (75%)	1	0.843697	3840.055	0	3842.553	0.8236	3788.772
	2	0.783794		0.7151		0.7622	
	3	0.000001		1		0	
	4	1		1		1	
	5	1		1		1	
	6	0.883049		0.7933		0.8805	
5334 (70%)	1	0.749969	3507.27	0	3546.438	0.7402	3464.713
	2	0.682477		0.5834		0.6698	
	3	0.000012		1		0	
	4	1		1		1	
	5	1		1		1	
	6	0.776363		0.6218		0.7572	

Table 3 Constant parameters of chillers in case 2

Chiller	a_i	b_i	c_i	d_i	RT_i
1	104.09	166.57	−430.13	512.53	450
2	−67.15	1177.79	−2174.53	1456.53	450
3	384.71	−779.13	1151.42	−63.2	1000
4	541.63	413.48	−3626.5	4021.41	1000

Table 4 (a) PLRs and objective functions found by GA, DE and PSO in case 2. (b) PLRs and objective functions found by DCSA, IWO and WGA in case 2

(a)

CL_t	Chiller	GA [11, 12]		DE [8]		PSO [16]	
	i	PLR_i^t	P_{total}^t	PLR_i^t	P_{total}^t	PLR_i^t	P_{total}^t
2610 (90%)	1	0.99	1862.18	0.99	1857.3	0.99	1857.3
	2	0.95		0.91		0.91	
	3	1		1		1	
	4	0.74		0.76		0.76	
2320 (80%)	1	0.86	1457.23	0.83	1455.66	0.83	1455.66
	2	0.81		0.81		0.81	
	3	0.88		0.90		0.90	
	4	0.69		0.69		0.69	
2030 (70%)	1	0.66	1183.8	0.73	1178.14	0.73	1178.14
	2	0.76		0.74		0.74	
	3	0.76		0.72		0.72	
	4	0.64		0.65		0.65	
1740 (60%)	1	0.6	1001.62	0.60	998.53	0.60	998.53
	2	0.7		0.66		0.66	
	3	0.57		0.56		0.56	
	4	0.59		0.61		0.61	
1450 (50%)	1	0.6	907.72	0.61	820.07	0.61	820.07
	2	0.36		0		0	
	3	0.44		0.57		0.57	
	4	0.58		0.61		0.61	
1160 (40%)	1	0.33	856.3	0	651.07	0	651.07
	2	0.32		0		0	
	3	0.32		0.56		0.56	
	4	0.54		0.6		0.6	

(continued)

Table 4 (continued)

(**b**)

CL_t	Chiller	DCSA [6]		IWO [14]		WGA	
	i	PLR_i^t	P_{total}^t	PLR_i^t	P_{total}^t	PLR_i^t	P_{total}^t
2610 (90%)	1	0.990988	1857.3	0.9913	1857.3	0.9930	1857.343
	2	0.905473		0.9059		0.9084	
	3	1		1		1	
	4	0.756593		0.7563		0.7544	
2320 (80%)	1	0.828756	1455.665	0.8287	1455.665	0.8290	1455.672
	2	0.805457		0.8054		0.8055	
	3	0.896722		0.8967		0.8965	
	4	0.687883		0.6880		0.6878	
2030 (70%)	1	0.773478	1178.137	0.7261	1178.14	0.7264	1178.139
	2	0.739801		0.7400		0.7402	
	3	0.721146		0.7217		0.7209	
	4	0.627878		0.6486		0.6490	
1740 (60%)	1	0.767678	942.183	0.6036	998.53	0.7412	942.110
	2	0.004531		0.6577		0.000039	
	3	0.746317		0.5647		0.7506	
	4	0.646189		0.6077		0.6558	
1450 (50%)	1	0.515832	753.004	0.6070	820.07	0.6067	752.928
	2	0.000001		0		0.00003	
	3	0.610547		0.5683		0.5683	
	4	0.607328		0.6086		0.6086	
1160 (40%)	1	0	583.923	0	651.07	0	583.922
	2	0.000014		0		0.000001	
	3	0.570369		0.5551		0.5551	
	4	0.589625		0.6049		0.6048	

Table 5 Constant factors of chillers in case 3

Chiller	a_i	b_i	c_i	d_i	RT_i
1	100.95	818.61	−973.43	788.55	800
2	66.598	606.34	−380.58	275.95	800
3	130.09	304.5	14.377	99.8	800

Table 6 (a) PLRs and objective functions found by GA, DE and PSO in case 3. (b) PLRs and objective functions found by DCSA, IWO and WGA in case 3

(**a**)

CL_t	Chiller	GA [11, 12]		DE [8]		PSO [16]	
	i	PLR_i^t	P_{total}^t	PLR_i^t	P_{total}^t	PLR_i^t	P_{total}^t
2160 (90%)	1	0.81	1590.96	0.73	1583.81	0.73	1583.81
	2	0.93		0.97		0.97	
	3	0.96		1		1	
1920 (80%)	1	0.7	1406.02	0.66	1403.2	0.66	1403.2
	2	0.8		0.86		0.86	
	3	0.9		0.88		0.88	
1680 (70%)	1	0.69	1250.06	0.6	1244.32	0.6	1244.32
	2	0.68		0.74		0.74	
	3	0.73		0.76		0.76	
1440 (60%)	1	0.52	1107.75	0	993.6	0	993.6
	2	0.74		0.89		0.89	
	3	0.54		0.91		0.91	
1200 (50%)	1	0.49	971.21	0	832.33	0	832.33
	2	0.44		0.74		0.74	
	3	0.57		0.76		0.76	
960 (40%)	1	0.31	842.18	0	692.25	0	692.25
	2	0.32		0.57		0.57	
	3	0.58		0.63		0.63	

(**b**)

CL_t	Chiller	DCSA [6]		IWO [14]		WGA	
	i	PLR_i^t	P_{total}^t	PLR_i^t	P_{total}^t	PLR_i^t	P_{total}^t
2160 (90%)	1	0.725258	1583.807	0.7254	1583.81	0.725093	1583.806
	2	0.974742		0.9746		0.974906	
	3	1		1		1	
1920 (80%)	1	0.659065	1403.196	0.6588	1403.20	0.658172	1403.196
	2	0.858458		0.8589		0.858771	
	3	0.882477		0.8823		0.883057	
1680 (70%)	1	0.6	1244.32	0.5959	1244.32	0.597557	1244.327

(continued)

Table 6 (continued)

(b)

CL_t	Chiller	DCSA [6]		IWO [14]		WGA	
	i	PLR_i^t	P_{total}^t	PLR_i^t	P_{total}^t	PLR_i^t	P_{total}^t
	2	0.74		0.7453		0.745451	
	3	0.76		0.7588		0.756992	
1440 (60%)	1	0	993.602	0	993.60	0	993.602
	2	0.896314		0.8854		0.885634	
	3	0.903686		0.9146		0.914366	
1200 (50%)	1	0	832.325	0	832.33	0	832.325
	2	0.743026		0.7431		0.743055	
	3	0.756974		0.7569		0.756945	
960 (40%)	1	0	692.251	0	692.25	0	692.251
	2	0.536846		0.57		0.570018	
	3	0.663154		0.63		0.629982	

4 Conclusion

Multiple-chiller systems are used more recently as a central part of heating, ventilating, and air-conditioning (HVAC) systems. Optimal combination of partial load ratio of each chiller can result the minimum energy consumption in air-conditioning units. By properly controlling of chiller load distribution, the multiple-chiller plants can provide more flexibility and standby capacity leading to less disruption and maintenance cost. In this chapter, Wild Goats algorithm was used to solve the OLC problem. The minimum energy consumption of the chillers is considered as the objective function and the partial load ratio of each chiller is considered as the decision variable. Simulation results, which obtained in three different case studies, compared with solutions of other algorithms and proved the competence of WGA in deal with the OLC problem. The obtained optimum results are better or equal to the other existing optimization methods. It is concluded that WGA has well convergence and is robust, stable and also fast enough, so it is an efficient algorithm to solve other optimal problems. Other trends which readers can concentrate on, is applying this algorithm on a multi-objective OLC problem in large-scale multiple-chiller systems.

MATLAB Codes

```
% WILD GOATS ALGORITHM
clc;
clear;
close all;
%% Problem Definition
model=CreateModel();
CostFunction=@(x) MyCost(x,model);      % Cost Function
nVar=model.nUnit;                   % Number of Decision Var-
iables
VarSize=[1 nVar]; % Size of Decision Variables Matrix
VarMin=0;         % Lower Bound of Variables
VarMax=1;         % Upper Bound of Variables
%% Algorithm Parameters
MaxIt=200;         % Maximum Number of Iterations
nPop=100;          % Population Size
nGrp=10;           % Number of Groups/Leaders
w=0.7298;          % Inertia Weight
c1=1.4962;         % Personal Learning Coefficient
m=0.95;            % Mutation Percentage
% Velocity Limits
VelMax=0.1*(VarMax-VarMin);
VelMin=-VelMax;
%% Globalization of Parameters and Settings
global ProblemSettings;
ProblemSettings.CostFunction=CostFunction;
ProblemSettings.nVar=nVar;
ProblemSettings.VarSize=VarSize;
ProblemSettings.VarMin=VarMin;
ProblemSettings.VarMax=VarMax;
global AlgorithmSettings;
AlgorithmSettings.MaxIt=MaxIt;
AlgorithmSettings.nPop=nPop;
AlgorithmSettings.nGrp=nGrp;
AlgorithmSettings.w=w;
AlgorithmSettings.c1=c1;
AlgorithmSettings.VelMin=VelMin;
AlgorithmSettings.VelMax=VelMax;
AlgorithmSettings.m=m;
%% Initialization
% Initialize Groups
grp=CreateInitialGroups();
% Array to Hold Best Cost Values
BestCost=zeros(MaxIt,1);
%% Algorithm Main Loop
```

```
for it=1:MaxIt
    % Leaders' Movement
    grp=LeaderMovement(grp);
    % Followers' Movement
    grp=FollowerMovement(grp);
    % Reevaluating
    grp=Reevaluating(grp);
    % Groups' Cooperation and Mutation of Young Goats
    grp=GroupsCooperation(grp);
    % Update Best Solution Ever Found
    led=[grp.Leader];
    [~, BestLeaderIndex]=min([led.BestCost]);
    BestSol=grp(BestLeaderIndex).Leader;
    % Update Best Cost
    BestCost(it)=BestSol.BestCost;
    % Show Iteration Information
    disp(['Iteration ' num2str(it) ': Best Cost = '
num2str(BestCost(it))]);
end
%% Results
figure;
semilogy(BestCost);
-------------------------------------------------------
MODEL CREATION
function model=CreateModel()
    model.CL=6858;
    model.Plants.Cmin=[0 0 0 0 0 0];
    model.Plants.Cmax=[1 1 1 1 1 1];
    model.Plants.alpha=[399.345 287.116 -120.505
-19.121 -95.029 191.750];
    model.Plants.beta=[-122.12 80.04 1525.99 898.76
1202.39 224.86];
    model.Plants.gamma=[770.46 700.48 -502.14 -98.15
-352.16 524.04];
    model.nUnit=numel(model.Plants.alpha);
end
-------------------------------------------------------
INITIALIZATION
function grp=CreateInitialGroups()
    global ProblemSettings;
    global AlgorithmSettings;
    CostFunction=ProblemSettings.CostFunction;
    nVar=ProblemSettings.nVar;
    VarSize=ProblemSettings.VarSize;
    VarMin=ProblemSettings.VarMin;
    VarMax=ProblemSettings.VarMax;
    nPop=AlgorithmSettings.nPop;
```

```
    nGrp=AlgorithmSettings.nGrp;
    empty_particle.Position=[];
    empty_particle.Cost=[];
    empty_particle.Out=[];
    empty_particle.Velocity=[];
    empty_particle.Weight=[];
    empty_particle.BestPosition=[];
    empty_particle.BestCost=[];
    empty_particle.BestOut=[];
    particle=repmat(empty_particle,nPop,1);
    GlobalBest.Cost=inf;
    for k=1:nPop
        % Initialize Position
        parti-
cle(k).Position=unifrnd(VarMin,VarMax,VarSize);
        % Initialize Velocity
        particle(k).Velocity=zeros(VarSize);
        % Evaluation
        [particle(k).Cost, parti-
cle(k).Out]=CostFunction(particle(k).Position);
        % Update Personal Best
        particle(k).BestPosition=particle(k).Position;
        particle(k).BestCost=particle(k).Cost;
        particle(k).BestOut=particle(k).Out;
        % Update Global Best
        if particle(k).BestCost<GlobalBest.Cost

GlobalBest.Position=particle(k).BestPosition;
            GlobalBest.Cost=particle(k).BestCost;
            GlobalBest.Out=particle(k).BestOut;
        end
    end
    for k=1:nPop
        % Initialize Weight
        particle(k).Weight=exp(-
nVar*((particle(k).Cost-
GlobalBest.Cost)/(sum([particle.Cost])-
nPop*GlobalBest.Cost)));
    end
    [~, SortOrder]=sort([particle.Weight]);
    particle=particle(SortOrder);
    led=particle(end:-1:end-nGrp+1);
    fol=particle(end-nGrp:-1:1);
    empty_group.Leader=[];
    empty_group.Follower=repmat(empty_particle,0,1);
    empty_group.nFol=0;
    empty_group.Weight=[];
```

```
    empty_group.alpha=[];
    grp=repmat(empty_group,nGrp,1);
    % Assign lederialists
    for k=1:nGrp
        grp(k).Leader=led(k);
        grp(k).Weight=led(k).Weight;
    end
    % Assign Colonies
    for k=1:nGrp
         grp(k).alpha=grp(k).Weight/sum([grp.Weight]);
         grp(k).nFol=round(grp(k).alpha*(numel(fol)));
    end
    grp = rmfield(grp,'alpha');
    if diff([sum([grp.nFol]) nPop-nGrp])>0
        for kk=1:abs(diff([sum([grp.nFol]) nPop-nGrp]))
            grp(kk).nFol=grp(kk).nFol+1;
        end
    elseif diff([sum([grp.nFol]) nPop-nGrp])<0
        for kk=nGrp:-1:nGrp+diff([sum([grp.nFol]) nPop-
nGrp])+1
            grp(kk).nFol=grp(kk).nFol-1;
        end
    end
    for k=1:nGrp
         for kk=1:grp(k).nFol
             if kk<nGrp
                 grp(k).Follower(kk)=fol((kk-1)*10+k);
             elseif kk==nGrp
                 grp(k).Follower(kk)=fol((kk-
2)*10+k+kk-
(sum([grp.nFol]>=nGrp)+sum([grp.nFol]>nGrp)));
             elseif kk==nGrp+1
                 grp(k).Follower(kk)=fol((kk-
3)*10+k+kk-(sum([grp.nFol]>=nGrp)));
             else
                 grp(k).Follower(kk)=fol((kk-
4)*10+k+kk-
(sum([grp.nFol]>=nGrp)+sum([grp.nFol]>nGrp)));
             end
         end
    end
end
-------------------------------------------------------
MODEL CALCULATIONS
function out=ModelCalculations(P,model)
    alpha=model.Plants.alpha;
    beta=model.Plants.beta;
```

```
        gamma=model.Plants.gamma;
        Cmax=model.Plants.Cmax;
        CL=model.CL;
         P(6)=(CL-1280*(sum(P(1:5))))/1280;
         CP=gamma.*P.*P+beta.*P+alpha;
         CPTotal=sum(CP);
         z=(CPTotal);
         out.P=P;
         out.CP=CP;
         out.CPTotal=CPTotal;
         out.z=z;
end
-------------------------------------------------------
COST EVALUATION
function [z, out]=MyCost(x,model)
    [P]=ParseSolution(x,model);
     out=ModelCalculations(P,model);
     z=out.z;
 end
-------------------------------------------------------
PARSE SOLUTION
function [P]=ParseSolution(x,model)
     nUnit=model.nUnit;
     Cmin=model.Plants.Cmin;
     Cmax=model.Plants.Cmax;
     P=x;
End
-------------------------------------------------------
REEVALUATION
function grp=Reevaluating(grp)
     global ProblemSettings;
     nVar=ProblemSettings.nVar;
     global AlgorithmSettings;
     nPop=AlgorithmSettings.nPop;
     nGrp=numel(grp);

led=[grp.Leader];fol=[grp.Follower];Cost=[led.BestCost
fol.BestCost];
     for k=1:nGrp
         grp(k).Leader.Weight=exp(-
nVar*((grp(k).Leader.BestCost-min(Cost))/(sum(Cost)-
nPop*min(Cost))));
         for kk=1:grp(k).nFol
         % Initialize Weight
         grp(k).Follower(kk).Weight=exp(-
nVar*((grp(k).Follower(kk).BestCost-
min(Cost))/(sum(Cost)-nPop*min(Cost))));
```

```
        end
        if
max([grp(k).Follower.Weight])>grp(k).Leader.Weight

[~,BestFollowerOrder]=max([grp(k).Follower.Weight]);

BestFollower=grp(k).Follower(BestFollowerOrder);

grp(k).Follower(BestFollowerOrder)=grp(k).Leader;
            grp(k).Leader=BestFollower;
        end

grp(k).Weight=(grp(k).Leader.Weight+sum([grp(k).Followe
r.Weight]))/(1+grp(k).nFol);
    end
%% First Method: Sum Is One.
for k=1:nGrp
   grp(k).Weight=((1+grp(k).nFol)/nPop)*grp(k).Weight;
end
Sum=sum([grp.Weight]);
for k=1:nGrp
    grp(k).Weight=(grp(k).Weight/Sum);
end
% % Second Method: Max Is One (Weighted).
for k=1:nGrp
   grp(k).Weight=((1+grp(k).nFol)/nPop)*grp(k).Weight;
   grp(k).Weight=1/grp(k).Weight;
end
Weights=[grp.Weight];
if nGrp>1
    for k=1:nGrp
        grp(k).Weight=exp(-nVar*((grp(k).Weight-
min(Weights))/(sum(Weights)-nGrp*min(Weights))));
    end
else
    grp(k).Weight=1;
end
end
--------------------------------------------------------
LEADERS' MOVEMENT ALGORITHM
function grp=LeaderMovement(grp)
    global ProblemSettings;
    CostFunction=ProblemSettings.CostFunction;
    VarSize=ProblemSettings.VarSize;
    VarMin=ProblemSettings.VarMin;
    VarMax=ProblemSettings.VarMax;
    global AlgorithmSettings;
```

```
    w=AlgorithmSettings.w;
    c1=AlgorithmSettings.c1;
    VelMin=AlgorithmSettings.VelMin;
    VelMax=AlgorithmSettings.VelMax;
    nGrp=numel(grp);
    for k=1:nGrp
        d=zeros(VarSize);
        for kk=1:nGrp
            if
grp(k).Leader.Weight<grp(kk).Leader.Weight

c=grp(kk).Leader.Weight*(grp(kk).Leader.BestPosition-
grp(k).Leader.Position);
            else
                c=zeros(VarSize);
            end
            d=d+c;
        end
        % Update Velocity
        grp(k).Leader.Velocity =
w*grp(k).Leader.Velocity ...
            +d/nGrp ...

+c1*grp(k).Leader.Weight*rand(VarSize).*(grp(k).Leader.
BestPosition-grp(k).Leader.Position);
     % Apply Velocity Limits
    grp(k).Leader.Velocity =
max(grp(k).Leader.Velocity,VelMin);
    grp(k).Leader.Velocity =
min(grp(k).Leader.Velocity,VelMax);
    % Update Position
   grp(k).Leader.Position = grp(k).Leader.Position +
grp(k).Leader.Velocity;
    % Velocity Mirror Effect
    IsOutside=(grp(k).Leader.Position<VarMin |
grp(k).Leader.Position>VarMax);
    grp(k).Leader.Velocity(IsOutside)=-
grp(k).Leader.Velocity(IsOutside);
    % Apply Position Limits
    grp(k).Leader.Position =
max(grp(k).Leader.Position,VarMin);
    grp(k).Leader.Position =
min(grp(k).Leader.Position,VarMax);
    % Evaluation
    [grp(k).Leader.Cost, grp(k).Leader.Out] =
CostFunction(grp(k).Leader.Position);
    if grp(k).Leader.Cost<grp(k).Leader.BestCost
```

```
grp(k).Leader.BestPosition=grp(k).Leader.Position;
            grp(k).Leader.BestCost=grp(k).Leader.Cost;
            grp(k).Leader.BestOut=grp(k).Leader.Out;
     end
    end
 end
```

FOLLOWERS' MOVEMNET ALGORITHM

```
function grp=FollowerMovement(grp)
    global ProblemSettings;
    CostFunction=ProblemSettings.CostFunction;
    VarSize=ProblemSettings.VarSize;
    VarMin=ProblemSettings.VarMin;
    VarMax=ProblemSettings.VarMax;
    global AlgorithmSettings;
    w=AlgorithmSettings.w;
    c1=AlgorithmSettings.c1;
    VelMin=AlgorithmSettings.VelMin;
    VelMax=AlgorithmSettings.VelMax;
    nGrp=numel(grp);
    for k=1:nGrp
        for kk=1:grp(k).nFol
            d=zeros(VarSize);
            for kkk=1:grp(k).nFol
                if
grp(k).Follower(kk).Weight<grp(k).Follower(kkk).Weight

c=grp(k).Follower(kkk).Weight*(grp(k).Follower(kkk).Bes
tPosition-grp(k).Follower(kk).Position);
                else
                    c=zeros(VarSize);
                end
                d=d+c;
            end
            % Update Velocity
            grp(k).Follower(kk).Velocity =
w*grp(k).Follower(kk).Velocity ...
                +d/grp(k).nFol ...

+grp(k).Leader.Weight*rand(VarSize).*(grp(k).Leader.Bes
tPosition-grp(k).Follower(kk).Position) ...

+c1*grp(k).Follower(kk).Weight*rand(VarSize).*(grp(k).F
ollower(kk).BestPosition-grp(k).Follower(kk).Position);
        % Apply Velocity Limits
```

```
        grp(k).Follower(kk).Velocity =
max(grp(k).Follower(kk).Velocity,VelMin);
        grp(k).Follower(kk).Velocity =
min(grp(k).Follower(kk).Velocity,VelMax);
        % Update Position
        grp(k).Follower(kk).Position =
grp(k).Follower(kk).Position +
grp(k).Follower(kk).Velocity;
        % Velocity Mirror Effect
        IsOutside=(grp(k).Follower(kk).Position<VarMin
| grp(k).Follower(kk).Position>VarMax);
        grp(k).Follower(kk).Velocity(IsOutside)=-
grp(k).Follower(kk).Velocity(IsOutside);
        % Apply Position Limits
        grp(k).Follower(kk).Position =
max(grp(k).Follower(kk).Position,VarMin);
        grp(k).Follower(kk).Position =
min(grp(k).Follower(kk).Position,VarMax);
        % Evaluation
        [grp(k).Follower(kk).Cost,
grp(k).Follower(kk).Out] =
CostFunction(grp(k).Follower(kk).Position);
        if
grp(k).Follower(kk).Cost<grp(k).Follower(kk).BestCost

grp(k).Follower(kk).BestPosition=grp(k).Follower(kk).Po
sition;

grp(k).Follower(kk).BestCost=grp(k).Follower(kk).Cost;

grp(k).Follower(kk).BestOut=grp(k).Follower(kk).Out;
        end
        end
    end
 end
-------------------------------------------------------
COOPERATION OF GROUPS
function grp=GroupsCooperation(grp)
    global ProblemSettings;
    CostFunction=ProblemSettings.CostFunction;
    VarSize=ProblemSettings.VarSize;
    VarMin=ProblemSettings.VarMin;
    VarMax=ProblemSettings.VarMax;
    global AlgorithmSettings;
    m=AlgorithmSettings.m;
    nGrp=numel(grp);
    % Groups' Cooperation
```

```
    k=1;kk=1;
    while k<=nGrp
        while kk<=nGrp
            if grp(k).Weight>grp(kk).Weight
                if isempty(grp(kk).Follower)==0

grp(k).Follower(end+1)=grp(kk).Follower(end);
                    grp(kk).Follower(end)=[];
                    grp(k).nFol=grp(k).nFol+1;
                    grp(kk).nFol=grp(kk).nFol-1;
                else

grp(k).Follower(end+1)=grp(kk).Leader;
                    grp(kk)=[];
                    grp(k).nFol=grp(k).nFol+1;
                    nGrp=nGrp-1;
                end
            end
            kk=kk+1;
        end
        k=k+1;
    end
    % Mutation of Young Goats
    MutLim=1;
    for k=1:nGrp
        if isempty(grp(k).Follower)==0
            for kk=1:grp(k).nFol
                if
rand*(1/grp(k).Weight)*(1/grp(k).Follower(kk).Weight)>m
                    mut(MutLim) =
setfield(grp(k).Follower(kk),'Address',[k,kk]);
                    MutLim=MutLim+1;
                end
            end
        end
    end
    [~,SortOrder]=sort([mut.Weight],'descend');
    if MutLim-1>nGrp
        mut=mut(SortOrder(1:nGrp));
    end
    for k=1:numel(mut)
        % Update Position
        mut(k).Position = mut(k).Position ...
            + rand(VarSize).*(rand(VarSize)-
mut(k).Position);
        % Velocity Mirror Effect
```

```
        IsOutside=(mut(k).Position<VarMin |
mut(k).Position>VarMax);
        mut(k).Velocity(IsOutside)=-
mut(k).Velocity(IsOutside);
        % Apply Position Limits
        mut(k).Position = max(mut(k).Position,VarMin);
        mut(k).Position = min(mut(k).Position,VarMax);
        % Evaluation
        [mut(k).Cost, mut(k).Out] =
CostFunction(mut(k).Position);
        if mut(k).Cost<mut(k).BestCost
            mut(k).BestPosition=mut(k).Position;
            mut(k).BestCost=mut(k).Cost;
            mut(k).BestOut=mut(k).Out;
        end
        MutLead=zeros(1,nGrp);
        for kk=1:nGrp
            if mut(k).BestCost<grp(kk).Leader.BestCost
& kk~=MutLead

grp(mut(k).Address(1)).Follower(mut(k).Address(2))=grp(
kk).Leader;

grp(kk).Leader=rmfield(mut(k),'Address');
                MutLead(kk)=kk;
                break
            elseif
mut(k).BestCost<grp(mut(k).Address(1)).Follower(mut(k).
Address(2)).BestCost

grp(mut(k).Address(1)).Follower(mut(k).Address(2))=rmfi
eld(mut(k),'Address');
            end
        end
    end
end
```

References

1. Ghazinoory, S., Huisingh, D.: National program for cleaner production (CP) in Iran: a framework and draft. J. Clean. Prod. **14**(2), 194–200 (2006)
2. Chang, Y.-C., Lin, F.-A., Lin, C.H.: Optimal chiller sequencing by branch and bound method for saving energy. Energy Convers. Manag. **46**(13), 2158–2172 (2005)
3. Chang, Y.-C., Chan, T.-S., Lee, W.-S.: Economic dispatch of chiller plant by gradient method for saving energy. Appl. Energy **87**(4), 1096–1101 (2010)
4. Chang, Y.-C.: An innovative approach for demand side management—optimal chiller loading by simulated annealing. Energy **31**(12), 1883–1896 (2006)

5. dos Santos Coelho, L., Mariani, V.C.: Improved firefly algorithm approach applied to chiller loading for energy conservation. Energy Build. **59**, 273–278 (2013)
6. Santos Coelho, L., Klein, C.E., Sabat, S.L., Mariani, V.C.: Optimal chiller loading for energy conservation using a new differential cuckoo search approach. Energy **75**, 237–243 (2014)
7. Chang, Y.-C., Lee, C.-Y., Chen, C.-R., Chou, C.-J., Chen, W.-H., Chen, W.-H.: Evolution strategy based optimal chiller loading for saving energy. Energy Convers. Manag. **50**(1), 132–139 (2009)
8. Lee, W.-S., Chen, Y.-T., Kao, Y.: Optimal chiller loading by differential evolution algorithm for reducing energy consumption. Energy Build. **43**(2), 599–604 (2011)
9. Jabari, F., Mohammadi-ivatloo, B.: Basic open-source nonlinear mixed integer programming based dynamic economic dispatch of multi-chiller plants. In: Operation, Planning, and Analysis of Energy Storage Systems in Smart Energy Hubs, pp. 121–127. Springer (2018)
10. Rao, R.V.: Optimization of multiple chiller systems using TLBO algorithm. In: Teaching Learning Based Optimization Algorithm, pp. 115–128. Springer (2016)
11. Chang, Y.-C., Lin, J.-K., Chuang, M.-H.: Optimal chiller loading by genetic algorithm for reducing energy consumption. Energy Build. **37**(2), 147–155 (2005)
12. Chang, Y.-C.: Genetic algorithm based optimal chiller loading for energy conservation. Appl. Therm. Eng. **25**(17), 2800–2815 (2005)
13. Lo, C.-C., Tsai, S.-H., Lin, B.-S.: Economic dispatch of chiller plant by improved ripple bee swarm optimization algorithm for saving energy. Appl. Therm. Eng. **100**, 1140–1148 (2016)
14. Zheng, Z.-X., Li, J.-Q.: Optimal chiller loading by improved invasive weed optimization algorithm for reducing energy consumption. Energy Build. **161**, 80–88 (2018)
15. Shefaei, A., Mohammadi-ivatloo, B.: Wild goats algorithm: an evolutionary algorithm to solve the real-world optimization problems. IEEE Trans. Industr. Inform. pp. 1–1 (2017)
16. Chen, C.-L., Chang, Y.-C., Chan, T.-S.: Applying smart models for energy saving in optimal chiller loading. Energy Build. **68**, 364–371 (2014)

Optimization of Tilt Angle for Intercepting Maximum Solar Radiation for Power Generation

Amit Kumar Yadav and Hasmat Malik

Abstract In this study the novelty is determination of optimum tilt angles (β_{opt}) for photovoltaic system at 11 different sites for Gujarat in India. The β_{opt} is searched for maximum incident solar radiation (SR). For calculation SR values given by National Aeronautics and Space Administration (NASA) is utilized. It was found that the optimum tilt angle varies between 1° and 57° throughout the year in Gujarat, India. The monthly optimum tilt angle is maximum in December for different sites in Gujarat India. This study is useful for industry and researcher to install PV system in India to generate maximum power.

Keywords Solar photovoltaic system · Optimum tilt angle · Power generation

1 Introduction

Due to depleting fossil fuels and environmental concern utilization of renewable energy increases during recent years [1]. Especially solar energy based photovoltaic (PV) generation is used to meet energy need [2].

For development and design of PV systems, incident SR over PV panel is needed. As measured SR data are not existing for most of the sites so it can be estimated [3]. On horizontal surface different methods are used to estimate SR [4]. Incident SR on PV panel is affected by tilt angle and orientation [5]. As per rule PV panel face towards north, south at southern, northern hemisphere respectively. The SR variation depending on meteorological conditions affect PV optimum tilt angles for the different sites.

A. K. Yadav (✉)
Electrical and Electronics Engineering Department, National Institute of Technology, Sikkim, Barfung Block, Ravangla, South Sikkim 737139, India
e-mail: aks.kings@gmail.com

H. Malik
Division of Instrumentation and Control Engineering, Netaji Subhas Institute of Technology Delhi, Delhi, India
e-mail: hasmat.malik@gmail.com

M. Pesaran Hajiabbas and B. Mohammadi-Ivatloo (eds.),
Optimization of Power System Problems, Studies in Systems, Decision and Control 262,
https://doi.org/10.1007/978-3-030-34050-6_9

The maximum power from PV system can be obtained by using tracking system that consist of mechanical part and follow sun trajectory continuously [6]. The trackers are costly, require power for operation and complex operation and maintenance procedures. Therefore orientation of PV system at optimum tilt angle (β_{opt}) is most easy implementation. For calculating β_{opt} several authors presented different studies [7]. In this study β_{opt} is calculated by varying tilt angle from 0° to 90° at steps of 1° and β_{opt} is selected at which incident SR on PV panel is maximum.

This chapter is organized as follows: methodology is presented in Sect. 2. The results and discussion are given in Sect. 3 and conclusion in Sect. 4.

2 Methodology

2.1 Optimum Tilt Angle Determination

The extraterrestrial radiation is the solar radiation received outside earth's surface and its intensity varies throughout the year. The average value of extraterrestrial radiation i.e. the solar constant is $S_0 = 1367\ \text{W/m}^2$. The extraterrestrial radiation H_0 for nth day of the year is given by following equation:

$$H_0 = \frac{24}{\pi} S_0 \left(1 + 0.033 \cos \frac{360n}{365}\right) \quad (1)$$

The polar axis of earth is inclined at an angle of 66.55° to the elliptical plane and by 23.45° from perpendicular to elliptical plane. The rotation of the earth on its inclined polar axis with respect to elliptic plane is responsible for different seasons on earth. It causes lengthy days in summer and shorter days in winter. The angle made by the lines joining the centre of the earth to centre of the sun with its projection on the equatorial plane of the earth is called the declination angle. It varies due to the inclination of the earth's polar axis and its revolution around the sun. It varies between −23.45° and 23.45°. The declination angle (δ) is as follows [8]:

$$\sigma = 23.45 \times \sin\left(\frac{2\pi(248 + n)}{365}\right) \quad (2)$$

The angular displacement of the sun about east or west of local meridian due to the rotary motion of earth on axis is called hour angle (w). It is an expression explaining the variation between local solar time and solar noon. After solar noon the hour angle measures time in duration of 1° for each 4 min or 15° per hour. The hour angle is positive after solar noon and it is negative before solar noon. The hour angle is given by following equation.

$$w = \cos^{-1}(-\tan\phi \tan\delta) \quad (3)$$

The incident radiation reaches the earth's surface without being absorbed or scattered, is called beam radiation. Some of the radiation from the sun is scattered back to the atmosphere and a part of it is scattered to earth, this scattered radiation reaching the earth's surface is called diffuse radiation (H_d). The H_d is given by Eqs. 4 and 5 for monthly average clearness index (K_t) and the K_t is $\frac{H_g}{H_0}$

$$H_d = H_g(1.391 - 3.560K_t + 4.189K_t^2 - 2.137K_t^3) \text{ if } \ w < 81.4^\circ \quad (4)$$

$$H_d = H_g(1.311 - 3.022K_t + 3.427K_t^2 - 1.821K_t^3) \text{ if } \ w > 81.4^\circ \quad (5)$$

When the solar radiation reaches the earth's surface, some of it is reflected by the ground and other objects on the ground. This radiation is called reflected radiation (H_r). The total solar radiation on a horizontal surface is called global radiation (H_g) which is the sum of the beam, diffused and reflected radiations. For maximum utilization of solar radiation on PV panel the tilt angle (β) is used which lies between 0° and 90° [8]. The angle between the plane of the PV panel surface and horizontal is called tilt angle (β). Thus for a surface tilted at a tilt angle (β) from the horizontal, the incident total global solar radiation on tilted surface (H_T) incorporating isotropic model [9] is given by the relation:

$$H_T = (H_g - H_d)R_b + H_g\rho\frac{(1 - \cos\beta)}{2} + H_d\frac{(1 + \cos\beta)}{2} \quad (6)$$

For the surface in the northern hemisphere sloped toward the equator Liu and Jordan [10] is used to calculate R_b given by following equation.

$$R_b = \frac{\cos(\phi - \beta)\cos\delta\sin w_{ss} + w_{ss}\sin(\phi - \beta)\sin\delta}{\cos\phi\cos\delta\sin w_{ss} + w_{ss}\sin\phi\sin\delta} \quad (7)$$

where sunset hour angle at tilted surface (w_{ss}) is given by following relation

$$w_{ss} = \min\left[\cos^{-1}(-\tan\phi\tan\delta), \cos^{-1}(-\tan(\phi + \beta)\tan\delta)\right] \quad (8)$$

The optimum tilt angle for maximizing incident solar radiation (β_{opt}) is calculated by changing β between 0° to 90° at step of 1° for which H_T is maximum. The optimum tilt angle is required to maximize photovoltaic (PV) array output and hence to minimize PV array capacity (C_{PV}) in SAPV system as shown below [11].

$$C_{PV} = \frac{\eta_{PV} A H_T}{L} \quad (9)$$

where A is PV array area (m^2), η_{PV} is efficiency of PV array, L is daily load consumption. Therefore C_{PV} is dependent on H_T and hence on (β_{opt}) i.e. $C_{PV} = f(\beta_{opt})$.

3 Results and Discussion

This section presents the results of the study. First, the optimum tilt angle (β_{opt}) are calculated using developed program in MATLAB (R2011a). This program utilizes monthly average solar radiation and for obtaining monthly optimum tilt angle, β is changed from 0° to 90° at interval of 1°. In order to obtain H_T for each β the program incorporates Eqs. (1)–(5) and (7)–(8) to calculate each term of Eq. 6. In the end β which cause the maximum value of H_T is selected by the program as the optimum tilt angle β_{opt} for that month. The monthly β_{opt} and its corresponding H_T for 11 sites of Gujarat namely Khambada (22.3° N, 72.62° E), Lamba (21.9° N, 69.31° E), Mahidad (22.27° N, 71.18° E), Roimal (22.0° N, 71.48° E), Sadodar (22.06° N, 70.21° E), Sangasar (22.19° N, 72.10° E), Sinugra (23.09° N, 69.96° E), Suvarda (22.38° N, 70.15° E), Vadgam (24.07° N, 72.48° E), Vandhya (23.24° N, 70.61° E), Lodhrani (23.88° N, 70.64° E) are shown Tables 1, 2, 3, 4, 5, 6, 7, 8, 9, 10 and 11 in which HL represents maximum SR at tilt angle equal to latitude and HY is yearly optimum tilt angle. Table 12 present increase in maximum solar radiation.

4 Conclusions

In this study optimum tilt angle (β_{opt}) of 26 different Indian cities are calculated by changing tilt angle (β) from 0° to 90° at step of 1°. Based on results and discussions the following conclusions are drawn:

- It is found that the optimum tilt angle changes between 1° (May, June, July) and 57° (December) throughout the year in Gujarat, India. The optimum tilt angle is maximum in December. The months in which β_{opt} is 1° for different locations in Gujarat, India are April, May, June, July and August (21.9° N $\leq \phi \leq$ 22.30° N); May, June, July (22.38° N $\leq \phi \leq$ 24.07° N). This angles is closed to angle suggested by Soulayman and Sabbagh [12].
- The average global solar radiation on monthly optimum tilted surface varies from 4.42 to 8.274 kWh/m^2/day throughout Gujarat India.
- The maximum incident solar radiation at monthly optimum tilt angle is more in comparison to yearly and latitude based tilt angle.
- The increase in maximum solar radiation at monthly optimum tilt angle in comparison to latitude based tilt angle and yearly optimum tilt angle varies from 7.13% to 7.30% and 4.60% to 5.51%, respectively, showing monthly optimum tilt angle is beneficial for maximum power generation for different sites in Gujarat, India.

Table 1 Extraterrestrial SR, clearness index, optimum tilt-angle and monthly average daily global radiation on optimum tilted surface for Khambada

	Jan	Feb	Mar	April	May	June	July	Aug	Sep	Oct	Nov	Dec
H_g	4.65	5.28	6.14	6.61	6.63	5.7	4.52	4.42	5.11	5.21	4.68	4.25
H_0	8.76	9.15	9.69	9.19	9.16	8.49	9.15	9.42	9.33	9.58	9.30	9.04
K_t	0.53	0.57	0.63	0.71	0.72	0.67	0.49	0.46	0.54	0.54	0.50	0.47
β_{opt}	54	42	27	8	1	1	1	1	20	38	50	56
H_T	7.28	6.66	6.69	6.64	6.63	5.7	4.52	4.42	5.34	6.25	6.63	6.93
HL	6.35	6.36	6.68	6.47	6.01	4.98	3.99	4.18	5.33	6.08	6.03	5.93
HY	6.70	6.56	6.80	6.50	5.99	4.95	3.97	4.18	5.40	6.26	6.54	6.28

Table 2 Extraterrestrial SR, clearness index, optimum tilt-angle and monthly average daily global radiation on optimum tilted surface for Lamba

	Jan	Feb	Mar	April	May	June	July	Aug	Sep	Oct	Nov	Dec
H_g	5.27	6.18	6.92	7.5	7.45	6.42	5.37	5.4	6.16	5.97	5.32	4.95
H_0	8.79	9.17	9.72	9.22	9.19	8.52	9.18	9.45	9.36	9.61	9.33	9.07
K_t	0.59	0.67	0.71	0.81	0.81	0.75	0.58	0.57	0.65	0.62	0.56	0.54
β_{opt}	54	41	26	8	1	1	1	1	19	37	49	56
H_T	8.27	7.69	7.49	7.53	7.45	6.42	5.37	5.4	6.41	7.11	7.47	8.13
HL	7.19	7.36	7.48	7.33	6.78	5.65	4.75	5.10	6.40	6.92	6.80	6.92
HY	7.59	7.59	7.61	7.36	6.76	5.61	4.73	5.11	6.48	7.11	7.43	7.32

Table 3 Extraterrestrial SR, clearness index, optimum tilt-angle and monthly average daily global radiation on optimum tilted surface for Mahidad

	Jan	Feb	Mar	April	May	June	July	Aug	Sep	Oct	Nov	Dec
H_g	4.65	5.3	6.19	6.65	6.68	5.77	4.71	4.5	5.24	5.21	4.68	4.31
H_0	8.77	9.15	9.70	9.20	9.16	8.50	9.15	9.43	9.33	9.58	9.31	9.04
K_t	0.53	0.57	0.63	0.72	0.72	0.67	0.51	0.47	0.56	0.54	0.50	0.47
β_{opt}	54	42	27	8	1	1	1	1	20	38	50	56
H_T	7.27	6.68	6.74	6.68	6.68	5.77	4.71	4.5	5.47	6.25	6.62	7.03
HL	6.35	6.38	6.73	6.51	6.06	5.05	4.16	4.25	5.47	6.08	6.02	6.02
HY	6.70	6.59	6.85	6.53	6.03	5.01	4.13	4.26	5.54	6.26	6.54	6.37

Table 4 Extraterrestrial SR, clearness index, optimum tilt-angle and monthly average daily global radiation on optimum tilted surface for Roimal 2

	Jan	Feb	Mar	April	May	June	July	Aug	Sep	Oct	Nov	Dec
H_g	4.65	5.3	6.19	6.65	6.68	5.77	4.71	4.5	5.24	5.21	4.68	4.31
H_0	8.78	9.17	9.71	9.21	9.18	8.51	9.17	9.44	9.35	9.60	9.33	9.06
K_t	0.52	0.57	0.63	0.72	0.72	0.67	0.51	0.47	0.56	0.54	0.50	0.47
β_{opt}	54	42	27	8	1	1	1	1	20	38	50	56
H_T	7.24	6.66	6.73	6.68	6.68	5.77	4.71	4.5	5.47	6.23	6.60	7.00
HL	6.32	6.36	6.72	6.50	6.06	5.05	4.16	4.25	5.46	6.07	6.00	5.99
HY	6.70	6.59	6.85	6.53	6.03	5.01	4.13	4.26	5.54	6.26	6.54	6.37

Table 5 Extraterrestrial SR, clearness index, optimum tilt-angle and monthly average daily global radiation on optimum tilted surface for Sadodar

	Jan	Feb	Mar	April	May	June	July	Aug	Sep	Oct	Nov	Dec
H_g	4.64	5.34	6.17	6.7	6.67	5.8	4.77	4.58	5.2	5.22	4.66	4.31
H_0	8.78	9.16	9.71	9.21	9.18	8.51	9.16	9.44	9.35	9.60	9.32	9.06
K_t	0.52	0.58	0.63	0.72	0.72	0.68	0.52	0.48	0.55	0.54	0.49	0.47
β_{opt}	54	42	27	8	1	1	1	1	20	38	50	56
H_T	7.23	6.71	6.71	6.73	6.67	5.8	4.77	4.58	5.43	6.25	6.58	7.00
HL	6.31	6.41	6.70	6.55	6.05	5.08	4.22	4.33	5.42	6.08	5.98	5.99
HY	6.69	6.64	6.83	6.58	6.03	5.04	4.19	4.33	5.50	6.27	6.51	6.37

Table 6 Extraterrestrial SR, clearness index, optimum tilt-angle and monthly average daily global radiation on optimum tilted surface for Sangasar

	Jan	Feb	Mar	April	May	June	July	Aug	Sep	Oct	Nov	Dec
H_g	4.65	5.28	6.14	6.61	6.63	5.7	4.52	4.42	5.11	5.21	4.68	4.25
H_0	8.77	9.15	9.70	9.20	9.17	8.50	9.15	9.43	9.34	9.59	9.31	9.05
K_t	0.52	0.57	0.63	0.71	0.72	0.67	0.49	0.46	0.54	0.54	0.50	0.46
β_{opt}	54	42	27	8	1	1	1	1	20	38	50	56
H_T	7.26	6.65	6.69	6.64	6.63	5.7	4.52	4.42	5.34	6.24	6.62	6.91
HL	6.34	6.35	6.678	6.47	6.01	4.98	3.99	4.18	5.33	6.08	6.02	5.92
HY	6.70	6.56	6.80	6.50	5.99	4.95	3.97	4.18	5.40	6.26	6.54	6.28

Table 7 Extraterrestrial SR, clearness index, optimum tilt-angle and monthly average daily global radiation on optimum tilted surface for Sinugra

	Jan	Feb	Mar	April	May	June	July	Aug	Sep	Oct	Nov	Dec
H_g	4.24	4.95	5.67	6.31	6.47	6.18	5.24	4.96	5.28	4.94	4.33	3.99
H_0	8.71	9.09	9.64	9.14	9.10	8.44	9.08	9.37	9.28	9.52	9.24	8.97
K_t	0.48	0.54	0.58	0.69	0.71	0.73	0.57	0.52	0.56	0.51	0.46	0.44
β_{opt}	54	43	28	9	1	1	1	2	21	39	50	56
H_T	6.67	6.31	6.22	6.35	6.47	6.18	5.24	4.96	5.54	5.98	6.20	6.55
HL	5.85	6.03	6.21	6.18	5.85	5.41	4.62	4.62	5.53	5.82	5.64	5.63
HY	6.14	6.20	6.31	6.21	5.84	5.38	4.60	4.69	5.59	5.97	6.08	5.93

Table 8 Extraterrestrial SR, clearness index, optimum tilt-angle and monthly average daily global radiation on optimum tilted surface for Suvarda

	Jan	Feb	Mar	April	May	June	July	Aug	Sep	Oct	Nov	Dec
H_g	4.64	5.34	6.17	6.7	6.67	5.8	4.77	4.58	5.2	5.22	4.66	4.31
H_0	8.76	9.14	9.69	9.19	9.15	8.49	9.14	9.42	9.32	9.57	9.30	9.03
K_t	0.52	0.58	0.63	0.72	0.72	0.68	0.52	0.48	0.55	0.54	0.50	0.47
β_{opt}	54	42	27	8	1	1	1	2	20	38	50	56
H_T	7.27	6.74	6.73	6.73	6.67	5.8	4.77	4.58	5.43	6.27	6.61	7.05
HL	6.35	6.43	6.71	6.56	6.05	5.07	4.21	4.33	5.43	6.10	6.01	6.03
HY	6.70	6.64	6.84	6.58	6.03	5.04	4.19	4.33	5.50	6.27	6.52	6.39

Table 9 Extraterrestrial SR, clearness index, optimum tilt-angle and monthly average daily global radiation on optimum tilted surface for Vadgam

	Jan	Feb	Mar	April	May	June	July	Aug	Sep	Oct	Nov	Dec
H_g	4.33	4.93	5.77	6.27	6.54	6.13	4.83	4.56	5.2	4.97	4.34	3.96
H_0	8.63	9.02	9.57	9.07	9.02	8.36	9.00	9.29	9.21	9.45	9.17	8.89
K_t	0.50	0.54	0.60	0.69	0.72	0.73	0.53	0.49	0.56	0.52	0.47	0.44
β_{opt}	55	44	29	10	1	1	1	3	22	39	53	57
H_T	6.96	6.36	6.37	6.32	6.54	6.13	4.83	4.56	5.48	6.08	6.63	6.63
HL	6.10	6.07	6.36	6.15	5.91	5.35	4.25	4.31	5.47	5.92	5.94	5.71
HY	6.39	6.25	6.46	6.18	5.90	5.32	4.23	4.32	5.54	6.06	6.20	6.00

Table 10 Extraterrestrial SR, clearness index, optimum tilt-angle and monthly average daily global radiation on optimum tilted surface for Vandhya

	Jan	Feb	Mar	April	May	June	July	Aug	Sep	Oct	Nov	Dec
H_g	4.22	4.88	5.63	6.35	6.67	6.27	5.16	4.88	5.18	4.81	4.23	3.96
H_0	8.69	9.08	9.63	9.13	9.09	8.42	9.07	9.35	9.27	9.51	9.23	8.96
K_t	0.48	0.53	0.58	0.69	0.73	0.74	0.56	0.52	0.55	0.50	0.45	0.44
β_{opt}	54	43	28	9	1	1	1	2	21	39	51	57
H_T	6.66	6.23	6.18	6.39	6.67	6.27	5.16	4.88	5.44	5.83	6.07	6.52
HL	5.84	5.95	6.17	6.22	6.04	5.49	4.55	4.61	5.43	5.67	5.52	5.60
HY	6.13	6.13	6.28	6.25	6.02	5.46	4.53	4.62	5.49	5.83	5.95	5.90

Table 11 Extraterrestrial SR, clearness index, optimum tilt-angle and monthly average daily global radiation on optimum tilted surface for Lodhrani

	Jan	Feb	Mar	April	May	June	July	Aug	Sep	Oct	Nov	Dec
H_g	4.22	4.88	5.63	6.35	6.67	6.27	5.16	4.88	5.18	4.81	4.23	3.96
H_O	8.64	9.03	9.58	9.08	9.04	8.37	9.02	9.31	9.22	9.46	9.18	8.91
K_t	0.48	0.53	0.58	0.69	0.73	0.74	0.57	0.52	0.56	0.50	0.46	0.44
β_{opt}	55	44	29	10	1	1	1	3	21	39	53	57
H_T	6.74	6.28	6.21	6.40	6.67	6.27	5.16	4.88	5.45	5.87	6.42	6.61
HL	5.91	6.00	6.20	6.23	6.03	5.48	4.54	4.61	5.45	5.71	5.75	5.68
HY	6.21	6.18	6.31	6.26	6.02	5.45	4.52	4.52	5.51	5.87	6.02	5.98

Table 12 Increase in SR at monthly OPTA in comparison to latitude and annual optimum based tilt angle

S. No.	Cities	Sum of SR at monthly optimum tilt angle	Sum of SR at yearly optimum tilt angle	Average SR at latitude based tilt angle	% increase in SR in comparison to yearly optimum	% increase in SR in comparison to latitude based tilt angle
1	Khambada	73.71	70.18	68.44	4.7890	7.1411
2	Lamba	84.78	80.73	78.63	4.7771	7.2511
3	Mahidad	74.45	70.88	69.12	4.7952	7.1530
4	Roimal	74.30	70.88	68.99	4.6030	7.1343
5	Sadodar	74.49	71.03	69.16	4.6449	7.1474
6	Sangasar	73.65	70.18	68.38	4.7115	7.1435
7	Sinugra	72.71	69.00	67.51	5.1025	7.1390
8	Suvarda	74.68	71.08	69.33	4.8206	7.1587
9	Vadgam	72.93	68.91	67.59	5.5121	7.3089
10	Vandhya	72.34	68.64	67.16	5.1147	7.1562
11	Lodhrani	73.00	69	67.66	5.4795	7.3033

MATLAB Code

```
clc
clear all; close all
for n=1:1:12
    % Latitude
    y=[22.3; 21.9; 22.27; 22.03; 22.06; 22.19; 23.09;
22.38; 24.07; 23.24; 23.88];
    % Global Solar Radiation
    Hg=[4.65 5.28 6.14 6.61 6.63 5.7 4.52 4.42 5.11
5.21 4.68 4.25;
        5.27 6.18 6.92 7.5 7.45 6.42 5.37 5.4 6.16 5.97
5.32 4.95;
        4.65 5.3 6.19 6.65 6.68 5.77 4.71 4.5 5.24 5.21
4.68 4.31;
        4.65 5.3 6.19 6.65 6.68 5.77 4.71 4.5 5.24 5.21
4.68 4.31;
        4.64 5.34 6.17 6.7 6.67 5.8 4.77 4.58 5.2 5.22
4.66 4.31;
        4.65 5.28 6.14 6.61 6.63 5.7 4.52 4.42 5.11
5.21 4.68 4.25;
```

```
        4.24 4.95 5.67 6.31 6.47 6.18 5.24 4.96 5.28
4.94 4.33 3.99;

        4.64 5.34 6.17 6.7 6.67 5.8 4.77 4.58 5.2 5.22
4.66 4.31;
        4.33 4.93 5.77 6.27 6.54 6.13 4.83 4.56 5.2
4.97 4.34 3.96;
        4.22 4.88 5.63 6.35 6.67 6.27 5.16 4.88 5.18
4.81 4.23 3.96;
        4.22 4.88 5.63 6.35 6.67 6.27 5.16 4.88 5.18
4.81 4.23 3.96];
    % Julian Day
    N=[17 47 75 105 135 163 198 228 258 288 318 344];
    i=1;
    x=11;
    % T=0:1:90;
    T=y(:,x);
    lat=y(:,x);
    % Declination Angle
    delta=[-20.92 -12.95 -2.42 9.41 18.79 23.15 21.18
13.45 2.22 -9.59 -18.91 -23.05];
    % Sunshine Hour Angle
    ws=(acosd(-tand(delta).*tand(lat)));
    % ws=[76.3550 81.8417 88.5055 95.8702 102.1205
105.2995 103.8347 98.4873 91.3708 84.0149 77.7949
74.7761]
    % Extraterrestial Radiation
Ho=[((24*3600)/pi)*1367.*(1+0.033.*cos((360.*N)/365)).*
cosd(lat).*cosd(delta).*sind(ws)+0.0175.*ws.*sind(lat).
*sind(delta)]./3600000;
    % Ho=[7.9366 8.3698 8.9185 8.4354 8.3233 7.6595
8.2782 8.6192 8.5849 8.7888 8.4533 8.1509]
    % H=[2.44 2.87 4.80 5.22 6.14 4.95 4.07 3.48 3.98
4.22 3.16 2.85]
    H=Hg(x,:);
    % Clearness Index
    Kt=H./Ho;
    % Kt=[0.3074 0.3429 0.5382 0.6188 0.7377 0.6463
0.4917 0.4037 0.4636 0.4802 0.3738 0.3497]
    % Kt=[0.3074 0.3738 0.3497]% for ?<81.40
    % Kt1=[0.3429 0.5382 0.6188 0.7377 0.6463 0.4917
0.4037 0.4636 0.4802]% for ?>81.4
    % Hd=1.391-3.560.*Kt+4.184.*(Kt.^2)-2.137.*(Kt.^3)
    % Hd1=1.311-
3.022.*Kt1+3.427.*(Kt1.^2)+1.821.*(Kt1.^3)
    % Hd=[0.6299 0.7511 0.9611 1.1847 1.6777 1.2810
0.8701 0.7693 0.8280 0.8517 0.5333 0.5663]
    % Calculation of Diffuse Radiation, Beam Radiation
    if ws(:,n)<81.4
```

```
        Hd=((1.391-3.560*(Kt(:,n))+4.189.*(Kt(:,n))^2)-
2.137*(Kt(:,n))^3);
    else
        Hd=((1.311-3.022*Kt(:,n)+ 3.427.* (Kt(:,n).^2)+
1.821*Kt(:,n).^3));
    end
    Hb=H(:,n)-Hd;
    % Hb=[1.8101 2.1189 3.8389 4.0353 4.4623 3.6690
3.1999 2.7107 3.1520 3.3683 2.6267 2.2837]
    wss=acosd(-tand(lat-T(i,:)).*tand(delta(:,n)));
    wts=min(ws(:,n),wss);
    Rb=((((wts.*pi)./180).*sind(delta(:,n)).*sind(lat-
T(i,:))+cosd(delta(:,n)).*sind(abs(wts)).* cosd(lat-
T(i,:)))./(((ws(:,n)*pi)./180).*
sind(delta(:,n))*sind(lat(i,:))+
cosd(delta(:,n)).*sind(ws(:,n))* cosd(lat(i,:))));
    Rd=(1+cosd(T))./2;
    ref=0.2;
    Rr=ref.*((1-cosd(T(i,:)))./2);
    Ht=Hb*Rb+Hd*Rd+H(:,n).*Rr;
    [m,y]=max(Ht);
end
```

References

1. Arikan, O., Isen, E., Durusu, A., Kekezoglu, B., Bozkurt, B., Erduman, A.: Introduction to hybrid systems—ildiz Technical University. In: Eurocon, pp. 1145–1149 (2013)
2. Kekezoglu, B., Arikan, O., Erduman, A., Isen, E., Durusu, A., Bozkurt, A.: Reliability analysis of hybrid energy systems: case study of Davutpasa Campus. Eurocon, pp. 1141–1144 (2013)
3. Yadav, A.K., Chandel, S.S.: Tilt angle optimization to maximize incident solar radiation: a review. Renew. Sustain. Energy Rev. **23**, 503–513 (2013)
4. Bakirci, K.: Models of solar radiation with hours of bright sunshine: a review. Renew. Sustain. Energy Rev. **13**(9), 2580–2588 (2009)
5. Chandel, S.S., Aggarwal, R.K.: Estimation of hourly solar radiation on horizontal and inclined surfaces in Western Himalayas. Smart Grid Renew. Energy **02**(01), 45–55 (2011)
6. Chandel, S.S., Aggarwal, R.K., Pandey, A.N.: New correlation to estimate global solar radiation on horizontal surfaces using sunshine hour and temperature data for indian sites. J. Sol. Energy Eng. **127**(3), 417 (2005)
7. El-Sebaii, A.A., Al-Hazmi, F.S., Al-Ghamdi, A.A., Yaghmour, S.J.: Global, direct and diffuse solar radiation on horizontal and tilted surfaces in Jeddah. Saudi Arabia. Appl. Energy **87**(2), 568–576 (2010)
8. Copper, P.I.: The absorption of solar radiation in solar stills. Sol. Energy **12**(3), 333–346 (1969)
9. Benghanem, M.: Optimization of tilt angle for solar panel: case study for Madinah. Saudi Arabia. Appl. Energy **88**, 1427–1433 (2011)

10. Liu, B.Y.H., Jordan, R.C.: Daily insolation on surfaces tilted toward the equator. ASHRAE Trans. **67**, 526–541 (1962)
11. Liu, B.Y.H., Jordan, R.C.: The long term average performance of flat plate solar energy collectors. Sol. Energy **7**, 53 (1963)
12. Soulayman, S., Sabbagh, M.: Optimum tilt angle at tropical region. Int. J. Renew. Energy Dev. **4**(1), 48–54 (2015). https://doi.org/10.14710/ijred.4.1.48-54

Probabilistic Power Flow Analysis of Distribution Systems Using Monte Carlo Simulations

Farkhondeh Jabari, Maryam Shamizadeh and Behnam Mohammadi-Ivatloo

Abstract Nowadays, population growth has led to increased electricity consumption in different residential, commercial and industrial district levels. This may leads to load-generation imbalance, voltage drop, cascaded failure and catastrophic blackout of interconnected power networks. To prevent from wide spread outages and uncontrolled islanding of large-scale and distributed grids, uncertainties associated with loads are considered in steady-state voltage stability analysis and reliability assessment. Therefore, this chapter aims to present a Monte Carlo simulations based probabilistic power flow method for finding all critical buses against variations of active and reactive loads. In this approach, backward-forward sweep based load flow is used to find optimal operating point of benchmark distribution grid in each scenario. Number of scenarios with bus voltage magnitude violation probability is used to cluster nodes into two critical and non-critical categories. Robustness and effectiveness of Monte Carlo based probabilistic power flow algorithm is revealed by simulations on 33-bus radial distribution system.

Keywords Optimal power flow · Monte Carlo simulation (MCS) · Backward-forward sweep (BFS) · Critical node · Bus voltage magnitude violation probability

---

F. Jabari · M. Shamizadeh (✉) · B. Mohammadi-Ivatloo
Faculty of Electrical and Computer Engineering, University of Tabriz, Tabriz, Iran
e-mail: m.shamizadeh94@ms.tabrizu.ac.ir

F. Jabari
e-mail: f.jabari@tabrizu.ac.ir

B. Mohammadi-Ivatloo
e-mail: bmohammadi@tabrizu.ac.ir

M. Pesaran Hajiabbas and B. Mohammadi-Ivatloo (eds.),
Optimization of Power System Problems, Studies in Systems, Decision and Control 262,
https://doi.org/10.1007/978-3-030-34050-6_10

Nomenclature

$\dot{J}_i^{k+1}$	Injected current to node i in $(k+1)$th iteration
$\dot{V}^{(k)}$	Voltage of node i in kth iteration
$\dot{S}$	Power injected to node i
$\dot{Y}$	Parallel admittance of node i
$\dot{I}_i^{k+1}$	Current of branch i in $(k+1)$th iteration
$\dot{I}_j^{k+1}$	Current of branch j in $(k+1)$th iteration
$\dot{Z}_j$	Impedance of branch j
$\dot{V}_j^{k+1}$	Voltage of bus i in $(k+1)$th iteration
$\dot{V}_j^{k+1}$	Voltage of bus j in $(k+1)$th iteration
F_{loss}	Total real power losses as objective function
$g_{i,j}$	Conductance of branch i to j
V_i	Voltage magnitude of bus i
V_j	Voltage magnitude of bus j
θ_i	Voltage angle of bus i
θ_j	Voltage angle of bus j
n_l	Total number of branches
$V_i^{\min}$, $V_i^{\max}$	Minimum and maximum values of voltage magnitude for bus i
I_b	Current of branch b
$I_b^{\max}$	Maximum current of branch b
$\mu_{P_{L_i}}, \sigma_{P_{L_i}}$	Mean and standard deviation of active power of ith load
$\mu_{Q_{L_i}}, \sigma_{Q_{L_i}}$	Mean and standard deviation of reactive power of ith load
$P_{L_i}(\mathrm{s}\vert\mu_{P_{L_i}}, \sigma_{P_{L_i}})$	Active power of ith load in sth scenario
$Q_{L_i}(\mathrm{s}\vert\mu_{Q_{L_i}}, \sigma_{Q_{L_i}})$	Reactive power of ith load in sth scenario

1 Introduction

In recent years, variable nature of electricity demand and uncertain productions of renewable energy resources based power generation facilities influence on stability and reliability of radial and meshed distribution systems. Hence, optimization of power system performance under uncertain operating conditions has attracted more attention.

Some scholars have focused on probabilistic optimal power flow (OPF) problem. Jabari et al. [1, 2] developed a forward-selection backward-elimination method based optimal power flow algorithm for defensive splitting of faulted power systems. Monte Carlo simulations (MCS) [3] and point estimation method (PEM) [4] based stochastic contingency analysis are applied on interconnected electricity systems for finding all critical buses while occurring different single transmission line outages. A wide variety of optimization techniques are implemented on power flow problems, such as nonlinear programming (NLP) [5–8], linear programming (LP)

[9–11] and quadratic programming (QP) [12]. Generally, nonlinear programming approaches have many drawbacks such as insecure convergence. Moreover, linear and quadratic programming methods have some disadvantages associated with cost approximation. OPF analysis uses different advanced optimization processes such as genetic algorithm, interior point method, simulated annealing, decomposition and Newton's method to find a good operating point for interconnected electricity grids and determine bus voltage angle and magnitude, output active and reactive powers of generator and transmission congestion. Interior point method is presented as computationally efficient scheme, but it suffers from bad initial, termination and in some cases is unable to solve nonlinear and quadratic objective function [13]. In [14], a simple genetic algorithm is used for OPF solution. Active power output of generators, voltages and transformer taps have been selected as control variables. Transmission capacity criterion, voltage permitted range and generation limits are considered as optimization constraints. Some scholars have considered Karush–Kuhn–Tucker (KKT) optimality conditions instead of solving OPF original problem. For equality-constrained optimization problems, KKT conditions are a set of nonlinear equations, which can be solved using a Newton algorithm. In Newton algorithm based OPF problem [15], inequality constraints are added as quadratic penalty terms to objective function. Probabilistic optimal power flow [16] is developed to estimate all steady state characteristics of power systems and solve optimal power flow problem with uncertain loads and variable renewable energy sources based power production processes. There is overwhelming agreement among scholars that Monte-Carlo based stochastic programming methods are more capable and accurate in comparison with other uncertainty modeling techniques. Besides, this approach requires much more deterministic Newton-Raphson power flow calculations to converge. To improve efficiency and speed of stochastic power system analysis, K-point estimation method [17–19] with computational burden less than Monte Carlo simulations is widely used to solve OPF problem under uncertain operating condition. Popular power flow algorithms such as Newton-Raphson and fast decoupled load flow method generally fail to converge when analyzing radial distribution systems with high R/X ratio [20]. In [21], an interval arithmetic based backward-forward sweep algorithm is used to model both lower and upper bounds of uncertain loads for balanced radial distribution system power flow analysis. However, upper and lower ranges estimated by interval arithmetic approach tend to be large especially in long iterative computations [22]. A quadratic convex approximation strategy is introduced in [23] for OPF in power distribution systems. This algorithm is based on linear formulation of power flow presented by Garces [24]. Different consideration are made ending at a non-iterative analytical solution for relaxed problem. Both, quadratic convex and analytical relaxed models are extensible to three-phase unbalanced distribution systems.

As reviewed, different probabilistic power flow algorithms have been implemented on distribution systems to model fluctuations of renewable energy resources based power generation processes and electrical loads. This chapter aims to present a novel backward-forward sweep (BFS) based probabilistic power flow strategy for radial distribution networks. In this method, number of Monte Carlo iterations is considered as known parameter. A random number between 0 and 1 is generated as probability of load value of each bus. Mean, standard deviation and probability of each load level are then used to determine an uncertain value for active or reactive power of demand of each bus. Afterward, BFS load flow is solved, total real power losses is calculated and voltage magnitude of each bus and current of each branch are obtained in each stochastic scenario. When all iterations of MCS are run, probability of voltage magnitude violation is computed for each node and critical buses are marked based on this index. To prove applicability of MCS based BFS load flow analysis in finding critical buses, a 33-bus radial distribution system is studied.

Other sections of this chapter are organized as follows: A comprehensive problem formulation on backward-forward sweep load flow analysis and Monte Carlo based uncertainty modeling technique is presented in Sect. 2. Simulation results and discussions are provided in Sect. 3. Finally, concluding remarks appear in Sect. 4.

2 Problem Formulation

2.1 Forward-Backward Sweep Algorithm

In this section, backward forward approach is introduced. By considering a sample distribution system as shown in Fig. 1, the injected current to the ith node can be calculated as Eq. (1).

$$\dot{J}_i^{k+1} = \left(\frac{\dot{S}_i}{\dot{V}_i^{(k)}}\right)^* - \dot{Y}_i \dot{V}_i^{(k)} \tag{1}$$

where,

$\dot{J}_i^{k+1}$: Injected current to node i in $(k + 1)$th iteration
$\dot{V}_i^{(k)}$: Voltage of node i in kth iteration
$\dot{S}_i$: Power injection of node i
$\dot{Y}_i$: Parallel admittance of node i
Backward sweep

In this step, sum of all currents injecting to branch i is obtained from Eq. (2).

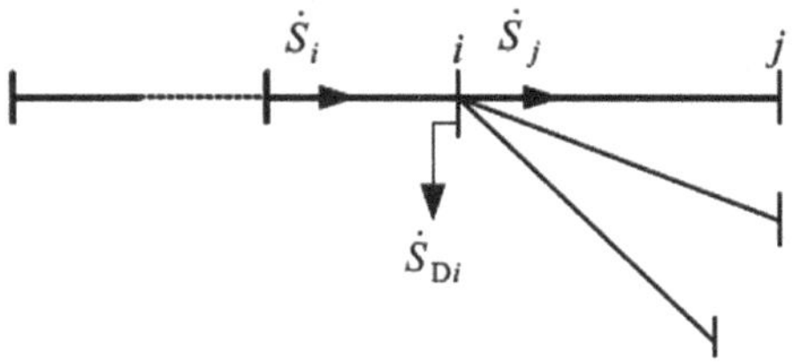

Fig. 1 A simple radial distribution network

$$\dot{I}_i^{k+1} = -\dot{J}_i^{k+1} + \sum_{j \in C_i} \dot{I}_j^{k+1} \quad (2)$$

where,

$\dot{I}_i^{k+1}$: Current of branch i in $(k + 1)$th iteration
$\dot{I}_j^{k+1}$: Current of branch j in $(k + 1)$th iteration
Forward sweep

In this process, according to obtained current in backward process, new value of bus voltage is formulated by Eq. (3).

$$\dot{V}_j^{k+1} = \dot{V}_i^{k+1} - \dot{I}_j^{k+1} \times Z_j \quad (3)$$

In which,

Z_j: Impedance of branch j
$\dot{V}_i^{k+1}$: Voltage of bus i in $(k + 1)$th iteration
$\dot{V}_j^{k+1}$: Voltage of bus j in $(k + 1)$th iteration

Finally, in order to finalize power flow calculations, limitation (4) should be satisfied; otherwise, Eqs. (2) and (3) will be repeated.

2.2 *Monte Carlo Simulations*

As mentioned in Sect. 1, Monte-Carlo simulation based stochastic programming methods are accurate in comparison with other uncertainty modeling techniques. In this research, it is used to generate numerous stochastic scenarios for modeling load uncertainties and finding all critical nodes of a radial distribution system. Let U_1, U_2, and U_3 denote three uniform random numbers on (0, 1). Hence, active and reactive powers (as two independent normal variables) of ith load in sth scenario can be obtained from Eqs. (4) and (5), respectively.

$$P_{L_i}(\mathrm{s}|\mu_{P_{L_i}}, \sigma_{P_{L_i}}) = F_{P_{L_i}}^{-1}(\text{`Normal'}, U_1(s), \mu_{P_{L_i}}, \sigma_{P_{L_i}}) \quad (4)$$

$$Q_{L_i}(\mathrm{s}|\mu_{Q_{L_i}}, \sigma_{Q_{L_i}}) = F_{Q_{L_i}}^{-1}(\text{`Normal'}, U_2(s), \mu_{Q_{L_i}}, \sigma_{Q_{L_i}}) \quad (5)$$

where,

$\mu_{P_{L_i}}, \sigma_{P_{L_i}}$: Mean and standard deviation of active power of ith load
$\mu_{Q_{L_i}}, \sigma_{Q_{L_i}}$: Mean and standard deviation of reactive power of ith load.

3 Proposed Approach and Case Study

In this section, a backward-forward sweep based probabilistic power flow method is proposed for finding all critical buses of IEEE 33-bus radial distribution system [25] against uncertainties of loads, as illustrated in Fig. 2. As shown in Matlab Code section, the bus data matrix is defined as "bdata.not.per.unit". The first column of this matrix refer to number of nodes. Active and reactive power consumptions in each bus are presented at second and third columns of bus data matrix in kW and kVAr, respectively. Similarly, "ldata.not.per.unit" is line data matrix of 33-bus radial benchmark network. In each row of branch information matrix, number of starting and ending point of each line is determined using bus numbers. The third and fourth columns of the line data matrix represent the resistance and reactance of each branch in Ohm. In this research, it is assumed that number of Monte Carlo iterations is known. As obvious from this figure, "MCS" and "mcs" refer to number of Monte Carlo simulations and iteration index, respectively. While, index "mcs" is not more than Monte Carlo iterations (MCS), both active and reactive powers of each bus are updated using two inverse transform functions (4) and (5) with average values equal to those of reported in primary bus data matrix and standard deviations 35 (for active power) and 15 (for reactive power). The probability of each scenario is randomly generated using variable "u(i) = rand". In this research, normal distribution function is used to generate stochastic values for active and reactive loads. Afterward, bus data matrix is updated according to random values generated for active and reactive loads. The backward-forward sweep based load flow analysis is then implemented using updated bus data matrix. In each iteration, if voltage magnitude of bus i is less than 0.9, it will be considered as an unstable state for this node. Finally, the probability of voltage drop for each bus can be calculated by dividing the number of unstable scenarios to total number of Monte Carlo iterations. Figure 3 illustrates the voltage profile in two cases "Deterministic backward-forward sweep based load flow" and "Monte Carlo based probabilistic load flow with 1000 iterations". As obvious from this figure, the voltage magnitude is less than 0.9 per unit in buses 13–18 and 31–33. Moreover, probability of voltage drop for mentioned buses is reported by Fig. 4. According to Fig. 4, buses 14–18 with higher probability of voltage instability are critical under uncertain operating conditions.

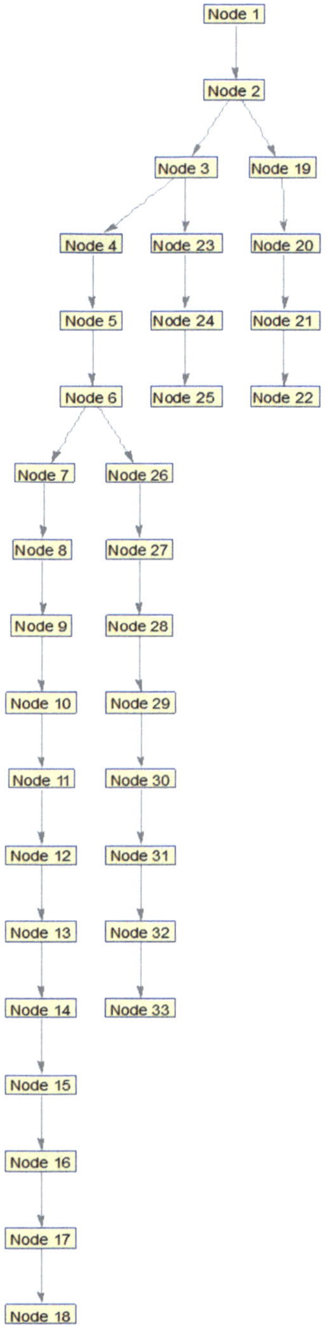

Fig. 2 Single line diagram of IEEE 33-bus radial distribution system

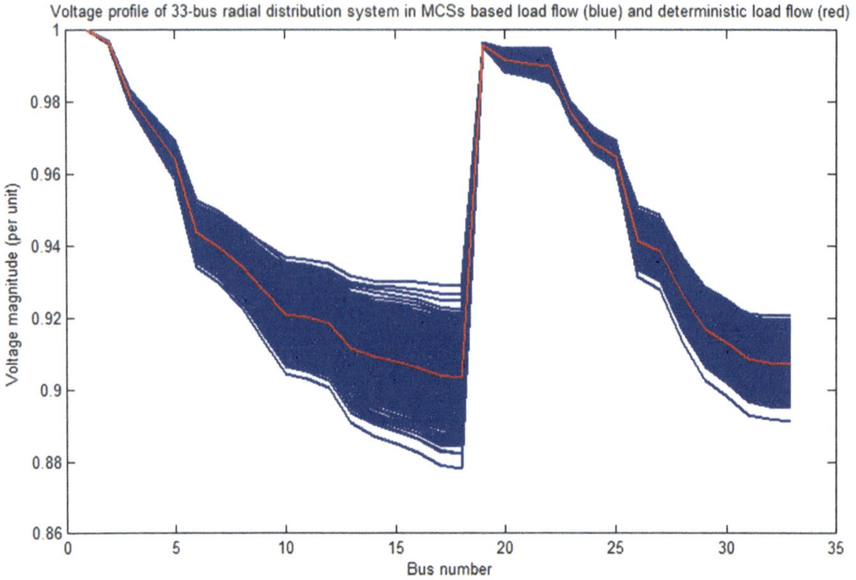

Fig. 3 Voltage profile in two cases: deterministic load flow (blue) and probabilistic load flow (red)

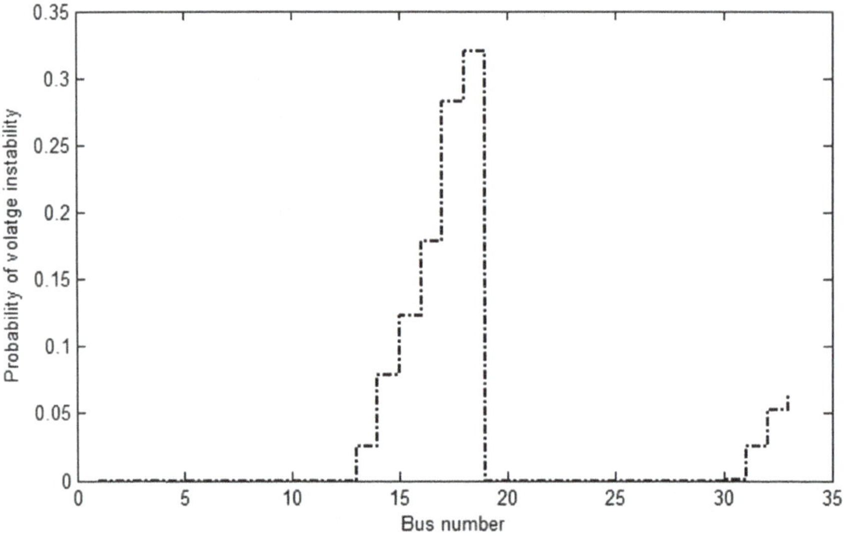

Fig. 4 Probability of voltage instability in 1000 Monte Carlo iterations

4 Conclusion

In this chapter, a novel Monte Carlo integrated backward-forward load flow algorithm was presented. Uncertain nature of active and reactive loads was modeled using inverse transform of normal distribution functions. By implementing 1000 iterations of MCS based BFS power flow scenarios on updated bus data matrix, voltage magnitude and probability of instability for each node were found. If voltage magnitude of bus i at scenario j is either smaller than 0.9 per unit or larger than 1.05 per unit, it will be marked as unstable node against given active and reactive powers. Moreover, probability of voltage instability is calculated based on number of unstable conditions of each bus and total number of Monte Carlo iterations. Simulations were carried out on IEEE 33-bus radial distribution system. Actual active and reactive powers reported in 2nd and 3rd columns of bus data matrix are used as mean values of inverse transform functions of Monte Carlo scenarios. Moreover, standard deviations of power variations are considered to be equal to 35 and 15 for active and reactive loads, respectively. It is found that buses 14–18 with higher probability of voltage instability are critical nodes against demand variations.

MATLAB Code

```
clear all
close all
clc
MCS=input(Enter number of Monte Carlo iterations:');
mcs=1;
while mcs<MCS+1
bdata.not.per.unit=[%Bus     P(Kw)   Q(Kvar)
                     1       0       0
                     2       120     72
                     3       108     48
                     4       144     96
                     5       72      36
                     6       72      24
                     7       240     120
                     8       240     120
                     9       72      24
                     10      72      24
                     11      54      36
                     12      72      42
                     13      72      42
                     14      144     96
                     15      72      12
                     16      72      24
                     17      72      24
                     18      108     48
                     19      108     48
                     20      108     48
                     21      108     48
                     22      108     48
                     23      108     60
                     24      504     240
                     25      504     240
                     26      72      30
                     27      72      30
                     28      72      24
                     29      144     84
                     30      240     720
                     31      180     84
                     32      252     120
                     33      72      48 ];
 for i=2:33
    u(i)=rand;
```

```
bdata.not.per.unit(i,2)=icdf('normal',u(i),bdata.not.pe
r.unit(i,2),35);

bdata.not.per.unit(i,3)=icdf('normal',u(i),bdata.not.pe
r.unit(i,3),15);
end
ldata.not.per.unit=[
%        Inbus  Outbus Resistance(ohm) Reactance(ohm)
          1       2       0.0922          0.0470
          2       3       0.4930          0.2511
          3       4       0.3660          0.1864
          4       5       0.3811          0.1941
          5       6       0.8191          0.7070
          6       7       0.1872          0.6188
          7       8       0.7114          0.2351
          8       9       1.0300          0.7400
          9       10      1.0440          0.7400
          10      11      0.1966          0.0650
          11      12      0.3744          0.1238
          12      13      1.4680          1.1550
          13      14      0.5416          0.7129
          14      15      0.5910          0.5260
          15      16      0.7463          0.5450
          16      17      1.2890          1.7210
          17      18      0.7320          0.5740
          2       19      0.1640          0.1565
          19      20      1.5042          1.3554
          20      21      0.4095          0.4784
          21      22      0.7089          0.9373
          3       23      0.4512          0.3083
          23      24      0.8980          0.7091
          24      25      0.8960          0.7011
          6       26      0.2030          0.1034
          26      27      0.2842          0.1447
          27      28      1.0590          0.9377
          28      29      0.8042          0.7006
          29      30      0.5075          0.2585
          30      31      0.9744          0.9630
          31      32      0.3105          0.3619
          32      33      0.3410          0.5302];
    sizbdata=size(bdata.not.per.unit);
    busnum=sizbdata(1,1);
    sizldata=size(ldata.not.per.unit);
    branchnum=sizldata(1,1);
    %per unit calculation:
    Sbase=10^3;
```

```
Vbase=12.66*10^3;
Zbase=Vbase^2/Sbase;
bdata=bdata.not.per.unit;
ldata=ldata.not.per.unit;
for n=1:busnum
    bdata(n,2)=(bdata(n,2)*1000)/Sbase;
    bdata(n,3)=(bdata(n,3)*1000)/Sbase;
end
for n=1:branchnum
    ldata(n,3)=ldata(n,3)/Zbase;
    ldata(n,4)=ldata(n,4)/Zbase;
end
%per unit calculation finished
terminatebus=zeros(busnum,1);
intermediatebus=zeros(busnum,1);
junctionbus=zeros(busnum,1);
junctionnum=zeros(busnum,1);
refbus=0;
busI=zeros(busnum,1);
v=ones(1,busnum);
I=zeros(busnum,busnum);
for k=1:busnum
    co=0;
    l=0;
    for n=1:branchnum
        if ldata(n,1)==k
            co=co+1;
        end
    end
    if co==0
        terminatebus(k,1)=k;
    elseif co>=2
        junctionbus(k,1)=k;
        junctionnum(k,1)=co;
    elseif co==1
        for m=1:branchnum
            l=l+1;
            if ldata(m,2)==k
                intermediatebus(k,1)=k;
                break
            elseif l==branchnum
                refbus=k;
            end
        end
    end
end
junctionbus;
```

```
    intermediatebus;
    terminatebus;
    refbus;
    junctionnum;
    tempterminatebus=terminatebus;
    controljunctionnum=zeros(busnum,1);
    k=0;
    c=0;
    itecount=0;
    for s=1:15 %iteration
        itecount=itecount+1;
        %backward sweep
        while c==0
            k=k+1;
            juncnum=0;
            n=0;
            stop=0;
            previousI=0;
            if tempterminatebus(k,1)==k

                while(n<branchnum)&&(stop==0)
                    n=n+1;
                    if ldata(n,2)==k
                        a=ldata(n,1);
                        if a==refbus
                            c=1;
                        end
           I(a,k)=busI(k)+(bdata(k,2)-
1i*bdata(k,3))/conj(v(k))+previousI;
                        previousI=I(a,k);
                        tempterminatebus(k,1)=0;
                        if junctionbus(a,1)==a
                            busI(a)=busI(a)+I(a,k);
controljunctionnum(a,1)=controljunctionnum(a,1)+1;
if controljunctionnum(a,1)==junctionnum(a,1)
tempterminatebus(a,1)=a;
                            end
                            break
                        end
                        k=a;
                        n=0;
                    end
                end
                k=0;
            end
        end
```

```
%end of backward sweep
%forward sweep
count=0;
beforev=v;
newldata=ldata;
stop1=0;
forwardbus=zeros(busnum,1);
c=0;
stopif=0;
while stop1==0
    for k=1:branchnum
        if(newldata(k,1)==refbus)&&(stopif==0)
            c=refbus;
            if junctionbus(refbus,1)==refbus
                forwardbus(refbus,1)=1;
            end
            stop2=0;
            while 1
                a=newldata(k,2);
              z=newldata(k,3)+1i*newldata(k,4);
                v(a)=v(c)-z*I(c,a);
                newldata(k,:)=0;
                if junctionbus(a,1)==a
                    forwardbus(a,1)=1;
                end
                if terminatebus(a,1)==a
                    stopif=1 ;
                    stop1=1;
                    break
                end
                for n=1:branchnum
                    if ldata(n,1)==a
                        c=a;
                        k=n;
                        break
                    end
                end
            end
        end
    end
end
stop3=0;
while stop3==0
    stopif1=0;
    for k=1:busnum
        if (forwardbus(k,1)==1)&& (stopif1==0)
            stop4=0;
```

```
            while stop4==0
                stopif2=0;
                hhh=0;
                counter=0;
                for n=1:branchnum
                    counter=counter+1;
           if (newldata(n,1)==k)&& (stopif2==0)
                        hhh=1;
                        c=k;
                        a=newldata(n,2);
              z=newldata(n,3)+1i*newldata(n,4);
                        v(a)=v(c)-z*I(c,a);
                        newldata(n,:)=0;
                        if junctionbus(a,1)==a
                            forwardbus(a,1)=1;
                        end
                        if terminatebus(a,1)==a
                            if newldata==0
                                stopif1=1;
                                stop3=1;
                                stopif2=1;
                                stop4=1;
                                break
                            end
                            stopif2=1;
                            stop4=1;
                        end
                        k=a;
                        stopif2=1;
                    end
            if (counter==branchnum) && (hhh==0)
                        stop4=1;
                    end
                end
            end
        end
    end
end
%end of forward sweep
for p=2:busnum
    if abs(v(p)-beforev(p))<=0.000001
        count=count+1;
    end
end
if count==(busnum-1)
    break
end
```

```
    end
    %data display
    voldisp=zeros(busnum,3);
    curdisp=zeros(branchnum,3);
    co=0;
    for n=1:busnum
        for k=1:busnum
            if I(n,k)~=0
                co=co+1;
                curdisp(co,1)=n;
                curdisp(co,2)=k;
                curdisp(co,3)=I(n,k);
            end
        end
    end
    for n=1:busnum
        voldisp(n,1)=n;
        voldisp(n,2)=abs(v(n));
        voldisp(n,3)=angle(v(n))*180;
    end
    co=0;
    W=zeros(1,branchnum);
    a=zeros(1,branchnum);
    b=zeros(1,branchnum);
    for n=1:busnum
        for k=1:busnum
            if I(n,k)~=0
                co=co+1;
                a(co)=n;
                b(co)=k;
            end
        end
    end
    DG = sparse(a,b,true,busnum,busnum);
    Ploss=0; Qloss=0;
    for k=1:branchnum
        z=curdisp(k,1);
        y=curdisp(k,2);
        for g=1:branchnum
            if ldata(g,1)==z && ldata(g,2)==y

Ploss=Ploss+abs(curdisp(k,3))^2*ldata(g,3);
%ploss=r*|I|^2
Qloss=Qloss+abs(curdisp(k,3))^2*ldata(g,4);
%qloss=x*|I|^2
                break
            end
```

```
            end
        end
        Inew=zeros(busnum,1);
        for l=1:busnum
          Inew(l,1)=(bdata(l,2)-1i*bdata(l,3))/conj(v(l));
        end
        %Outputs
        P_loss=abs(Ploss);
        Q_loss=abs(Qloss);
        Loss(mcs)=P_loss;
        for i=1:33
        voltage(i,mcs)=abs(v(i));
        end
    for i=1:33
        n(i)=0;
    end
    for i=1:33
        for j=1:MCS
            if  voltage(i,j)<0.9
              n(i)=n(i)+1;
            end
        end
    end
    P_instability=n/MCS %Probability of voltage instability
```

References

1. Jabari, F., Seyedi, H., Ravadanegh, S.N.: Large-scale power system controlled islanding based on backward elimination method and primary maximum expansion areas considering static voltage stability. Int. J. Electr. Power Energy Syst. **67**, 368–380 (2015)
2. Jabari, F., Mohammadi-Ivatloo, B.: Backward-forward sweep-based islanding scenario generation algorithm for defensive splitting of radial distribution systems. In: Power Quality in Future Electrical Power Systems. 2017, Institution of Engineering and Technology, pp. 283–304
3. Jabbari, F., Mohammadi-Ivatloo, B.: Static voltage stability assessment using probabilistic power flow to determine the critical PQ buses **8**(4), 9 (2014)
4. Jabari, F., et al.: Stochastic contingency analysis based on voltage stability assessment in islanded power system considering load uncertainty using MCS and k-PEM. In: Handbook of Research on Emerging Technologies for Electrical Power Planning, Analysis, and Optimization, IGI Global, pp. 12–36 (2016)
5. Alsac, O., Stott, B.: Optimal load flow with steady-state security. IEEE Trans. Power Appar. Syst. **3**, 745–751 (1974)
6. Shoults, R.R., Sun, D.: Optimal power flow based upon PQ decomposition. IEEE Trans. Power Appar. Syst. **2**, 397–405 (1982)
7. Bottero, M., Galiana, F., Fahmideh-Vojdani, A.: Economic dispatch using the reduced hessian. IEEE Trans. Power Appar. Syst. **10**, 3679–3688 (1982)

8. Momoh, J.A.: A generalized quadratic-based model for optimal power flow. In: IEEE International Conference on Systems, Man and Cybernetics, 1989. Conference Proceedings. IEEE, New York (1989)
9. Stott, B., Hobson, E.: Power system security control calculations using linear programming, Part I. IEEE Trans. Power Appar. Syst. **5**, 1713–1720 (1978)
10. Stott, B., Marinho, J.: Linear programming for power-system network security applications. IEEE Trans. Power Appar. Syst. **3**, 837–848 (1979)
11. Mota-Palomino, R., Quintana, V.: A penalty function-linear programming method for solving power system constrained economic operation problems. IEEE Trans. Power Appar. Syst. **6**, 1414–1422 (1984)
12. Reid, G.F., Hasdorff, L.: Economic dispatch using quadratic programming. IEEE Trans. Power Appar. Syst. **6**, 2015–2023 (1973)
13. Momoh, J.A., Zhu, J.: Improved interior point method for OPF problems. IEEE Trans. Power Syst. **14**(3), 1114–1120 (1999)
14. Lai, L.L., et al.: Improved genetic algorithms for optimal power flow under both normal and contingent operation states. Int. J. Electr. Power Energy Syst. **19**(5), 287–292 (1997)
15. Sun, D.I., et al.: Optimal power flow by Newton approach. IEEE Trans. Power Appar. Syst. **10**, 2864–2880 (1984)
16. Aien, M., Fotuhi-Firuzabad, M., Rashidinejad, M.: Probabilistic optimal power flow in correlated hybrid wind–photovoltaic power systems. IEEE Trans. Smart Grid **5**(1), 130–138 (2014)
17. Morales, J.M., Perez-Ruiz, J.: Point estimate schemes to solve the probabilistic power flow. IEEE Trans. Power Syst. **22**(4), 1594–1601 (2007)
18. Su, C.-L.: Probabilistic load-flow computation using point estimate method. IEEE Trans. Power Syst. **20**(4), 1843–1851 (2005)
19. Caramia, P., Carpinelli, G., Varilone, P.: Point estimate schemes for probabilistic three-phase load flow. Electr. Power Syst. Res. **80**(2), 168–175 (2010)
20. Singh, S., Ghose, T.: Improved radial load flow method. Int. J. Electr. Power Energy Syst. **44**(1), 721–727 (2013)
21. Das, B.: Radial distribution system power flow using interval arithmetic. Int. J. Electr. Power Energy Syst. **24**(10), 827–836 (2002)
22. Vaccaro, A., Cañizares, C.A., Bhattacharya, K.: A range arithmetic-based optimization model for power flow analysis under interval uncertainty. IEEE Trans. Power Syst. **28**(2), 1179–1186 (2013)
23. Garces, A.: A quadratic approximation for the optimal power flow in power distribution systems. Electr. Power Syst. Res. **130**, 222–229 (2016)
24. Garces, A.: A linear three-phase load flow for power distribution systems. IEEE Trans. Power Syst. **31**(1), 827–828 (2016)
25. Venkatesh, B., Ranjan, R., Gooi, H.: Optimal reconfiguration of radial distribution systems to maximize loadability. IEEE Trans. Power Syst. **19**(1), 260–266 (2004)

Long-Term Load Forecasting Approach Using Dynamic Feed-Forward Back-Propagation Artificial Neural Network

Amin Masoumi, Farkhondeh Jabari, Saeid Ghassem Zadeh and Behnam Mohammadi-Ivatloo

Abstract In recent years, due to increasing rate of electricity demand and power system restructuring with a limited investment in transmission expansion, large power systems may closely be operated at their stability margins. Meanwhile, uncertain and intermittent nature of electricity demand with traditional load forecasting error seriously effects on operation and planning of bulk power grids. Hence, this chapter aims to present a novel approach based on dynamic feed-forward back-propagation artificial neural network (FBP-ANN) for long-term forecasting of total electricity demand. A feed-forward back-propagation time series neural network consists of an input layer, hidden layers, and an output layer and is trained in three steps: (a) Forward the input load data, (b) Compute and propagate the error backward, (c) Update the weights. First, all examples of the training set are entered into the input nodes. The activation values of the input nodes are weighted and accumulated at each node in the hidden layer and transformed by an activation function into the node's activation value. Hence, it becomes an input into the nodes in the next layer, until eventually the output activation values are found. The training algorithm is used to find the weights that minimize mean squared error. The main characteristics of FBP-TSNN are the self-learning and self-organizing. The proposed algorithm is implemented on Canada's power network to prove its accuracy along with effectiveness, and then compared with real historical data.

Keywords Feed-forward back-propagation artificial neural network (FBP-ANN) · Long-term forecasting · Electricity demand

A. Masoumi (✉) · F. Jabari · S. Ghassem Zadeh · B. Mohammadi-Ivatloo
Faculty of Electrical and Computer Engineering, University of Tabriz, Tabriz, Iran
e-mail: a.masoumi95@ms.tabrizu.ac.ir

F. Jabari
e-mail: f.jabari@tabrizu.ac.ir

S. Ghassem Zadeh
e-mail: g_zadeh@tabrizu.ac.ir

B. Mohammadi-Ivatloo
e-mail: bmohammadi@tabrizu.ac.ir

M. Pesaran Hajiabbas and B. Mohammadi-Ivatloo (eds.),
Optimization of Power System Problems, Studies in Systems, Decision and Control 262,
https://doi.org/10.1007/978-3-030-34050-6_11

1 Introduction

Nowadays, interconnected power systems have been developed from reliability and stability aspects. Studies indicate that increasing rate of population all around the world leads to a growth in electricity consumption [1]. Meanwhile, uncertain nature of electrical demand and renewable energy sources such as solar and wind adds some limitations to dynamic stability of large electricity grids [2]. Hence, forecasting of electrical demand becomes more and more critical for assisting power system operators in electricity grid management, whether for short-term analysis or long-term applications such as economic emission dispatch, unit commitment, optimal scheduling, etc. [3–5].

Recently, many scholars have proposed different short and long-term load forecasting algorithms. In this context, authors of [6] have simulated a specific aggregated state prediction for electrical consumption of interconnected power networks with 1% error in 700 h. In [7], combination of genetic algorithm and neural network is an illustrative example with accuracy of 98.95% for expanding a feed backward neural network for forecasting of heterogeneous demand time series in very-short and short time intervals. Application of support vector machine (SVM) has been presented in [8] for one hour ahead demand forecasting. In addition, this two-phase technique consisting of artificial neural network (ANN) and SVM has demonstrated the resolution of speed and accuracy through precise experiment on real historical data of 4th July 2012. Son et al. [9] has evaluated the application of support vector regression (SVR), fuzzy logic, and particle swarm optimization (PSO) with mean demand scaling of 149.28754 kW for short-term electrical demand forecasting. Guo et al. [10] has introduced a self-learning algorithm for load forecasting process which benefits from economic factors. This approach inserts some economic elements in searching process to reduce computational error. According to overall implementation of automation system in residential consumption as demand response strategies, it is found that modified algorithms, which made up SVR, can fix the intermittent nature of internal loads (i.e. cooling, heating, and ventilation). Thereby, poly-phase prediction can practically actualize the demand response strategies in such programming. In addition, Le Cam et al. [11] has aimed to forecast total electricity cost of automation system in a benchmark building by providing a poly-stage prediction model that 14.2–22.5% optimum absolute error has been observed. Li et al. [12] has combined a wavelet decomposition technique with ANN to diminish the negative impacts of volatile load data. Plus, in the noted study, it can be given as advanced intelligent algorithm with 2.4% mean absolute percentage error (MAPE). Effectiveness of ANNs in Poland's natural gas consumption forecasting has been described in [13]. This approach has been investigated on historical time series of Szczecin city with characteristics of 22 input, 36 hidden layer and 1 output and 8% MAPE. Authors of [14] have used ANNs for hour ahead prediction of universal solar radiation with 7.86% RMSE through filtering the low frequency of input data set. Reference [15] runs two different learning rules on ANNs. Therefore, it is observed that integration of back propagation (BP) and extreme learning method (ELM) for decoding the fluctuation of wind speed prediction leads to 1.33% and 1.1965% RMSE, respectively. In a similar standpoint, conducting PSO and genetic algorithm for optimal selecting of weight vectors of ANN in solar irradiance estimation system can be found in [16]. It indicates that the combination of BP-ANN and PSO technique has demonstrated correlated results as 0.78 RMSE and 0.685 mean absolute error (MAE). Authors of [17] have aimed to apply a proper orthogonal decomposition (POD) to ANNs for wind and demand forecasting of high altitude towers. It is found that such complex algorithm can reaches RMSE and mean error of 4% and 0.98%, respectively. As mentioned, ANNs support both regression based and computational methods under various prediction scales. Ramasamy et al. [18] has formed a unique wind power forecasting with respect to ANNs through speed estimation experiment in western Himalayan. To prove its robustness, output series have compared with real historical set considering some environmental and geographical factors such as temperature, air pressure, latitude, and longitude. Resiliency of this method against the time variant nature of ANN's input parameters has been revealed by 6.489% as MAPE. Yadav and Chandel [19] have identified relevant input variables for predicting of 1-min time-step photovoltaic module power using ANNs and multiple linear regression models with 2.15–2.55% MAPE. da Silva et al. [20] has reached to important point that using Bayesian Regularization (BR) and Levenberg Marquardt (LM) as training of ANNs has real time result in comparison with others for solar power estimation in a way that MAPE and RMSE are equal to 0.02%, 0.11% (BR), and 0.31%, 0.74% (LM), respectively.

This chapter aims to present a dynamic feed-forward back-propagation ANNs based method for long-term forecasting of electrical demand. In addition, the highly features of compatibility and accuracy of the proposed algorithm is revealed using a comparison between the forecasted and the actual electrical demands of Canada's, Ontario independent electricity operator system (IESO), low voltage grid. The remainder of this chapter can be organized as follows: Sect. 2 presents the problem formulation. Simulation result and discussions are provided in Sect. 3. Finally, concluding remarks appear in Sect. 4.

2 Problem Formulation

2.1 Artificial Neural Network (ANN)

Artificial neural network (ANN) is a powerful and extensive tool for engineering applications such as fitting, pattern recognition, clustering, and prediction. The performance of this algorithm is that, by applying weigh matrix and bias vectors to the input vectors and mathematical operations, we can reach the desired output which is demand consumption as in Fig. 1. The aim is to learn our ANN for obtaining the desired output, and also updating the error vectors in any steps. Then, after several iterations, we can regain the optimal weight values for the input vectors. The type of learning in our ANN is supervised learning rule which it benefits from three dynamic training techniques. Hence, the learning rule will be discussed in the next section. During learning, we update the weight matrix to determine the best error as well as observing a tolerable output dependency. Then, ANN will initiate with parameter load and time series for the network as an input, and then variables would have multiplied by weight matrix. Finally, they would have added by bias vectors. The next step begins in such order that, a specific mathematical function is presented to start the calculation as in Fig. 2.

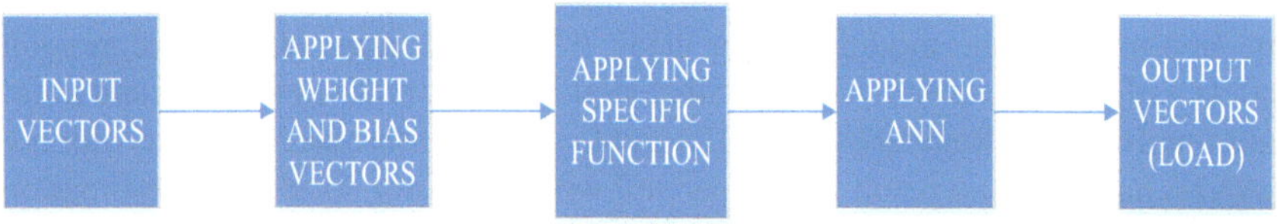

Fig. 1 The block diagram of proposed algorithm

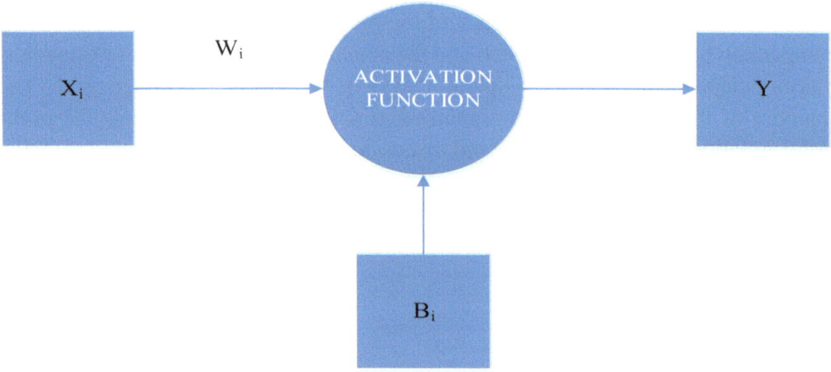

Fig. 2 The operation of the proposed algorithm

Where, x_i, w_i, B_i are the input vectors, weight matrixes and bias vectors respectively and Y is the output of neural network as is depicted in Eq. (1).

$$y_i = F(x_i \times w_i + b_i) \tag{1}$$

In order to satisfy the convergence condition, the algorithm is constructed based on supervised learning rule. In supervised learning at any moment in time K input $x(k)$ is applied to the network. Network desired response $\hat{Y}(\mathrm{k})$ is given and Couples $(x(k), \hat{Y}(\mathrm{k}))$ belong to a given set of learning that are pre-selected. The $x(i)$ and $\hat{Y}(i), i = 1 \ldots, \mathrm{N}$ (N is number of neurons) are used in supervised learning rule when $\hat{Y}(\mathrm{k}) = \hat{Y}(i), x(k) = x(i)$. Our desired network is Multi-Layer Perceptron (MLP) which has a group of vectors like input, output (validation), and network response $(Y(\mathrm{k}))$. MLP is a computational unit in the ANN architecture that is consisted of input layer, hidden layer, and one output layer. After the combination of this input, calculation process will begun as in Fig. 3.

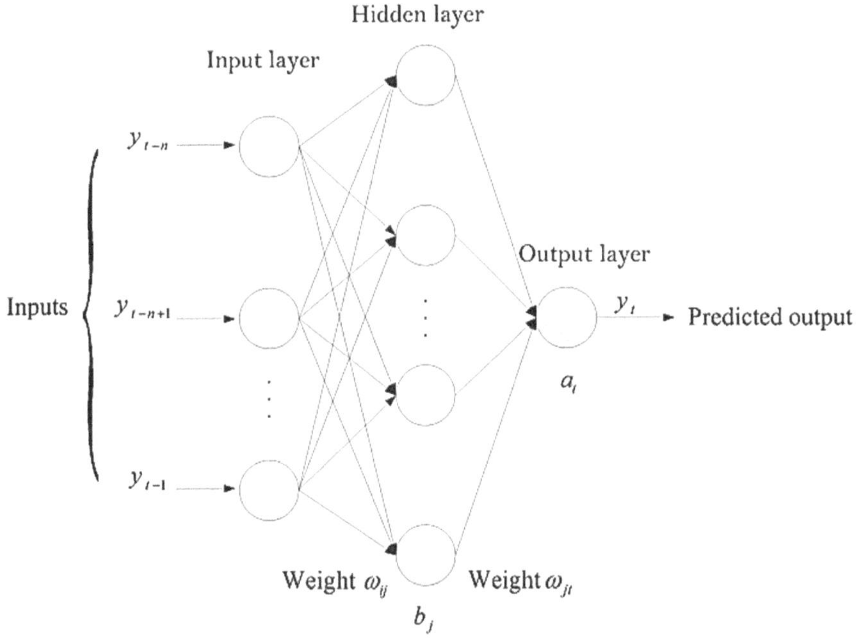

Fig. 3 Different layers of MLP

2.2 Dynamic Artificial Neural Network (DANN)

Dynamic artificial neural network is a computational operating system in which the continuous integration between its elements and the training processes improves the power of prediction. As it is depicted in Fig. 3, the conventional neural network has been made up of one input layer, one hidden layer and, one output layer. In addition, the relation among the parameters, layers, and also the training steps, leads to the prediction purpose. According to the nonlinear feature of input data set, using effective learning rules which they will define in the following section, can help the propagation procedure to be more dependable in comparison with individual learning network. Plus, the volume of calculation is a very critical point that must be considered if the number of sample data is notable for the convergence application. From this view point, overall conformity of forecasting algorithm will lead to the better understanding of estimated output. As it is clear in Eq. (1), the weight matrix and bias vectors are the stimulation parameters of ANN that are needed to generalize the diversity of input vectors. They simply interconnect the intermittent based inputs which are electric demand in this case, into the correcting steps of predicting. The activation function F is the operator of the net that is accommodated with both correcting and training processes. In another word, the F will demonstrate the predicting based on the termination of gradient process. The connectors, which are neurons, transfer the optimized values of weight and bias to the output layer in specific order. Moreover, dynamic neural network is defined as the combination of three regression based learning method which are Levenberg Marquardt (LM), Bayesian Regulation (BR), and Scaled Conjugated Gradient (SCG). First of all, the learning system operator will initiate with LM to train the portion of input data set (70%) and will allocate primary weight and bias vectors. After the generalization, the output trained is employed to the second propagation network (BR) to be normalize and filter the white noises of set with respect to the error performance (MSE) of the early neural network with 70% as training ratio. The next view is to conjugate the performance of two aforementioned learnings techniques to scale the searching process of optimal weight and bias vectors selecting in the hypothesis space. In another word, scaled conjugated will find the specific vectors for minimizing the error (MSE). Plus, this algorithm uses desired values of weight matrix at each stage to change them so that, the downward slope of the error curve is going to be descent. The flowchart of proposed algorithm for load forecasting is depicted as in Fig. 4.

2.3 Back Propagation Technique (BP)

Back propagation is a learning and adjusting method which conveys several partial derivatives from the basic parameter of neural network. In this method, we try to minimize the objective function and obtain mean square error (MSE) between the output of net and the desired output of electrical demand using dynamic algorithm. It

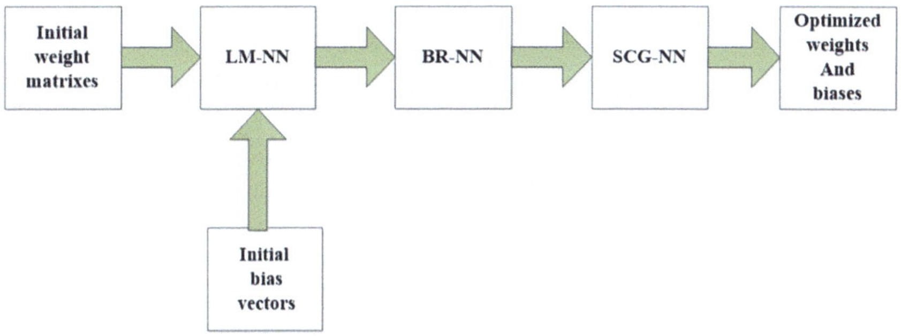

Fig. 4 The proposed algorithm

is observed that the hypothesis searching space is a large space that defines all possible values of the weights. The ANN tries to optimize the error to achieve reasonable states. However, there is no guarantee that it will reach to the absolute minimum. Therefore, the training algorithm (DANN) is used to find the weights that minimize mean squared error (MSE). Broadly, the mechanism of BP is based on the operating of tansigmoid as a sigmoid function for the hidden layer as well as pure linear function for the output layer. In this context, Eqs. (1) and (2) are the illustrative of hidden and output structure, respectively. As the same manner, the predicted output of the net can be achieved through the Eqs. (3) and (4) and algorithms were trained in three steps:

1. Forward the input data
2. Compute and propagate error backward
3. Update the weights

$$x_j = \sum_{i=t-1}^{t-n} \sum_{j=1}^{h} \omega_{ij} \times y_i + b_j \tag{2}$$

$$y_j = \frac{1}{1 + \exp(-x_j)} \quad j = 1, 2, \ldots, h \tag{3}$$

$$x_t = \sum_{j=1}^{h} \omega_{jt} \times y_j + a_t \quad t = 1, \ldots, T \tag{4}$$

$$y_t = x_t \quad t = 1, \ldots, T \tag{5}$$

where,

x_j, y_j	Input and output of the *j*th node of the hidden layer
ω_{ij}	Weight between *i*th input layer neuron and *j*th hidden layer neuron
b_j, a_j	Bias of the input and the hidden layers which are within the range of [−1, 1]
n, h, T	Number of input, hidden, and output layer nodes
x_t, y_t	Input and output values of the output layer at time horizon *t*
ω_{jt}	Connection weights of the *j*th hidden and output layers.

The mean square error of per cycle or epoch (Total square error for all learning models) and the norm of the gradient error is less than a predetermined value. The BP's view rest on the assumption of error gradient technique in the weight space. Hence, there is possibility to catch in Local minimum. To avoid such obstacle, we can use stochastic gradient with different values for weights. Considering the aforementioned concept, the weight adjustment rule in *i*th iteration depends on the size of the weight in the previous iteration as in Eq. (6).

$$\text{MSE} = \frac{1}{2}\sum_{t=1}^{T} (\hat{y}_t - y_t)^2 \tag{6}$$

where, $\hat{y}_t$ and y_t are the predicted results and expected output of the neural network, respectively. As a result, trapping in local minimum and placing on flat surfaces can be avoided, however, the search speed increases with gradual increase of step modification. It is observed from BP's properties that it can show undetected features of input data in hidden layer of network. Hence, the adjusting procedure is initiated by Eqs. (7) and (8) to propagate the weights of hidden and input neurons as follows:

$$\Delta\omega_{jt} \propto -\frac{\partial \text{MSE}}{\partial \omega_{jt}} \tag{7}$$

$$\begin{aligned} \Delta\omega_{jt} &= -\eta\left(\frac{\partial \text{MSE}}{\partial y_t}\right)\left(\frac{\partial y_t}{\partial x_t}\right)\left(\frac{\partial x_t}{\partial \omega_{jt}}\right) = \eta(\hat{y}_t - y_t)\left(\frac{\partial\left((1+\exp(-x_t))^{-1}\right)}{\partial x_t}\right)y_j \\ &= \eta(\hat{y}_t - y_t)y_t(1-y_t)y_j \\ &\text{for } j = 1, \ldots, h, t = 1, \ldots, T \end{aligned} \tag{8}$$

In which,

$\Delta\omega_{jt}$	Weights of hidden neurons
η	Learning rate
$\frac{\partial \text{MSE}}{\partial y_t}$	Derivative of the error with respect to the activation
$\frac{\partial y_t}{\partial x_t}$	Derivative of the activation with respect to the net input

$\frac{\partial x_t}{\partial \omega_{jt}}$ Derivative of the net input with respect to a weight.

According to the association of error (MSE) in each step of iteration, if the algorithm continues until the error is less than a certain amount, BP will terminate which can lead us to over-fitting. Over-fitting is caused by the weight adjusting that may've not a conformity with overall distribution data. By increasing the number of iteration, the complexity of the hypothesis space learned by the algorithm becomes more and more comprehendible until it can able to evaluate noise and rare example of a network in the training set properly. The solution is to import approved collection called validation set to stop learning when the error is small enough in this series and also to network for simpler hypothesis spaces. Then, the amount of weight in each iteration can be reduced. After determining the optimized values of weights, the error in all nodes can participate as follows:

$$\Delta\omega_{ij} \propto -\frac{\partial \mathrm{MSE}}{\partial \omega_{ij}} \tag{9}$$

Consequently,

$$\begin{aligned}\Delta\omega_{ij} &= -\sum_{t=1}^{T}\left[\left(\frac{\partial \mathrm{MSE}}{\partial y_t}\right)\left(\frac{\partial y_t}{\partial x_t}\right)\left(\frac{\partial x_t}{\partial y_j}\right)\right]\left(\frac{\partial y_j}{\partial x_j}\right)\left(\frac{\partial x_j}{\partial \omega_{ij}}\right)\\ &= \eta\sum_{t=1}^{T}\left[(\hat{y}_t - y_t)y_t(1-y_t)\omega_{it}\right]y_j(1-y_j)y_i\\ i &= t-n,\ldots,t-1 \quad j = 1,\ldots,h\end{aligned} \tag{10}$$

2.4 Levenberg Marquardt Algorithm (LM)

The LM is a computational approach for data mining problems of NN which include uncertain parameter structure. In this premise, LM categorize the input data set by learning the NN algorithm to adapt with the previous state of parameter through the error expected (MSE). This method is basically drowned out by the popular Gaussi-Newton technique [21] in non-singularity functions (tansig) as in Eqs. (11) and (12):

$$x^{k+1} = x^k + \Delta x \quad k = 1,\ldots,N \tag{11}$$

In which, x^{k+1}, x^k and Δx represent the current state, historical recent state, and the deviation with time step of time series, respectively. The deviation can be modeled in the LM concept in which the Jacobians of errors train each node of neural network as follows:

$$\Delta x = \left[J^T J + \eta I\right]^{-1} J^T MSE \tag{12}$$

where, the J, η and MSE represent the first derivative of errors with respect to the back propagation process, learning rate (70%) and the mean square error, respectively. The merit of LM is the speed of convergence that aims to escape from the local minima for the sake of prediction [22]. According to the abovementioned equation, LM method has been conducted a correcting system on error (MSE) instead of using Hessian matrix. It is noted that the main point in the weight adjusting of NN is the propagation of neurons of hidden layer which over fitting may have been occurred, if the covariance of data set is contaminated with heterogeneous pattern [23]. Hence, the propagation search is described as Eqs. (13) and (14):

$$\omega_{ij}^{k+1} = \omega_{ij}^{k} + \Delta\omega_{ij} \quad k = 1, \ldots, N \tag{13}$$

$$\Delta\omega_{ij} = \left[J^T J + \eta I\right]^{-1} J^T MSE \tag{14}$$

2.5 Bayesian Regularization (BR)

After considering the standardizing steps of LM in the propagation process, Bayesian Regularization (BR) is applied for the over-fitting problem of weight allocation in NN [24, 25]. Meanwhile, BR detects the unregulated weights considering their error (MSE) as well as accelerates the search speed for classifying the weights by the help of reducing their possibility from the state space. In another word, BR filters out the unbiased weights which are selected randomly. Plus, by determining such weight that are the white noises of NN, the optimum values can be more achievable than its former state. Then, by adding the extra term to the propagation equations as the sum of all weights of net, the decision function for the learning is described as follows [26]:

$$\begin{gathered} \text{Min } \Delta\omega_{ij} = \alpha E(w) + \beta MSE \\ \alpha, \beta > 0 \end{gathered} \tag{15}$$

$$E(\mathrm{w}) = \frac{1}{2}\sum_{i=1}^{N} \omega_i^2 \tag{16}$$

In which, MSE, $E(\mathrm{w})$, α and β are the mean square error of NN, total sum of all weights, and filtering variables, respectively [27, 28]. Hence, when the possibility of unbiased weights decreases, the convergence of forecasting increases till it is turned to a resistant computational unit against the local optimums. According to the volume of input data as well as the learning interactions, the training rate of BR technique has been set to 70%. As it can be inferred from the Eq. (15), the propagation procedure is converted to the quadratic equation optimization which the filtering variables play a regularization role in this problem. By solving this equation and finding the minimum point for the feasible solution of variables, the propagation process will be improved as follows [29]:

$$\begin{cases} \alpha^{op} = \frac{\gamma}{2E_{\omega}(\omega^{op})} \\ \beta^{op} = \frac{N-\gamma}{2MSE} \\ \gamma = K - \alpha \mathrm{Trace}(\mathrm{A})^{-1} \\ \alpha = \frac{1}{\sigma_{\omega}^{2}} \end{cases} \tag{17}$$

The cooperation of γ that is the optimum diagnostic number of regularized weights with the covariance factor of input data set leads to the refinement of feasible solutions of quadratic problem. In this equation, K is the weight matrix of neural network and A is the Hessian matrix of the quadratic problem which acts as variance operator to determine the error deviation as well as α has an inverse relation with diversity of weights. It is noted that, the effective number of γ can vary from 0 to K because of the priority of input data set. Hence, the suitable set of solutions which are the best-fitted in the quadratic problem enhance the propagation modelling by diminishing the noises from the Eq. (15).

2.6 Scaled Conjugated Gradient (SCG)

In this step, the conjugate scope is used to maximize the optimization feature of dynamic technique. The concept of SCG is based on the arrangement of overall minima of quadratic problem which aims to decrease the slop of errors. In this category, we consider a gradient operator for both errors and gradient of errors. After conducting the two aforementioned techniques, SCG initialize with x_0 as the primary point of linear searching algorithm for weights in accordance to Eq. (18) [30]. The combinatory gradient can be checked with Eq. (19).

$$x_0 \in R,\ f(\mathrm{x}) \le f(\mathrm{x}_0) \tag{18}$$

$$\begin{aligned} &\|\nabla f(x) - \nabla f(y)\| \le \mathrm{L}\|x - y\| \\ &y = g_{k+1} - g_k \\ &g_k = \nabla f(x_k) \end{aligned} \tag{19}$$

In which $\nabla f(x)$, $\nabla f(y)$ and L are the error gradient, gradient of error gradient of weight matrixes as well as x and y are the demonstration of input weights, and error gradient, respectively [31]. It has to be noted that the procedure can be achieved under the differentiability of objective function (15). In this assumptions, the propagating process of SCG can be expressed as follows (20) [32]:

$$
\begin{aligned}
x_{k+1} &= x_k + \alpha_k d_k \rightarrow (\omega_{k+1} = \omega_k + \alpha_k d_k) \\
d_{k+1} &= -\theta_{k+1} g_{k+1} + \beta_k s_k \\
\theta_{k+1} &= \frac{s_k^T s_k}{y_k s_k} \\
s_k &= x_{k+1} - x_k
\end{aligned} \tag{20}
$$

In which d_k and α_k are the direction and step counter of searching technique [33]. According to the quasi-newton theorem, if $\beta_k = 1$ then the possibility of θ_k for being a positive definite matrix increases. Therefore, we can call the first step to be innervated as:

$$
g_0 = \nabla f(x_0), d_0 = -g_0, \alpha_0 = \frac{1}{\|g_0\|} \tag{21}
$$

In addition, the searching algorithm updates every iteration until the Eq. (19) can be satisfied. Hence, the Eq. (20) indicates us that the propagating process of SCG is completely depends on the optimal selecting of d_k land α_k [34]. This premises imposes us that the step counter (α_k) must be determined originally for accelerating the computation search. Thereby, the Wolf condition is implemented on the objective function for this specific purpose as in Eq. (22) [35]:

$$
\begin{aligned}
f(x_k + \alpha_k d_k) - f(x_k) &\leq \sigma_1 \alpha_k g_k^T d_k \\
\nabla f(x_k + \alpha_k d_k)^T &\geq \sigma_2 g_k^T d_k
\end{aligned} \tag{22}
$$

where σ_1 and σ_2 are the positive constant considering $0 < \sigma_1 \leq \sigma_2 < 1$. At last, the configuration of three strong computational units which compensate the propagating search that is depicted in the following section.

3 Numerical Result and Discussions

3.1 Resiliency of Hybrid Proposed Strategy

In the conducted survey, the set of electrical load demand are assembled by three steps as: the training, the validation, and the testing that are valued by a tentative options 70%, 15%, and 15%, respectively. In order to fit the assumption of proposed technique by NN, the MSE criterion serves as the best identification of error distance. This criteria is defined for each stages of DANN to reach the constraints satisfaction. Moreover, after aforementioned scaling standardize the output, the analogy between the real historical demand data set and the linearized prediction set is obtained to verify the compatibility of algorithm as shown in Fig. 5.

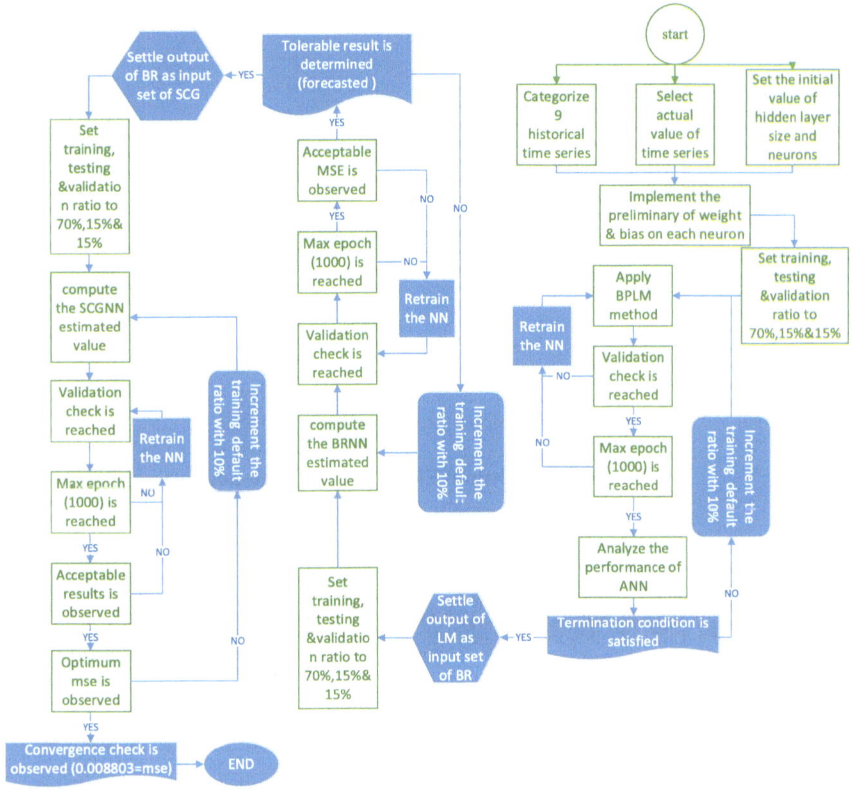

Fig. 5 The flowchart of the proposed strategy

3.2 *Robustness and Scalability*

The forecasting operatory system is made by main body of NN which benefits from three learning technique that are LM, BR, and SCG. The selection of 4344 net input is allocated from the Canada, Ontario independent electricity operator system (IESO), during 1/1/2001 to 6/30/2001 till 1/1/2009 to 6/30/2009 in six month time horizons as well as 9 years which have been imported to DANN. The configuration of abovementioned structure are set as 10 hidden layer within 24 hidden neurons for each stage as well as 4320 output set iteratively. Furthermore, the comparison of predetermined and forecasted outputs in association with error functionality (MSE) are denoted as in Figs. 6, 7, 8, and 9. According to the Figs. 8 and 9, the result is accommodated with the actual historical data set. To qualify the

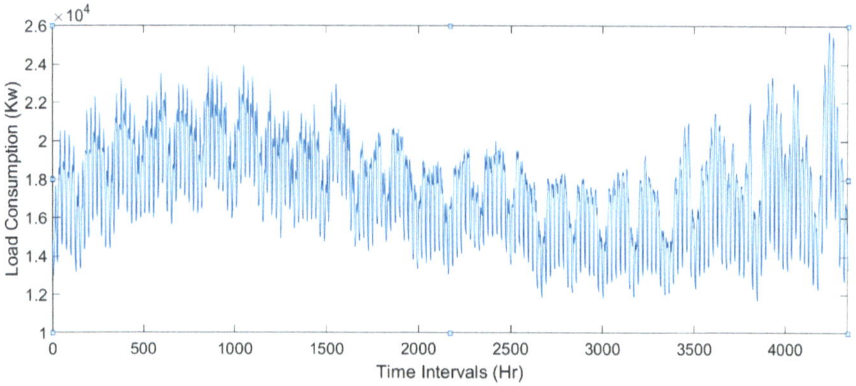

Fig. 6 Actual value of the power consumption from 1/1/2010 to 6/30/2010

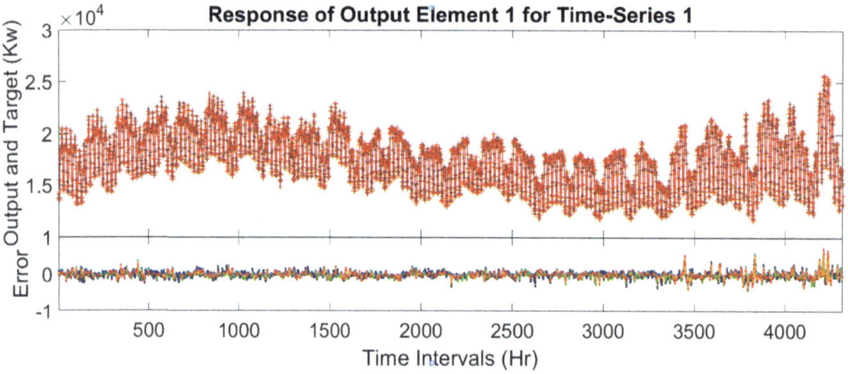

Fig. 7 The demonstration of predicted set of input

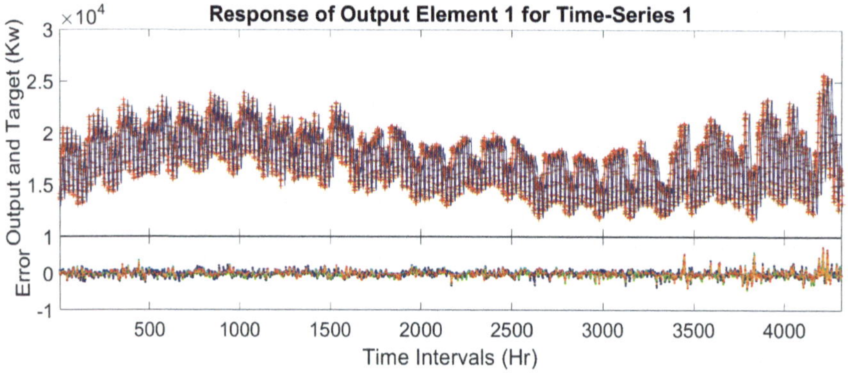

Fig. 8 The comparison of actual and estimated power consumption from 1/1/2010 to 6/30/2010

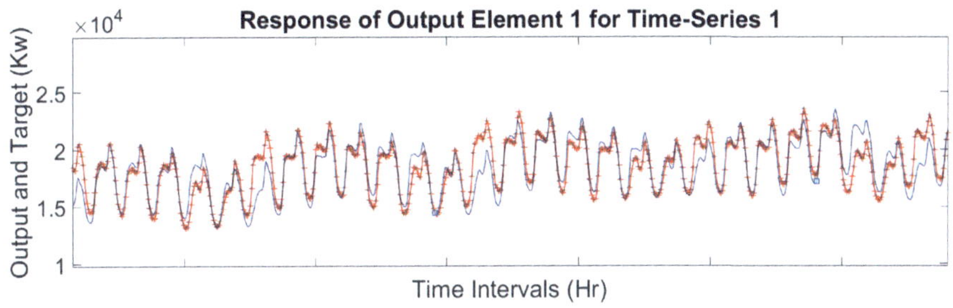

Fig. 9 The comparison of actual and estimated power consumption from 1/1/2010 to 2/1/2010

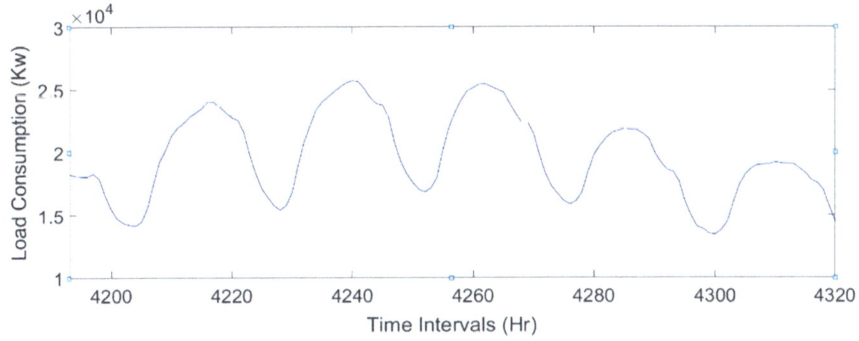

Fig. 10 The actual value of input data set for the last week during 6/24/2010 to 6/30/2010

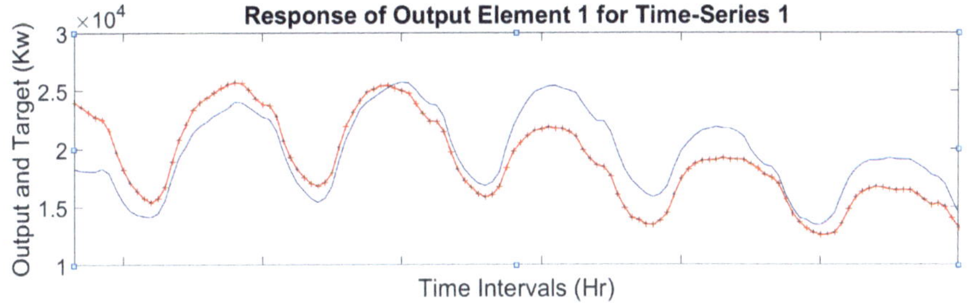

Fig. 11 The comparison of actual and estimated power consumption 6/24/2010 to 6/30/2010

contribution of simulation, the symmetric resolution of compared output of network are presented in Figs. 10 and 11 which convey the participation of measured dataset with forecasted output, respectively. As it is obvious, the blue line considered as the actual selection of measured data from IESO. In addition the red curve is defined to output value of simulated approach coherently. According to the identification of error trials of our correlated method, after 1000 epoch, the MSE is decreased to 8.803×10^{-3} (ε) which enables the conformity of strategy. The appraisal of NN construction incorporated with performance of NN are depicted in Figs. 12 and 13, respectively. Moreover Figs. 14 and 15 attached to clarify the feasibility of algorithm substantially. It is worth mentioning that, Fig. 16 is represented as linear regression view of simulated structure to fit all data set simultaneously.

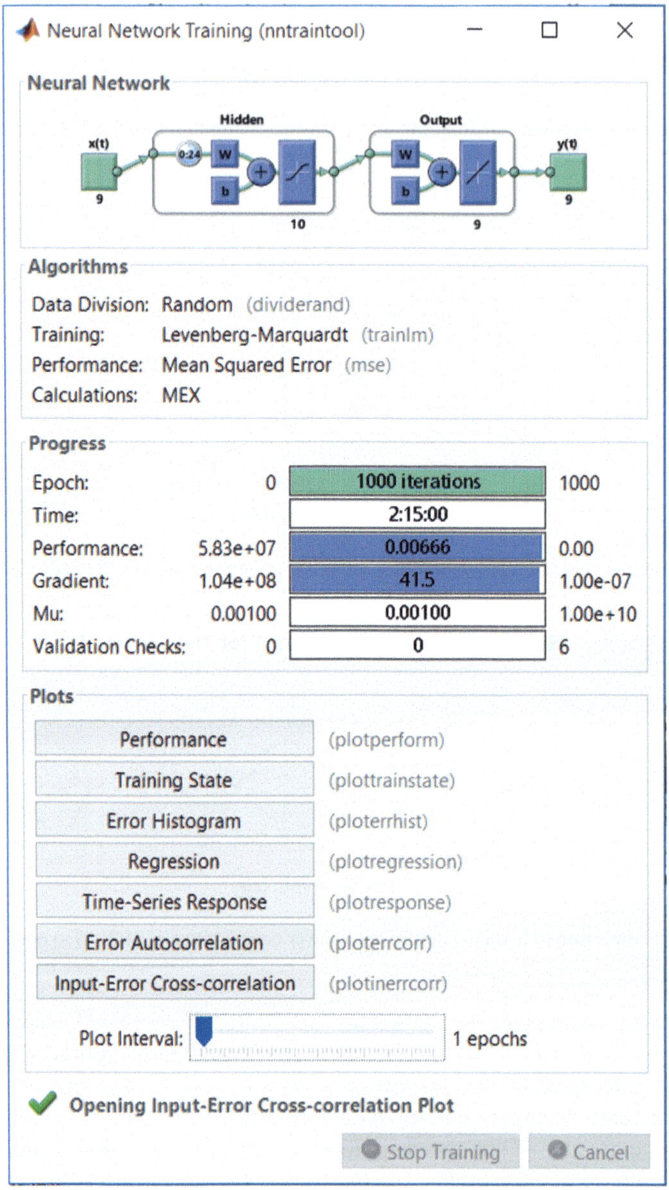

Fig. 12 The configuration of LM method

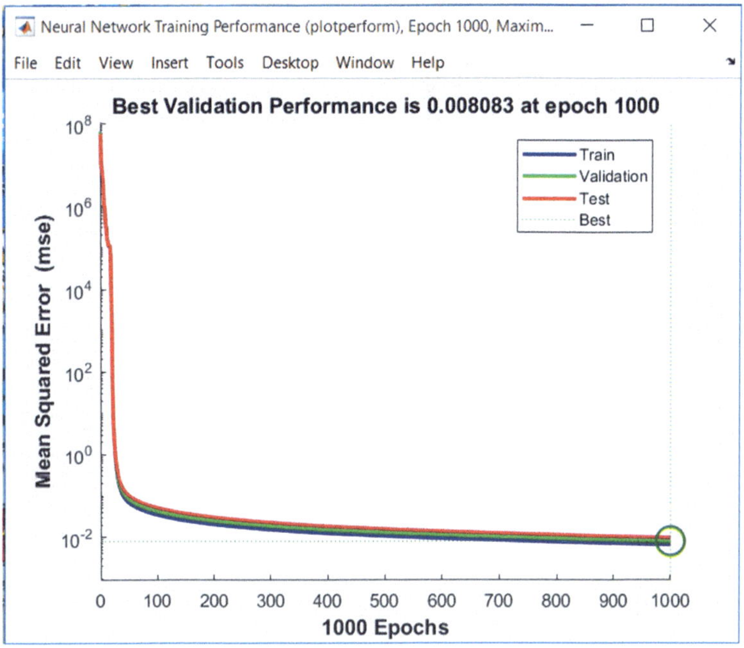

Fig. 13 The performance of LM-DANN

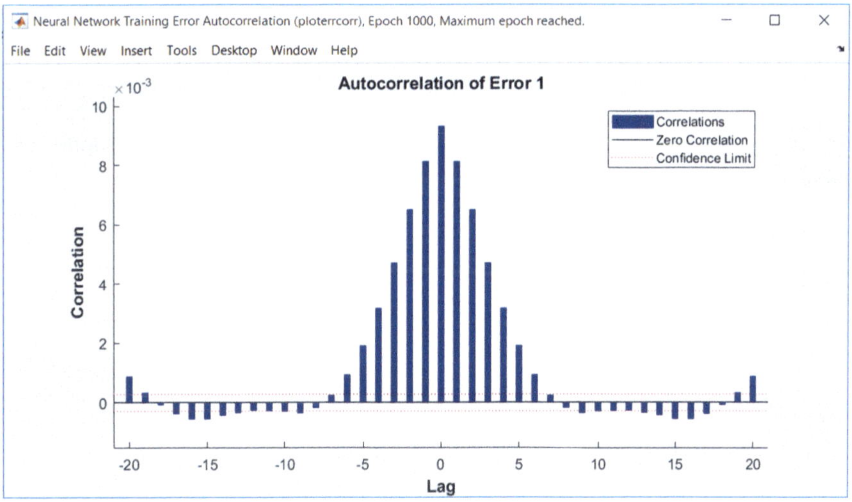

Fig. 14 The autocorrelation of error

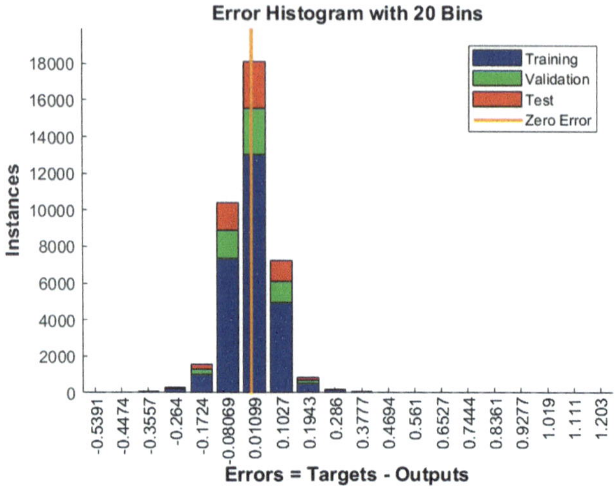

Fig. 15 The error histogram of proposed architecture

4 Conclusion

All in all, we have assumed the advantage of ANN for the long term forecasting of electrical consumption as well as to predict the desired data set using the error criterion. In this paper, the intermittent nature of our problem has been depicted us that implementing proposed method is applicable for the uncertain frequency data sets. Thereby, the historical sets is reported by IESO, Canada's power network, for the purpose of estimating . Plus, after determining the composition of DANN, the regulating steps which is guided by training progress of demand curve are applied to gain dependable results. Consequently, the simulation performance of DANN covers the sensitivity and practicable operating of proposed architecture which is obtained as tolerable minimum MSE.

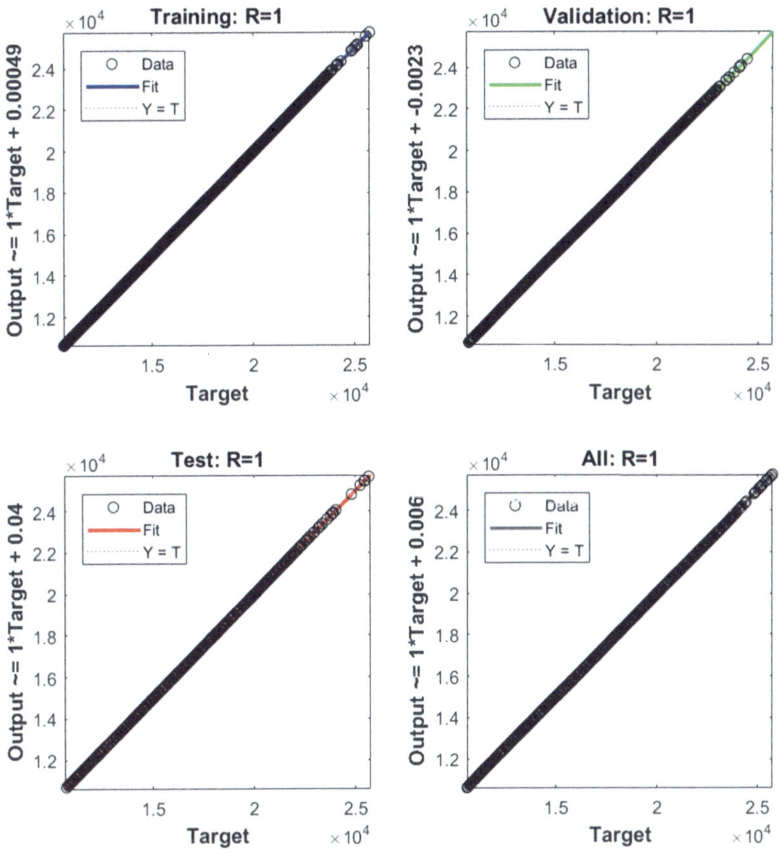

Fig. 16 The regression criterion for proposed architecture

MATLAB Code

Note that sheets 1 and 2 are the historical data and the time series data set of input, respectively. In addition, the training progress should be applied after each stages met their termination condition and gained acceptable performance. The dash sign is separated all training slops in order.

```
% Import the file which is the historical data set with
respect to month/day/year in Excel format

function [newData1] = importfile3(fileToRead1)
sheetName='Sheet1';
[numbers, strings] = xlsread(fileToRead1, sheetName);
if ~isempty(numbers)
    newData1.data =  numbers;
end
if ~isempty(strings)
    newData1.textdata =  strings;
end
function [newData1] = importfile3(fileToRead1)

% Import the file which is the actual data in Excel
format.

sheetName='Sheet2';
[numbers, strings] = xlsread(fileToRead1, sheetName);
if ~isempty(numbers)
    newData1.data =  numbers;
end
if ~isempty(strings)
    newData1.textdata =  strings;
end

% Construct the main body of NN and adjust the hidden
layer size and its neurons in another Matlab file.

inputSeries = tonndata(data,false,false);
targetSeries = tonndata(data,false,false);
inputDelays = 0:24;
hiddenLayerSize = 10;
net = timedelaynet(inputDelays,hiddenLayerSize);
[inputs,inputStates,layerStates,targets] = pre-
parets(net,inputSeries,targetSeries);

% set the primary values of training, validation and
testing stages.
```

```
net.divideParam.trainRatio = 70/100;
net.divideParam.valRatio = 15/100;
net.divideParam.testRatio = 15/100;

% Determine the training approach.
net.trainFcn = 'trainlm'
[net,tr] =
train(net,inputs,targets,inputStates,layerStates);

% Check the validation in accord with the flowchart.

outputs = net(inputs,inputStates,layerStates);
errors = gsubtract(targets,outputs);
performance = perform(net,targets,outputs)

% Plot the forecasted output.

figure, plotperform(tr)
figure, plottrainstate(tr)
figure, plotresponse(targets,outputs)
figure, ploterrcorr(errors)
figure, plotinerrcorr(inputs,errors)

nets = removedelay(net);
[xs,xis,ais,ts] = pre-
parets(nets,inputSeries,targetSeries);
ys = nets(xs,xis,ais);
earlyPredictPerformance = perform(net,tc,yc)
% Make another Matlab files and apply the code for both
BRNN and SCGNN.

% Construct the main body of NN and adjust the hidden
layer size and its neurons in another Matlab file.

inputSeries = tonndata(data,false,false);
targetSeries = tonndata(data,false,false);
inputDelays = 0:24;
hiddenLayerSize = 10;
net = timedelaynet(inputDelays,hiddenLayerSize);
[inputs,inputStates,layerStates,targets] =
preparets(net,inputSeries,targetSeries);

% set the primary values of training, validation and
testing stages.

net.divideParam.trainRatio = 70/100;
```

```
net.divideParam.valRatio = 15/100;
net.divideParam.testRatio = 15/100;

% Determine the training approach.
net.trainFcn = 'trainbr'
[net,tr] =
train(net,inputs,targets,inputStates,layerStates);

% Check the validation in accord with the flowchart.
outputs = net(inputs,inputStates,layerStates);
errors = gsubtract(targets,outputs);
performance = perform(net,targets,outputs)

% Plot the forecasted output.

figure, plotperform(tr)
figure, plottrainstate(tr)
figure, plotresponse(targets,outputs)
figure, ploterrcorr(errors)
figure, plotinerrcorr(inputs,errors)

nets = removedelay(net);
[xs,xis,ais,ts] = pre-
parets(nets,inputSeries,targetSeries);
ys = nets(xs,xis,ais);
earlyPredictPerformance = perform(net,tc,yc)

% Construct the main body of NN and adjust the hidden
layer size and its neurons in another Matlab file.

inputSeries = tonndata(data,false,false);
targetSeries = tonndata(data,false,false);
inputDelays = 0:24;
hiddenLayerSize = 10;
net = timedelaynet(inputDelays,hiddenLayerSize);
[inputs,inputStates,layerStates,targets] =
 preparets(net,inputSeries,targetSeries);

% set the primary values of training, validation and
testing stages.

net.divideParam.trainRatio = 70/100;
net.divideParam.valRatio = 15/100;
net.divideParam.testRatio = 15/100;

% Determine the training approach.
net.trainFcn = 'trainscg'
```

```
[net,tr] =
train(net,inputs,targets,inputStates,layerStates);

% Check the validation in accord with the flowchart.
outputs = net(inputs,inputStates,layerStates);
errors = gsubtract(targets,outputs);
performance = perform(net,targets,outputs)

% Plot the forecasted output.

figure, plotperform(tr)
figure, plottrainstate(tr)
figure, plotresponse(targets,outputs)
figure, ploterrcorr(errors)
figure, plotinerrcorr(inputs,errors)

nets = removedelay(net);
[xs,xis,ais,ts] = pre-
parets(nets,inputSeries,targetSeries);
ys = nets(xs,xis,ais);
earlyPredictPerformance = perform(net,tc,yc)
```

References

1. Jabari, F., Nojavan, S., Ivatloo, B.M., Sharifian, M.B.: Optimal short-term scheduling of a novel tri-generation system in the presence of demand response programs and battery storage system. Energy Convers. Manag. **122**, 95–108 (2016)
2. Jabari, F., Nojavan, S., Ivatloo, B.M.: Designing and optimizing a novel advanced adiabatic compressed air energy storage and air source heat pump based μ-Combined Cooling, heating and power system. Energy **116**, 64–77 (2016)
3. Jabari, F., Shamizadeh, M., Mohammadi-ivatloo, B.: Dynamic economic generation dispatch of thermal units incorporating aggregated plug-in electric vehicles. In: 3rd International Conference of IEA Technology and Energy Management, pp. 28–29 (2017)
4. Jabari, F., Seyedia, H., Ravadanegh, S.N., Ivatloo, B.M.: Stochastic contingency analysis based on voltage stability assessment in islanded power system considering load uncertainty using MCS and k-PEM. In: Handbook of Research on Emerging Technologies for Electrical Power Planning, Analysis, and Optimization, pp. 12–36. IGI Global (2016)
5. Jabari, F., Masoumi, A., Mohammadi-ivatloo, B.: Long-term solar irradiance forecasting using feed-forward back-propagation neural network. In: 3rd International Conference of IEA, Tehran, Iran, 28th Feb–1st Mar 2017
6. Tajer, A.: Load forecasting via diversified state prediction in multi-area power networks. IEEE Transactions on Smart Grid **8**(6), 2675–2684 (2016)
7. Khan, G.M., Zafari, F.: Dynamic feedback neuro-evolutionary networks for forecasting the highly fluctuating electrical loads. Genet. Program Evolvable Mach. **17**, 391–408 (2016)
8. Mitchell, G., Bahadoorsingh, S., Ramsamooj, N., Sharma, C.: A comparison of artificial neural networks and support vector machines for short-term load forecasting using various load types. In: 2017 IEEE Manchester PowerTech, pp. 1–4 (2017)

9. Son, H., Kim, C.: Short-term forecasting of electricity demand for the residential sector using weather and social variables. Resour. Conserv. Recycl. **123**, 200–207 (2016)
10. Guo, H., Chen, Q., Xia, Q., Kang, C., Zhang, X.: A monthly electricity consumption forecasting method based on vector error correction model and self-adaptive screening method. Int. J. Electr. Power Energy Syst. **95**, 427–439 (2018)
11. Le Cam, M., Zmeureanu, R., Daoud, A.: Cascade-based short-term forecasting method of the electric demand of HVAC system. Energy **119**, 1098–1107 (2017)
12. Li, B., Zhang, J., He, Y., Wang, Y.: Short-term load-forecasting method based on wavelet decomposition with second-order gray neural network model combined with ADF test. IEEE Access **5**, 16324–16331 (2017)
13. Szoplik, J.: Forecasting of natural gas consumption with artificial neural networks. Energy **85**, 208–220 (2015)
14. Monjoly, S., André, M., Calif, R., Soubdhan, T.: Hourly forecasting of global solar radiation based on multiscale decomposition methods: a hybrid approach. Energy **119**, 288–298 (2017)
15. Petković, D., Nikolić, V., Mitić, V.V., Kocić, L.: Estimation of fractal representation of wind speed fluctuation by artificial neural network with different training algorothms. Flow Meas. Instrum. **54**, 172–176 (2017)
16. Xue, X.: Prediction of daily diffuse solar radiation using artificial neural networks. Int. J. Hydrogen Energy **42**, 28214–28221 (2017)
17. Dongmei, H., Shiqing, H., Xuhui, H., Xue, Z.: Prediction of wind loads on high-rise building using a BP neural network combined with POD. J. Wind Eng. Ind. Aerodyn. **170**, 1–17 (2017)
18. Ramasamy, P., Chandel, S., Yadav, A.K.: Wind speed prediction in the mountainous region of India using an artificial neural network model. Renew. Energy **80**, 338–347 (2015)
19. Yadav, A.K., Chandel, S.: Identification of relevant input variables for prediction of 1-minute time-step photovoltaic module power using artificial neural network and multiple linear regression models. Renew. Sustain. Energy Rev. **77**, 955–969 (2017)
20. da Silva, T.V., Monteiro, R.V.A., Moura, F.A.M., Albertini, M.R.M.C., Tamashiro, M.A., Guimaraes, G.C.: Performance analysis of neural network training algorithms and support vector machine for power generation forecast of photovoltaic panel. IEEE Latin Am. Trans. **15**, 1091–1100 (2017)
21. Lourakis, M., Argyros, A.: The Design and Implementation of a Generic Sparse Bundle Adjustment Software Package Based on the Levenberg-Marquardt Algorithm. Technical Report 340. Institute of Computer Science-FORTH, Heraklion, Crete, Greece (2004)
22. Lera, G., Pinzolas, M.: Neighborhood based Levenberg-Marquardt algorithm for neural network training. IEEE Trans. Neural Networks **13**, 1200–1203 (2002)
23. Wilamowski, B.M., Yu, H.: Improved computation for Levenberg–Marquardt training. IEEE Trans. Neural Networks **21**, 930–937 (2010)
24. Giovanis, D.G., Papaioannou, I., Straub, D., Papadopoulos, V.: Bayesian updating with subset simulation using artificial neural networks. Comput. Methods Appl. Mech. Eng. **319**, 124–145 (2017)
25. Kumar, P., Merchant, S., Desai, U.B.: Improving performance in pulse radar detection using Bayesian regularization for neural network training. Digit. Signal Proc. **14**, 438–448 (2004)
26. Sun, Z., Chen, Y., Li, X., Qin, X., Wang, H.: A Bayesian regularized artificial neural network for adaptive optics forecasting. Opt. Commun. **382**, 519–527 (2017)
27. Wan, L., Zeiler, M., Zhang, S., Le Cun, Y., Fergus, R.: Regularization of neural networks using dropconnect. In International Conference on Machine Learning, pp. 1058–1066 (2013)
28. MacKay, D.J.: Probable networks and plausible predictions—a review of practical Bayesian methods for supervised neural networks. Netw. Comput. Neural Syst. **6**, 469–505 (1995)
29. Williams, P.M.: Bayesian regularization and pruning using a Laplace prior. Neural Comput. **7**, 117–143 (1995)
30. Cheng, C., Chau, K., Sun, Y., Lin, J.: Long-term prediction of discharges in Manwan Reservoir using artificial neural network models. In: International Symposium on Neural Networks, pp. 1040–1045 (2005)

31. Møller, M.F.: A scaled conjugate gradient algorithm for fast supervised learning. Neural Netw. **6**, 525–533 (1993)
32. Sözen, A., Arcaklioğlu, E., Özalp, M.: A new approach to thermodynamic analysis of ejector–absorption cycle: artificial neural networks. Appl. Therm. Eng. **23**, 937–952 (2003)
33. Estevez, P.A., Vera, P., Saito, K.: Selecting the most influential nodes in social networks. In: 2007 International Joint Conference on Neural Networks, IJCNN 2007, pp. 2397–2402 (2007)
34. Sözen, A., Özalp, M., Arcaklioğlu, E.: Investigation of thermodynamic properties of refrigerant/absorbent couples using artificial neural networks. Chem. Eng. Process. **43**, 1253–1264 (2004)
35. Babaie-Kafaki, S.: Two modified scaled nonlinear conjugate gradient methods. J. Comput. Appl. Math. **261**, 172–182 (2014)

Multi-objective Economic and Emission Dispatch Using MOICA: A Competitive Study

Soheil Dolatabadi and Saeid Ghassem Zadeh

Abstract In this chapter, the application of multi-objective imperialist competitive algorithm is investigated for solving economic and emission dispatch problem. It is aimed to minimize two conflicting objectives, economic and environmental, while satisfying the problem constraints. In addition, nonlinear characteristics of generators such as prohibited zone and ramp up/down limits are considered. To check applicability of the MOICA, it is applied to 12 h of IEEE 30-bus test system. Then, results of MOICA are compared with those derived by non-dominated sorting genetic algorithm and multi-objective particle swarm optimizer. The finding indicates that MOICA exhibits better performance.

Keywords Multi-objective imperialist competitive algorithm · Economic and emission dispatch · Non-convex optimization problem

Nomenclature

$C(P_G)$	Total cost of power generation
$E(P_G)$	Total emission
P_{Loss}	Total network loss
P_i	Power generated at *i*th unit
$P_{L,i}$	Power flow of *i*th line
P_D	Total load demand
P_i^0	Output power of *i*th unit in previous dispatch interval
a_i, b_i, c_i, e_i, f_i	Fuel cost coefficients of *i*th unit
$\alpha_i, \beta_i, \gamma_i, \xi_i, \lambda_i$	Emission coefficients of *i*th unit
UR_i/DR_i	Up-ramp/down-ramp limits of *i*th unit
f_i	*i*th objective function
U(.)	Uniform distribution function

S. Dolatabadi (✉) · S. Ghassem Zadeh
Faculty of Electrical and Computer Engineering, University of Tabriz, Tabriz, Iran
e-mail: soheildolat@gmail.com

M. Pesaran Hajiabbas and B. Mohammadi-Ivatloo (eds.),
Optimization of Power System Problems, Studies in Systems, Decision and Control 262,
https://doi.org/10.1007/978-3-030-34050-6_12

n_g	Number of units
n_L	Number of transmission lines
n_{Obj}	Number of objective functions
n_{pop}	Number of population

Symbols

$X \preccurlyeq y$	x weakly dominates y
$x \prec y$	x strictly dominates y

1 Introduction

Economic dispatch (ED) program is used to define the optimum power output of generators in order to minimize total generation cost of units. Many constraints should be satisfied in ED problem, which make it a complex optimization problem. In this regard, many traditional methods are suggested for ED problem such as lambda iteration, interior point and linear programing methods [1]. In addition, nonlinear characteristics of generator such as prohibited zones and valve-point loading change ED problem to a non-convex optimization problem. For non-convex ED, dynamic programing is proposed, however it produces extremely large dimension which needs huge computation [2].

In recent years, new methods based on artificial intelligence are proposed to solve non-convex ED problems which show promising results. In [3] genetic algorithm (GA) is successfully applied to ED problem with considering valve point discontinuities and the results are verified by comparing with DP's results. Particle swarm optimizer (PSO) is proposed for solving ED problems in [4] which shows better performances compare to traditional methods and GA. After introduction of evolutionary multi-objective optimization algorithms, many researchers studied the application of these algorithms for solving economic and emission dispatch (EED) problem, where both pollution and fuel cost are considered for minimization. For instance, in [5] the performance of second version of non-dominated sorting genetic algorithm (NSGA-II) has been investigated for solving EED problem and it was successfully tested on four IEEE standard test systems, and the result indicates satisfying fitness values for both pollution and fuel cost. In [6] multi-objective gravitational search algorithm (MOGSA) is used and its results are compared with biography-based optimization (BBO). The multi-objective PSO (MOPSO) is proposed for EED problem and is tested on IEEE 30-bus test system [7]. In [8], the application of multi-objective harmony search algorithms has been investigated for EED and its performance is validated in three test systems.

The single-objective ICA is introduced by Atashpaz et al. in 2007 [9]. This algorithm has been successfully applied to many power engineering problem, such as machine design [10], distributed generation sizing and placement [11], FACTS devices allocating [12] and PID controller tuning [13]. The multi-objective version of ICA (MOICA) is presented in [14] and is tested on different standard benchmarks which the results show superiority of MOICA over MOPSO and NSGA-II. Despite of good performance of MOICA, still it has been not used for EED problem. Only in [15], The performance of MOICA in EED problem has been studied, but the researchers didn't consider nonlinear characteristics of generators such as valve-point loading and prohibited zones. In addition, some important constraints like up/down ramping rate limits are ignored which makes the solution impractical. Thus, for the first time, this paper is going to investigate application of MOICA on a non-convex EED problem. To evaluate the results of MOICA, a comparison is done between MOICA and two other popular multi-objective algorithms namely NSGA-II and MOPSO. The study is done on 12 h of IEEE 30-bus test system and the simulation is carried out in MATLAB. The rest of paper is organized as follows. The EED problem description and formulation are presented in Sect. 2. Section 3 briefly explains the multi-objective algorithm and the MOICA. The results of simulations are discussed in Sect. 4 and finally Sect. 5 concludes the paper.

2 Problem Description

As mentioned earlier, in EED problem, it is aimed to optimize both economic and environmental objectives simultaneous. This problem is formed from objective functions along with a number of equality and inequality constraints which make the EED problem a complex optimization problem. The whole problem can be described briefly as follow:

$$\begin{aligned} &Minimize(C(P_G), E(P_G)) \\ &s.t.: g(P_G) = 0, h(P_G) \geq 0 \end{aligned} \tag{1}$$

The fuel cost is formed from a quadratic term and a sinusoidal term which is related to valve-point loading. As mentioned in [16] considering valve-point loading makes EED solution more accurate and practical.

$$C(P_G) = \sum_{j=1}^{n_g} a_j + b_j P_j + c_j P_j^2 + \left| e_j \times \sin\left(f_j \times \left(P_j^{\min} - P_j\right)\right)\right| \tag{2}$$

The total ton/h gas emission can be calculated by

$$E(P_G) = \sum_{j=1}^{n_g} 10^{-2}\left(\alpha_j + \beta_j P_j + \gamma_j P_j^2\right) + \xi_j \exp\left(\lambda_j P_j\right) \tag{3}$$

2.1 Equalities and Inequalities Constraints

1. Active power balance: the produced active power by generators should meet the total load demand plus total network transmission loss:

$$\sum_{j=1}^{n_g} P_j = P_D + P_{Loss} \tag{4}$$

Which total network loss (P_{Loss}) is function of generator power output (P) and can be calculated using B coefficient [17]:

$$P_{Loss} = \sum_{i=1}^{n_g} \sum_{j=1}^{n_g} P_i B_{ij} P_j + \sum_{j=1}^{n_g} B_{0j} P_j + B_{00} \tag{5}$$

2. Generator power output limit: the output power of a generator is limited to a maximum and minimum value and should be considered.

$$P_i^{\min} < P_i < P_i^{\max} \tag{6}$$

3. Ramp rate limit: the generators are not capable to increase or decrease the output power instantly.

$$P_i - P_i^0 \le UR_i \quad and \quad P_i - P_i^0 \le DR_i \tag{7}$$

4. Prohibited zone: according to [18], some thermal units with many valve points are not capable to operate in some specific ranges of output power. Therefore, the feasible operation range can be defined as follow:

$$\begin{aligned} &P_i^{\min} \le P_i \le P_{i,1}^l \\ &P_{i,j-1}^u \le P_i \le P_{i,j}^l, \quad j = 2, 3, \ldots, n_g \\ &P_{i,n_g}^u \le P_i \le P_i^{\max} \end{aligned} \tag{8}$$

5. Line flow limit: the power flow through each transmission line should not exceed from its nominal value.

$$P_{L,i} \le P_{L,i}^{\max}, i = 1, 2, \ldots, n_l \tag{9}$$

3 Multi-objective Optimization Algorithm

The evolutionary multi-objective optimization (EMO) algorithms aim to find solution for minimization/maximization of a series objective functions [19]:

$$\begin{cases} Minimize/Maximize\ f_m(x) & m = 1, 2, \ldots, M \\ subject\ to: g_i(x) \geq 0, h_j(x) = 0 & i = 1, 2, \ldots, I, \quad j = 1, 2, \ldots, J \\ x_k^{\min} \leq x_k < x_k^{\max} & k = 1, 2, \ldots, n \end{cases} \tag{10}$$

where the optimization solution (X) is a vector of n decision variables $\boldsymbol{X} = (x_1, x_2, \ldots, x_n)$. Contrary to single-objective, multi-objective optimization should search throughout the multi-dimensional search space and therefore to distinguish the best solution from others, the *domination* concept was introduced. According to [20] the x dominates y if and only if: (1) the solution x is better or equal to y in all objectives and (2) x is strictly better than y at least in one objective. The mathematical expression of domination is given in (11).

$$\begin{cases} \forall i \in I, x_i \preccurlyeq y_i \\ \exists i \in I, x_i \prec y_i \end{cases} \Leftrightarrow x \prec y \tag{11}$$

The solutions which are not dominated by any other solutions are called non-dominated solutions which are near to Pareto-front (Fig. 1). A proper set of non-dominated solutions, in addition to be located near to Pareto-front, should also have diversity to represent the whole range of Pareto-front (*niching* strategy [21]).

In this regard, different strategies are proposed for finding Pareto-front but non-dominated sorting method is the most popular and efficient one. This method which proposed by Deb et al. in 2002 [22], at first was used in NSGA-II, but then many researchers apply the concept of non-dominated sorting method on various type of optimization algorithms to create new MOA such as multi-objective ant colony [23], multi-objective bat algorithm [24] and multi-objective imperialistic competitive

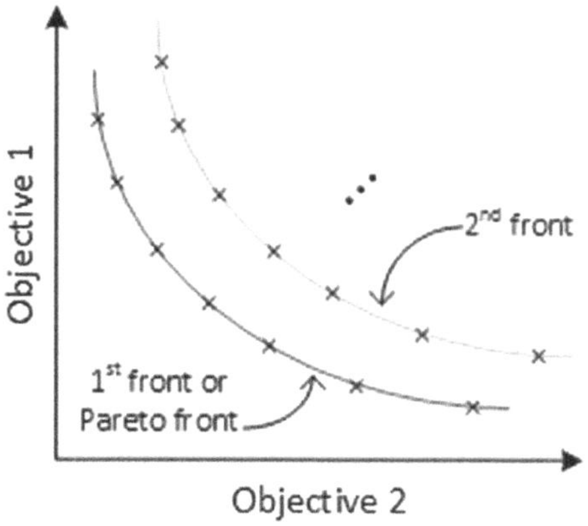

Fig. 1 Depiction of Pareto-front and other fronts

algorithm [14]. In other words, an evolutionary algorithm is responsible for exploration and exploitation the multi-dimensional search space and NS is responsible for ranking the solutions in multi-objective space to find the Pareto-front.

3.1 The Imperialist Competitive Algorithm

The ICA algorithm is introduced by Atashpaz and Lucas in 2007 which is inspired from imperialistic competition [9]. In the first step, the algorithm is initialized with random population or countries and based on their power or inverse of their fitness values, these countries are divided to imperialists (powerful countries) and colonies (rest of countries). Each imperialist, along with its colonies form an empire. The algorithm tries to find the best solution by eliminating imperialists via competition where each imperialist tries to extend its territory and possesses colonies of other empires. Therefore, a weak empire with worse fitness value will lose all of its colonies to more powerful empires.

In each iteration of ICA, the colonies of an empire move toward their imperialist. The step size of this movement is calculated by using a uniform distribution function.

$$x \sim U(0, \beta \times d) \tag{12}$$

where β is a number greater than 1 and d is the distance between the colony and imperialist. By moving colonies, it is possible that a colony gain a better fitness value compare to the imperialist. This issue leads to position exchange between the best colony and the imperialist. Therefore, in addition to the competition among empires, an internal competition exists between the imperialist and the colonies (Fig. 2).

The total power of nth empire ($T\,.\,C_n$) in each iteration is evaluated as follow:

$$T\,.\,C_n = Cost(imperilist_n) + \zeta mean\{Cost(Colonies\ Of\ Empire_n)\} \tag{13}$$

At the final stage of ICA, only one imperialist will be remained which is the solution of the optimization problem.

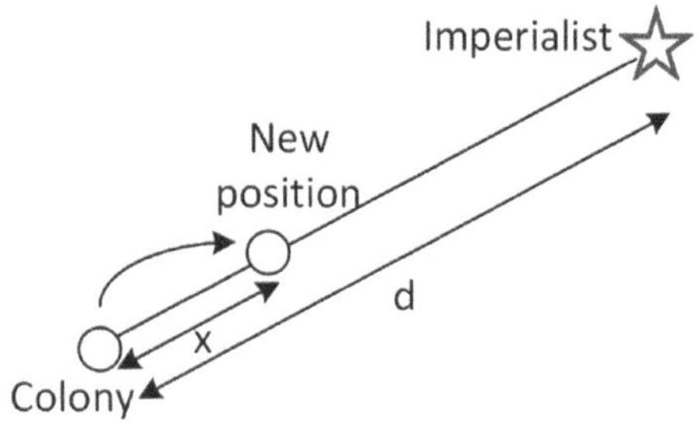

Fig. 2 The movement of colony toward the imperialist [9]

3.2 The MOICA

In each iteration, the MOICA uses ICA for generating new solutions and then non-dominated sorting method to rank them. This process can be explained in three important steps. In beginning of each iteration, the ICA generates a new set of population (P) using the previous generation. In next step, set P goes to NS algorithm for ranking purpose. The NS works in two phases, first finding the non-dominated solutions (rank one) and then ranking other solutions. Two entities are attributed to each solution: the number of solutions which dominate the solution i (n_i) and set of solutions which solution i dominates (S_i). The solutions with $n_i = 0$ are in first front (F_i) or current front. Then, for other solutions (j) which is placed in set S_i of first-front members, the n_j count will be reduced by one. By doing so, any solutions with $n_j = 0$ will be placed in set H. After checking all solutions in set S_i of current front, the solutions in set H will become new current front and the same process will be repeated. The pseud code of NS is given in Table 1.

And finally, the solutions of each fronts are sorted by using crowding distance method.

$$d(x_j) = \sum_{i=1}^{n_{obj}} \frac{|f_i(x_{j+1}) - f_i(x_{j-1})|}{f_i^{\max} - f_i^{\min}} \tag{14}$$

Having sorted solutions, the algorithm picks out N_{pop} number of high ranked solutions to produce next generation or to deliver it as the final result. This process is shown in Fig. 3.

Figure 4, presents the process of using MOICA for solving the non-convex multi-objective EED problem.

Table 1 The pseud code on non-dominated sorting for a given population (P)

Phase one:	**Phase two:**
for each p • P	i=1
for each q • P	while F_i • Ø
if(p<q) then	H • Ø
$S_P = S_P \cup \{q\}$	for each p • F_i
else if(q<p) then	for each q • S_p
$n_p=n_p+1$	$n_q=n_q-1$
if $n_p = 0$ then	if $n_q=0$ then H= H ∪
$F_1=F_1 \cup \{p\}$	{q}
	i=i+1
	F_i=H

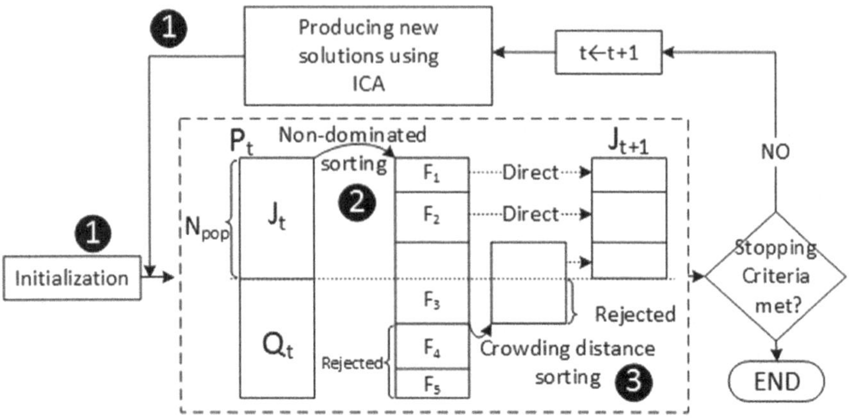

Fig. 3 The whole process of MOICA

Fig. 4 Process of using MOICA for solving EED problem

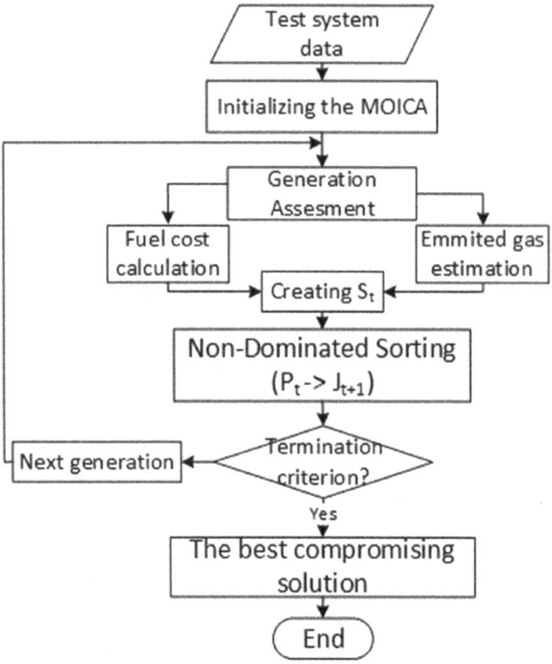

3.3 Selecting the Best Compromise Solution

In this study, the fuzzy-based method is used to find the best compromising solution by evaluating each non-dominated solutions of Pareto-front [25]. In this regard, the value of member function of each solution should be evaluated using (15), and the best compromising solution will be the one with the highest value of member function.

$$\mu_{f_i} = \begin{cases} 1 & f_i^{\min} \geq f_i \\ \frac{f_i^{\max} - f_i}{f_i^{\max} - f_i^{\min}} & f_i^{\min} < f_i < f_i^{\max} \\ 0 & f_i^{\max} \leq f_i \end{cases} \tag{15}$$

4 The Numeric Results

The IEEE 30-bus test system is used to test the performance of MOICA for solving EEC problem. This system has 6 thermal units which the detail data of this system are presented in Table 2 [26]. The prohibited zones and emission coefficients of thermal units are given in Tables 3 and 4 respectively [27]. A 12-h load profile is considered for this study (Table 5).

Table 2 Generating unit capacity, fuel cost coefficients and ramp rate limits

Unit	P_i^{min} (MW)	P_i^{max} (MW)	a ($)	b ($/MW)	c ($/MW^2)	UR_i (MW/h)	DR_i (MW/h)	P_i^0 (MW)
1	100	500	240	7.0	0.0070	80	120	340.00
2	50	200	200	10.0	0.0095	50	90	134.00
3	80	300	220	8.5	0.0090	65	100	240.00
4	50	150	200	11.0	0.0090	50	90	90.00
5	50	200	220	10.5	0.0080	50	90	110.00
6	50	120	190	12.0	0.0075	50	90	52.00

Table 3 The prohibited zones of units

Unit	Prohibited zone (MW)
1	[210 240]–[350 380]
2	[90 110]–[140 160]
3	[150 170]–[210 240]
4	[80 90]–[110 120]
5	[90 110]–[140 150]
6	[75 85]–[100 105]

Table 4 The emission coefficients of thermal units

Unit	α	β	γ	ξ	λ
1	4.091	−5.554	6.490	2.0E−4	2.857
2	2.543	−6.047	5.638	5.0E−4	3.333
3	4.258	−5.094	4.586	1.0E−6	8.000
4	5.326	−3.550	3.380	2.0E−3	2.000
5	4.258	−5.094	4.586	1.0E−6	8.000
6	6.131	−5.555	5.151	1.0E−5	6.667

Table 5 A 12-h load demand (MW)

Hour	1	2	3	4	5	6	7	8	9	10	11	12
Load	955	942	935	930	935	963	989	1023	1126	1150	1201	1235

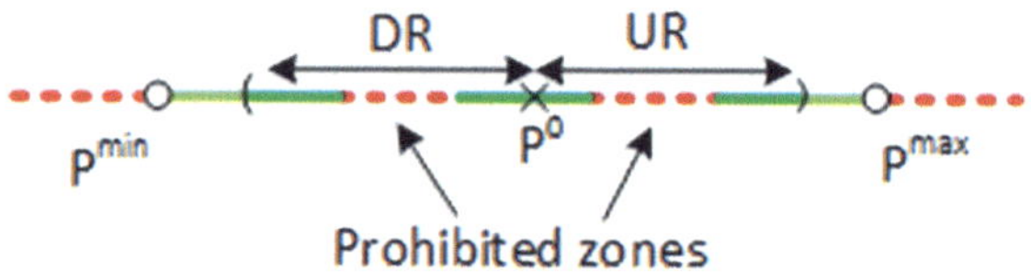

Fig. 5 Possible values for intializing the MOICA

4.1 *Initialization of Algorithm*

To initialize the MOICA, random values are chosen by considering maximum/minimum capacity of units, ramp rating limits and prohibited zones. This process increases accuracy of algorithm while reduces number of fitness function evaluations. Same method is applied to produce new generation of solutions in each iteration. Figure 5 shows the feasible values for a given thermal units by considering the constrains (dark green areas).

4.2 *The Simulation Results*

The Pareto-fronts of MOICA, MOPSO and NSGA-II for 7 different hours of power system are shown in Fig. 6. According to these graphs, the MOICA shows better performance compare to other ones. Figure 7 shows the relative generation costs and gas emissions for solutions of Pareto-front. The best compromising solution of MOICA for each hour is calculated using the previously presented fuzzy-based method. Using these compromising solutions, the values of generators output power, fuel cost and pollution for each thermal unit are calculated which due to lack of space only the MOICA's results are presented in Table 6.

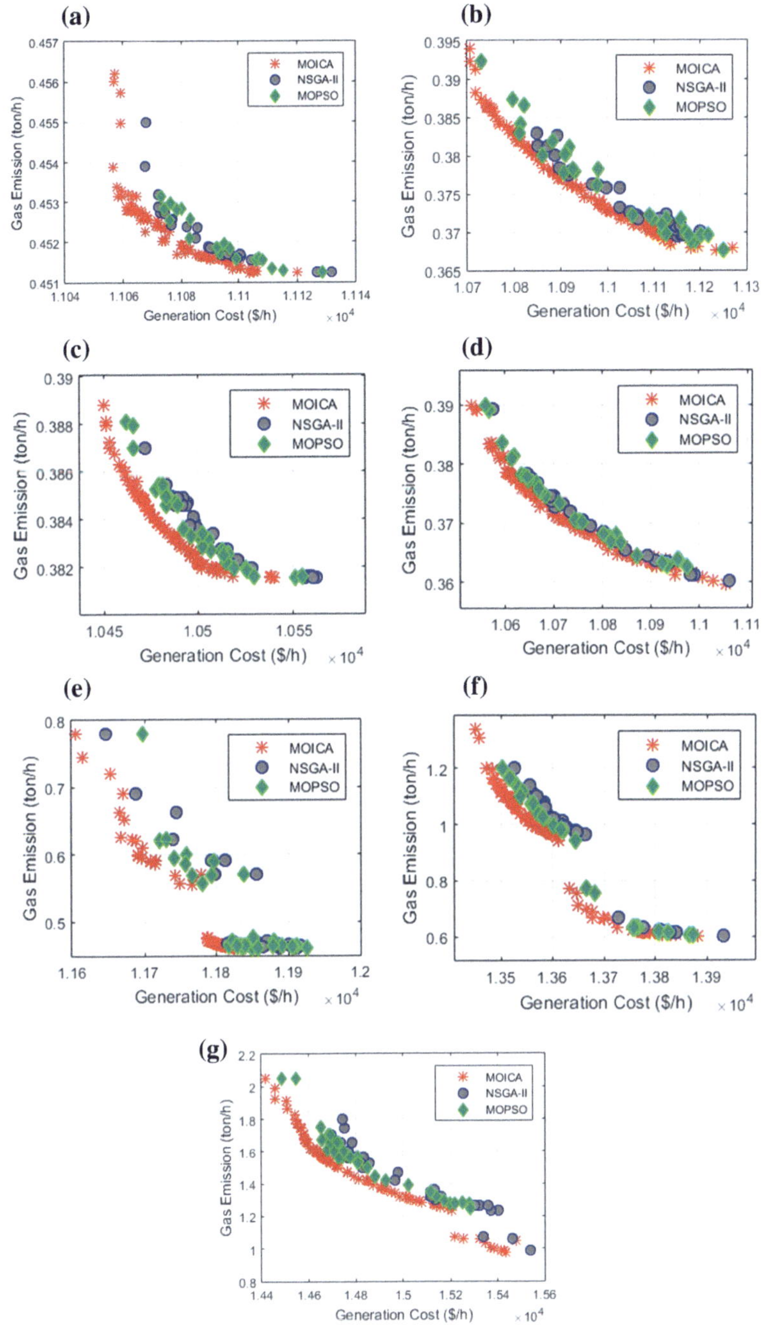

Fig. 6 Comparison among resulted Pareto-fronts of MOICA, MOPSO and NSGA-II: **a** 1st hour, **b** 2nd hour **c** 4th hour, **d** 6th hour, **e** 8th hour, **f** 10th hour, **g** 12th hour

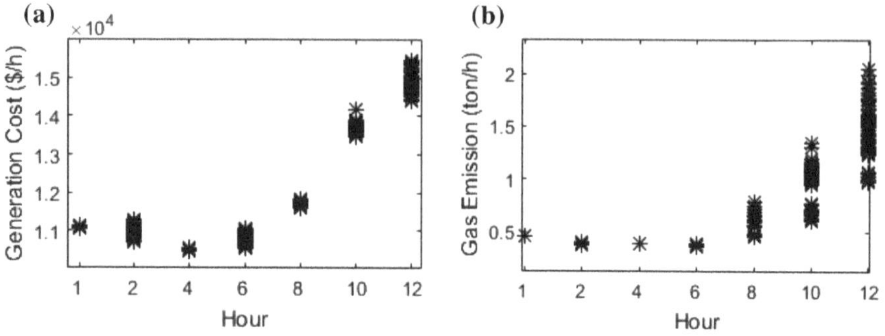

Fig. 7 The distribution of Pareto-front solutions for **a** generation cost, **b** gas emission

Figure 8 compares the results of MOICA with results which is derived from NSGA-II and MOPSO. The MOICA has better performance in both environmental and economic objectives compare to the two other algorithms.

5 Conclusion

This paper has studied the application of multi-objective imperialist completive algorithm (MOICA) to solve emission and economic dispatch (EED) problems. In this regard, the algorithm has been applied for solving EED problem of a IEEE 30-bus test system. the non-linear characteristics of generators such as valve-point loading and prohibited zone along with other constraints of the system are considered. For comparison purpose, the results of MOICA are compared with results of NSGA-II and MOPSO, which indicate better performance of MOICA.

Table 6 The result of compromising solution in each hour

Hour		Unit						Total
		1	2	3	4	5	6	
1	Generation	281.73	154.10	183.24	106.68	138.48	100.77	965.01
	Cost	2586.24	1778.08	2112.67	1413.49	1729.87	1466.85	11,087.22
	Pollution	0.2083	0.0382	0.0663	0.0544	0.0363	0.0481	0.4516
2	Generation	264.78	150.01	220.59	103.25	123.43	90.48	952.54
	Cost	2307.36	1739.52	2613.61	1399.98	1488.41	1420.79	10,969.68
	Pollution	0.1761	0.0354	0.1955	0.0538	0.0324	0.0471	0.5403
3	Generation	223.50	166.13	182.56	128.94	145.38	98.54	945.04
	Cost	1963.93	1926.14	2098.33	1622.97	1874.87	1457.83	10,944.08
	Pollution	0.1164	0.0477	0.0654	0.0599	0.0386	0.0478	0.3758
4	Generation	245.19	166.98	166.78	117.63	136.55	106.97	940.09
	Cost	2083.80	1938.64	1778.73	1483.48	1692.16	1494.56	10,471.38
	Pollution	0.1449	0.0484	0.0499	0.0568	0.0357	0.0491	0.3847
5	Generation	226.87	150.82	172.80	127.12	157.39	110.11	945.11
	Cost	1974.99	1746.74	1894.08	1595.32	2140.31	1512.04	10,863.48
	Pollution	0.1204	0.0359	0.0548	0.0593	0.0440	0.0496	0.3641
6	Generation	243.45	162.20	177.33	122.35	153.59	114.17	973.08
	Cost	2069.53	1871.92	1987.72	1532.36	2057.42	1540.16	11,059.11
	Pollution	0.1424	0.0443	0.0593	0.0580	0.0421	0.0505	0.3966
7	Generation	255.89	154.27	207.48	124.46	135.37	121.71	999.17
	Cost	2191.77	1779.75	2518.66	1558.45	1669.95	1614.46	11,333.04
	Pollution	0.1612	0.0383	0.1238	0.0586	0.0353	0.0524	0.4696
8	Generation	259.76	173.03	224.72	119.00	136.92	120.05	1033.48
	Cost	2239.31	2035.04	2630.01	1496.31	1699.28	1595.39	11,695.34
	Pollution	0.1675	0.0541	0.2296	0.0571	0.0358	0.0519	0.5960
9	Generation	253.79	162.96	236.44	97.14	244.19	143.99	1138.51
	Cost	2167.81	1882.04	2670.23	1379.10	3212.02	2020.31	13,331.50
	Pollution	0.1579	0.0450	0.3759	0.0527	0.5341	0.0615	1.2269
10	Generation	296.53	162.50	260.28	132.44	195.73	112.53	1160.01
	Cost	2874.33	1875.87	2901.88	1681.81	2617.83	1527.98	13,479.70
	Pollution	0.2415	0.0446	1.1608	0.0610	0.0883	0.0501	1.6463
11	Generation	389.64	181.66	229.17	118.94	184.76	107.08	1211.24
	Cost	4159.59	2193.18	2644.69	1495.78	2553.72	1495.10	14,542.04
	Pollution	0.6688	0.0632	0.2751	0.0571	0.0684	0.0491	1.1817
12	Generation	412.13	182.97	229.98	136.60	179.97	103.46	1245.12
	Cost	4239.28	2219.19	2647.25	1760.20	2510.52	1478.14	14,854.57
	Pollution	0.8912	0.0647	0.2845	0.0623	0.0622	0.0485	1.4135

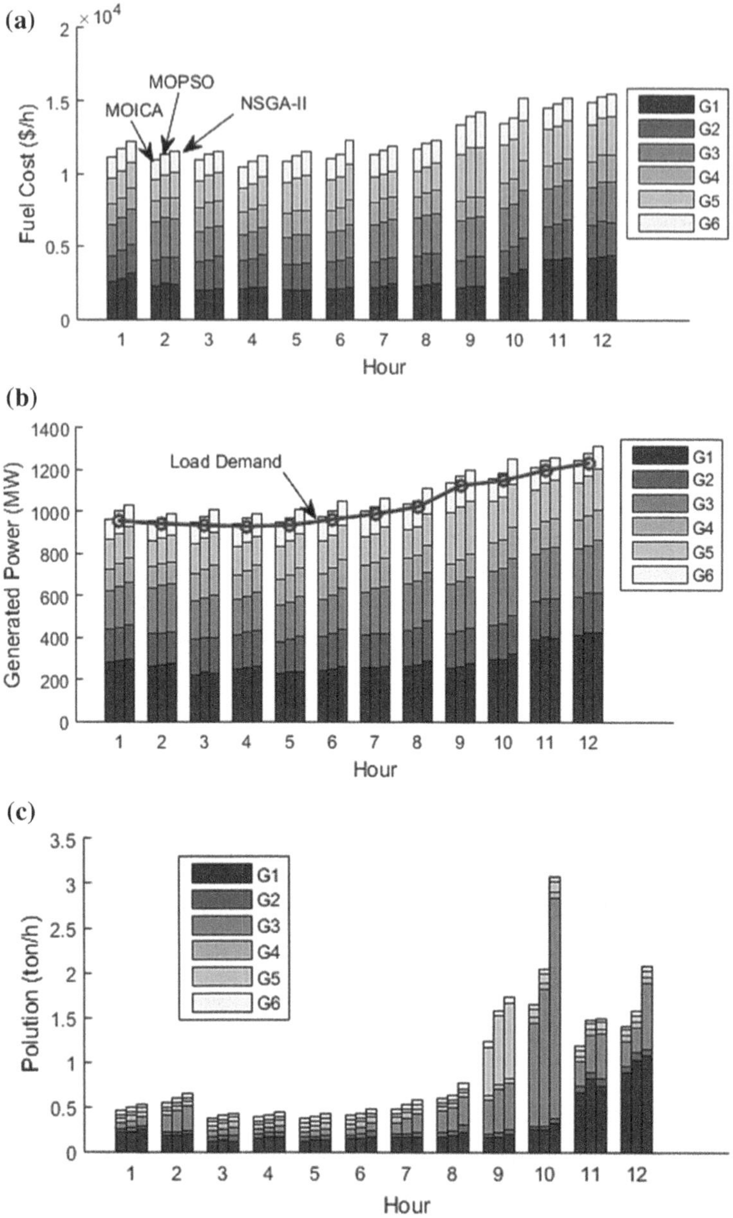

Fig. 8 The optimization results of MOICA, MOPSO and NSGA-II **a** fuel cost, **b** generated power, **c** emitted gas

MATLAB Codes

Genetic algorithm script and function.

```
clc;
clear;
close all;
global FE;
FE=0;

%% Problem Definition
CostFunction=@(x) CostF(x);      % Cost Function
nVar=6;                % Number of Decision Variables
VarSize=[1 nVar];      % Decision Variables Matrix Size
VarMin-0;             % Lower Bound of Variables
VarMax= 1;             % Upper Bound of Variables

%% GA Parameters
MaxIt=200;      % Maximum Number of Iterations
nPop=1000;        % Population Size
pc=0.7;                     % Crossover Percentage
nc=2*round(pc*nPop/2);  % Number of Offsprings (also
Parnets)
gamma=0.4;                  % Extra Range Factor for Cross-
over
pm=0.3;                     % Mutation Percentage
nm=round(pm*nPop);          % Number of Mutants
mu=0.1;           % Mutation Rate

%% Initialization
empty_individual.Position=[];
empty_individual.Cost=[];
pop=repmat(empty_individual,nPop,1);
for i=1:nPop
    % Initialize Position
    pop(i).Position=unifrnd(VarMin,VarMax,VarSize);
    % Evaluation
    pop(i).Cost=CostFunction(pop(i).Position);
end
```

```
% Sort Population
Costs=[pop.Cost];
[Costs, SortOrder]=sort(Costs);
pop=pop(SortOrder);
% Store Best Solution
BestSol=pop(1);
% Array to Hold Best Cost Values
BestCost=zeros(MaxIt,1);
% Store Cost
WorstCost=pop(end).Cost;
%% Main Loop
for it=1:MaxIt
    if FE>100000
        break;
    end
    % Crossover
    popc=repmat(empty_individual,nc/2,2);
    for k=1:nc/2
        i1=randi([1 nPop]);
        i2=randi([1 nPop]);
        % Select Parents
        p1=pop(i1);
        p2=pop(i2);
        % Apply Crossover
        [popc(k,1).Position, popc(k,2).Position]=
Crossover(p1.Position,p2.Position,gamma,VarMin,VarMax);
        % Evaluate Offsprings
        popc(k,1).Cost= CostFunc-
tion(popc(k,1).Position);
        popc(k,2).Cost= CostFunc-
tion(popc(k,2).Position);
    end
    popc=popc(:);
    % Mutation
    popm=repmat(empty_individual,nm,1);
    for k=1:nm
        % Select Parent
        i=randi([1 nPop]);
        p=pop(i);
        % Apply Mutation
        popm(k).Position= Mu-
tate(p.Position,mu,VarMin,VarMax);
        % Evaluate Mutant
        popm(k).Cost=CostFunction(popm(k).Position);
    end
    % Create Merged Population
    pop=[pop
```

```
        popc
        popm]; %#ok
    % Sort Population
    Costs=[pop.Cost];
    [Costs, SortOrder]=sort(Costs);
    pop=pop(SortOrder);
    % Update Worst Cost
    WorstCost=max(WorstCost,pop(end).Cost);
    % Truncation
    pop=pop(1:nPop);
    Costs=Costs(1:nPop);
    % Store Best Solution Ever Found
    BestSol=pop(1);
    % Store Best Cost Ever Found
    BestCost(it)=BestSol.Cost;
    % Show Iteration Information
    disp(['Iteration ' num2str(it) ': Best Cost = '
num2str(BestCost(it))]);
end

%% Results
printresult(BestSol.Position)
figure;
semilogy(BestCost,'LineWidth',2);
% plot(BestCost,'LineWidth',2);
xlabel('Iteration');
ylabel('Best Cost');



function fitness_value=CostF(x)
global FE; % Number of function evaluations
FE=FE+1;
P_base=100; % base power
n=size(x,2); % number of units;
%Fuel coost coefficients
cost_c(:,1)=[0.0070;0.0095;0.0090;0.0090;0.0080;0.0075]
; % for unit 1, 2, ..., n
cost_c(:,2)=[7.0;10.0;8.5;11.0;10.5;12.0]; % for unit
1, 2, ..., n
cost_c(:,3)=[240;200;220;200;220;190]; % for unit 1, 2,
..., n
% Valve-Point Loading Effects Coefficients
cost_c(:,4)=[300;200;150;150;150;150]; % for unit 1, 2,
..., n
cost_c(:,5)=[0.031;0.042;0.063;0.063;0.063;0.063]; %
for unit 1, 2, ..., n
% B-Coefficients
```

```
B_c=[0.0017,0.0012,0.0007,-0.0001,-0.0005,-0.0002;...
    0.0012,0.0014,0.0009,0.0001,-0.0006,-0.0001;...
    0.0007,0.0009,0.0031,0.0000,-0.0010,-0.0006;...
    -0.0001,0.0001,0.0000,0.0024,-0.0006,-0.0008;...
    -0.0005,-0.0006,-0.0010,-0.0006,0.0129,-0.0002;...
    -0.0002,-0.0001,-0.0006,-0.0008,-0.0002,0.0150];
B0_c=0.001*[-0.3908,-0.1297,0.7047,0.0591,0.2161,-
0.6635];
B00_c=0.056;
% Emission-Rates Coefficients
Alpha=[4.091;2.543;4.258;5.426;4.258;6.131]; % for unit
1, 2, ..., n
Beta=-1*[5.554;6.047;5.094;3.550;5.094;5.555]; % for
unit 1, 2, ..., n
Gamma=[6.490;5.638;4.586;3.380;4.586;5.151]; % for unit
1, 2, ..., n
Xi=[0.000200;0.000500;0.000001;0.002000;0.000001;0.0000
10]; % for unit 1, 2, ..., n
Delta=[2.857;3.333;8.000;2.000;8.000;6.667]; % for unit
1, 2, ..., n
% Ramp-Rate Limits
Pnow=[440;170;200;150;190;110]/P_base; % for unit 1, 2,
..., n
UR=[80;50;65;50;50;50]/P_base; % for unit 1, 2, ..., n
DR=[120;90;100;90;90;90]/P_base; % for unit 1, 2, ...,
n
% Prohibited Zones
PZL1=[210;90;150;80;90;75]/P_base; % 1st Lower Limit of
Prohibited Zone 1
PZU1=[240;110;170;90;110;85]/P_base; % 1st Upper Limit
of Prohibited Zone 1
PZL2=[350;140;210;110;140;100]/P_base; % 2nd Lower Lim-
it of Prohibited Zone 2
PZU2=[380;160;240;120;150;105]/P_base; % 2nd Upper Lim-
it of Prohibited Zone 2
P_D=1263/P_base; % Load value in p.u.
P_G_min=[100;50;80;50;50;50]/P_base; % for P1, P2, ...,
Pn (in MW)
P_G_max=[500;200;300;150;200;120]/P_base; % for P1, P2,
..., Pn (in MW)
% Calculating generator output power by considering
generator limits
for i=1:n
    P_G(i)=P_G_min(i)+(P_G_max(i)-P_G_min(i))*x(i);
end
% Calculating transmission loss
P_loss=P_G*B_c*P_G'+B0_c*P_G'+B00_c;
```

```
% Calculating power unblance penalty
err_unblance=abs(sum(P_G)-P_loss-P_D);
% Checking prohibited zone
err_PZ1=(P_G'<PZL1 .* P_G'>P_G_min + P_G'>PZU1 .*
P_G'<PZL2 +...
    P_G'>PZU2 .* P_G'<P_G_max);
% Cheking ramp limit
err_RL=~((P_G'>Pnow).*((P_G'-Pnow)<UR) +
(P_G'<=Pnow).*((Pnow-P_G')<DR));
% calculating fuel cost
fuel_cost=(P_G.^2)*cost_c(:,1)+P_G*cost_c(:,2)+sum(cost
_c(:,3))+abs(cost_c(:,4)'*sin(cost_c(:,5).*(P_G_min-
P_G')));
% Calculating emmision
emmision=1e-
2*(Alpha+P_G'.*Beta+P_G'.^2.*Gamma)+sum(Xi.*exp(P_G'.*D
elta));
% FINALLY, Calculating fitness function
fit-
ness_value=sum(emmision)+fuel_cost+1e6*(sum(~err_PZ1)+e
rr_unblance+sum(err_RL));
end
```

```
function [y1, y2]=Crossover(x1,x2,gamma,VarMin,VarMax)
alpha=unifrnd(-gamma,1+gamma,size(x1));
y1=alpha.*x1+(1-alpha).*x2;
y2=alpha.*x2+(1-alpha).*x1;
y1=max(y1,VarMin);
y1=min(y1,VarMax);
y2=max(y2,VarMin);
y2=min(y2,VarMax);
end
```

```
function y=Mutate(x,mu,VarMin,VarMax)
nVar=numel(x);
nmu=ceil(mu*nVar);
j=randsample(nVar,nmu);
sigma=0.1*(VarMax-VarMin);
y=x;
y(j)=x(j)+sigma*randn(size(j));
y=max(y,VarMin);
y=min(y,VarMax);
end
```

```
function z=printresult(x)
P_base=100;
n=6; % number of generators;
cost_c(:,3)=[240;200;220;200;220;190]; % for unit 1, 2,
..., n
cost_c(:,2)=[7.0;10.0;8.5;11.0;10.5;12.0]; % for unit
1, 2, ..., n
cost_c(:,1)=[0.0070;0.0095;0.0090;0.0090;0.0080;0.0075]
; % for unit 1, 2, ..., n
% Emission-Rates Coefficients
Alpha=[4.091;2.543;4.258;5.426;4.258;6.131]; % for unit
1, 2, ..., n
Beta=-1*[5.554;6.047;5.094;3.550;5.094;5.555]; % for
unit 1, 2, ..., n
Gamma=[6.490;5.638;4.586;3.380;4.586;5.151]; % for unit
1, 2, ..., n
Xi=[0.000200;0.000500;0.000001;0.002000;0.000001;0.0000
10]; % for unit 1, 2, ..., n
Delta=[2.857;3.333;8.000;2.000;8.000;6.667]; % for unit
1, 2, ..., n
% B-Coefficients
B_c=[0.0017,0.0012,0.0007,-0.0001,-0.0005,-0.0002;...
     0.0012,0.0014,0.0009,0.0001,-0.0006,-0.0001;...
     0.0007,0.0009,0.0031,0.0000,-0.0010,-0.0006;...
     -0.0001,0.0001,0.0000,0.0024,-0.0006,-0.0008;...
     -0.0005,-0.0006,-0.0010,-0.0006,0.0129,-0.0002;...
     -0.0002,-0.0001,-0.0006,-0.0008,-0.0002,0.0150];
B0_c=0.001*[-0.3908,-0.1297,0.7047,0.0591,0.2161,-
0.6635];
B00_c=0.056;
P_D=1263/P_base;
% Prohibited Zones
PZL1=[210;90;150;80;90;75]/P_base; % 1st Lower Limit of
Prohibited Zone 1
PZU1=[240;110;170;90;110;85]/P_base; % 1st Upper Limit
of Prohibited Zone 1
PZL2=[350;140;210;110;140;100]/P_base; % 2nd Lower Lim-
it of Prohibited Zone 2
PZU2=[380;160;240;120;150;105]/P_base; % 2nd Upper Lim-
it of Prohibited Zone 2
Pnow=[440;170;200;150;190;110]/P_base; % for unit 1, 2,
..., n
UR=[80;50;65;50;50;50]/P_base; % for unit 1, 2, ..., n
DR=[120;90;100;90;90;90]/P_base; % for unit 1, 2, ...,
n
P_G_min=[100;50;80;50;50;50]/P_base; % for P1, P2, ...,
Pn (in MW)
P_G_max=[500;200;300;150;200;120]/P_base; % for P1, P2,
..., Pn (in MW)
```

```
for i=1:n
    P_G(i)=P_G_min(i)+(P_G_max(i)-P_G_min(i))*x(i);
end
P_loss=P_G*B_c*P_G'+B0_c*P_G'+B00_c;
for i=1:n
    cost1(i)= P_G(i)^2*cost_c(i,1)+P_G(i)*cost_c(i,2)+
cost_c(i,3);
end
%      cost1
pout=[P_G'*100]
cost1'
sum(P_G*100)-P_D*100-P_loss*100
err_PZ1=(P_G'<PZL1 .* P_G'>P_G_min + P_G'>PZU1 .*
P_G'<PZL2 +...
    P_G'>PZU2 .* P_G'<P_G_max)
err_RL=(P_G'>Pnow).*((P_G'-Pnow)<UR) +
(P_G'<=Pnow).*((Pnow-P_G')<DR)
emmision=1e-2*(Alpha+P_G'.*Beta+P_G'.^2.*Gamma)+
Xi.*exp(P_G'.*Delta)
P_loss*P_base
sum(cost1)
sum(emmision)
end
```

MATLAB Code for Imperialist Competitive Algorithm

```
clc;
clear;
close all;
global FE;
FE=0;
%% Problem Definition
CostFunction=@(x) CostF(x);          % Cost Function
nVar=6;                  % Number of Decision Variables
VarSize=[1 nVar];        % Decision Variables Matrix Size
VarMin= 0;              % Lower Bound of Variables
VarMax= 1;              % Upper Bound of Variables
%% ICA Parameters
MaxIt=200;              % Maximum Number of Iterations
nPop=1000;                 % Population Size
nEmp=10;                 % Number of Empires/Imperialists
alpha=1;                 % Selection Pressure
beta=1.5;                % Assimilation Coefficient
pRevolution=0.05;        % Revolution Probability
mu=0.1;                  % Revolution Rate
```

```
zeta=0.2;            % Colonies Mean Cost Coefficient
%% Globalization of Parameters and Settings
global ProblemSettings;
ProblemSettings.CostFunction=CostFunction;
ProblemSettings.nVar=nVar;
ProblemSettings.VarSize=VarSize;
ProblemSettings.VarMin=VarMin;
ProblemSettings.VarMax=VarMax;
global ICASettings;
ICASettings.MaxIt=MaxIt;
ICASettings.nPop=nPop;
ICASettings.nEmp=nEmp;
ICASettings.alpha=alpha;
ICASettings.beta=beta;
ICASettings.pRevolution=pRevolution;
ICASettings.mu=mu;
ICASettings.zeta=zeta;
%% Initialization
% Initialize Empires
emp=CreateInitialEmpires();
% Array to Hold Best Cost Values
BestCost=zeros(MaxIt,1);
%% ICA Main Loop
for it=1:MaxIt
    if FE>100000
        break;
    end
    % Assimilation
    emp=AssimilateColonies(emp);
    % Revolution
    emp=DoRevolution(emp);
    % Intra-Empire Competition
    emp=IntraEmpireCompetition(emp);
    % Update Total Cost of Empires
    emp=UpdateTotalCost(emp);
    % Inter-Empire Competition
    emp=InterEmpireCompetition(emp);
    % Update Best Solution Ever Found
    imp=[emp.Imp];
    [~, BestImpIndex]=min([imp.Cost]);
    BestSol=imp(BestImpIndex);
    % Update Best Cost
    BestCost(it)=BestSol.Cost;
    % Show Iteration Information
    disp(['Iteration ' num2str(it) ': Best Cost = '
num2str(BestCost(it))]);
end
```

```
%% Results
figure;
plot(BestCost,'LineWidth',2);
% semilogy(BestCost,'LineWidth',2);
xlabel('Iteration');
ylabel('Best Cost');



function fitness_value=CostF(x)
global FE;
FE=FE+1;
P_base=100; % base power
n=6; % number of units;

%Fuel coost coefficients
cost_c(:,1)=[0.0070;0.0095;0.0090;0.0090;0.0080;0.0075]
; % for unit 1, 2, ..., n
cost_c(:,2)=[7.0;10.0;8.5;11.0;10.5;12.0]; % for unit
1, 2, ..., n
cost_c(:,3)=[240;200;220;200;220;190]; % for unit 1, 2,
..., n

% Valve-Point Loading Effects Coefficients
cost_c(:,4)=[300;200;150;150;150;150]; % for unit 1, 2,
..., n
cost_c(:,5)=[0.031;0.042;0.063;0.063;0.063;0.063]; %
for unit 1, 2, ..., n
% B-Coefficients
B_c=[0.0017,0.0012,0.0007,-0.0001,-0.0005,-0.0002;...
    0.0012,0.0014,0.0009,0.0001,-0.0006,-0.0001;...
    0.0007,0.0009,0.0031,0.0000,-0.0010,-0.0006;...
    -0.0001,0.0001,0.0000,0.0024,-0.0006,-0.0008;...
    -0.0005,-0.0006,-0.0010,-0.0006,0.0129,-0.0002;...
    -0.0002,-0.0001,-0.0006,-0.0008,-0.0002,0.0150];
B0_c=0.001*[-0.3908,-0.1297,0.7047,0.0591,0.2161,-
0.6635];
B00_c=0.056;
```

```
% Emission-Rates Coefficients
Alpha=[4.091;2.543;4.258;5.426;4.258;6.131]; % for unit
1, 2, ..., n
Beta=-1*[5.554;6.047;5.094;3.550;5.094;5.555]; % for
unit 1, 2, ..., n
Gamma=[6.490;5.638;4.586;3.380;4.586;5.151]; % for unit
1, 2, ..., n
Xi=[0.000200;0.000500;0.000001;0.002000;0.000001;0.0000
10]; % for unit 1, 2, ..., n
Delta=[2.857;3.333;8.000;2.000;8.000;6.667]; % for unit
1, 2, ..., n

% Ramp-Rate Limits
Pnow=[440;170;200;150;190;110]/P_base; % for unit 1, 2,
..., n
UR=[80;50;65;50;50;50]/P_base; % for unit 1, 2, ..., n
DR=[120;90;100;90;90;90]/P_base; % for unit 1, 2, ...,
n

% Prohibited Zones
PZL1=[210;90;150;80;90;75]/P_base; % 1st Lower Limit of
Prohibited Zone 1
PZU1=[240;110;170;90;110;85]/P_base; % 1st Upper Limit
of Prohibited Zone 1
PZL2=[350;140;210;110;140;100]/P_base; % 2nd Lower Lim-
it of Prohibited Zone 2
PZU2=[380;160;240;120;150;105]/P_base; % 2nd Upper Lim-
it of Prohibited Zone 2
P_D=1263/P_base;
P_G_min=[100;50;80;50;50;50]/P_base; % for P1, P2, ...,
Pn (in MW)
P_G_max=[500;200;300;150;200;120]/P_base; % for P1, P2,
..., Pn (in MW)
for i=1:n
    P_G(i)=P_G_min(i)+(P_G_max(i)-P_G_min(i))*x(i);
end
P_loss=P_G*B_c*P_G'+B0_c*P_G'+B00_c;
if sum(P_G)>=P_loss+P_D
    err_unblance=abs(sum(P_G)-P_loss-P_D);
else
    err_unblance=1e6;
end
```

```
err_PZ1=(P_G'<PZL1 .* P_G'>P_G_min + P_G'>PZU1 .*
P_G'<PZL2 +...
    P_G'>PZU2 .* P_G'<P_G_max);
err_RL=~((P_G'>Pnow).*((P_G'-Pnow)<UR) +
(P_G'<=Pnow).*((Pnow-P_G')<DR));
fuel_cost=(P_G.^2)*cost_c(:,1)+P_G*cost_c(:,2)+sum(cost
_c(:,3))+abs(cost_c(:,4)'*sin(cost_c(:,5).*(P_G_min-
P_G')));
emmision=1e-2*(Alpha+P_G'.*Beta+P_G'.^2.*Gamma)+
sum(Xi.*exp(P_G'.*Delta));
fit-
ness_value=sum(emmision)+fuel_cost+1e6*(sum(~err_PZ1)+e
rr_unblance+sum(err_RL));
end
function emp=CreateInitialEmpires()
global ProblemSettings;
global ICASettings;
CostFunction=ProblemSettings.CostFunction;
nVar=ProblemSettings.nVar;
VarSize=ProblemSettings.VarSize;
VarMin=ProblemSettings.VarMin;
VarMax=ProblemSettings.VarMax;
nPop=ICASettings.nPop;
nEmp=ICASettings.nEmp;
nCol=nPop-nEmp;
alpha=ICASettings.alpha;
empty_country.Position=[];
empty_country.Cost=[];
country=repmat(empty_country,nPop,1);
for i=1:nPop
    country(i).Position=unifrnd(VarMin,VarMax,VarSize);
    country(i).Cost=CostFunction(country(i).Position);
end
costs=[country.Cost];
[~, SortOrder]=sort(costs);
country=country(SortOrder);
imp=country(1:nEmp);
col=country(nEmp+1:end);
empty_empire.Imp=[];
empty_empire.Col=repmat(empty_country,0,1);
empty_empire.nCol=0;
empty_empire.TotalCost=[];
emp=repmat(empty_empire,nEmp,1);

% Assign Imperialists
for k=1:nEmp
    emp(k).Imp=imp(k);
end
```

```
% Assign Colonies
P=exp(-alpha*[imp.Cost]/max([imp.Cost]));
P=P/sum(P);
for j=1:nCol
    k=RouletteWheelSelection(P);
    emp(k).Col=[emp(k).Col
        col(j)];
    emp(k).nCol=emp(k).nCol+1;
end
emp=UpdateTotalCost(emp);
end
function emp=AssimilateColonies(emp)
global ProblemSettings;
CostFunction=ProblemSettings.CostFunction;
VarSize=ProblemSettings.VarSize;
VarMin=ProblemSettings.VarMin;
VarMax=ProblemSettings.VarMax;
global ICASettings;
beta=ICASettings.beta;
nEmp=numel(emp);
for k=1:nEmp
    for i=1:emp(k).nCol
        emp(k).Col(i).Position = emp(k).Col(i).Position
...
            + beta*rand(VarSize).*(emp(k).Imp.Position-
emp(k).Col(i).Position);
        emp(k).Col(i).Position=
max(emp(k).Col(i).Position,VarMin);
        emp(k).Col(i).Position =
min(emp(k).Col(i).Position,VarMax);
        emp(k).Col(i).Cost = CostFunc-
tion(emp(k).Col(i).Position);
    end
end
end
```

```
function emp=DoRevolution(emp)
global ProblemSettings;
CostFunction=ProblemSettings.CostFunction;
nVar=ProblemSettings.nVar;
VarSize=ProblemSettings.VarSize;
VarMin=ProblemSettings.VarMin;
VarMax=ProblemSettings.VarMax;
global ICASettings;
pRevolution=ICASettings.pRevolution;
mu=ICASettings.mu;
nmu=ceil(mu*nVar);
sigma=0.1*(VarMax-VarMin);
nEmp=numel(emp);
for k=1:nEmp
    NewPos = emp(k).Imp.Position + sig-
ma*randn(VarSize);
    jj=randsample(nVar,nmu)';
    NewImp=emp(k).Imp;
    NewImp.Position(jj)=NewPos(jj);
    NewImp.Cost=CostFunction(NewImp.Position);
    if NewImp.Cost<emp(k).Imp.Cost
        emp(k).Imp = NewImp;
    end
    for i=1:emp(k).nCol
        if rand<=pRevolution
            NewPos = emp(k).Col(i).Position + sig-
ma*randn(VarSize);
            jj=randsample(nVar,nmu)';
            emp(k).Col(i).Position(jj) = NewPos(jj);
            emp(k).Col(i).Position =
max(emp(k).Col(i).Position,VarMin);
            emp(k).Col(i).Position =
min(emp(k).Col(i).Position,VarMax);
            emp(k).Col(i).Cost = CostFunc-
tion(emp(k).Col(i).Position);
        end
    end
end
end
```

```
function emp=IntraEmpireCompetition(emp)
nEmp=numel(emp);
for k=1:nEmp
    for i=1:emp(k).nCol
        if emp(k).Col(i).Cost<emp(k).Imp.Cost
            imp=emp(k).Imp;
            col=emp(k).Col(i);

            emp(k).Imp=col;
            emp(k).Col(i)=imp;
        end
    end
end
end



function emp=UpdateTotalCost(emp)
global ICASettings;
zeta=ICASettings.zeta;
nEmp=numel(emp);
for k=1:nEmp
    if emp(k).nCol>0
        emp(k).TotalCost=
emp(k).Imp.Cost+zeta*mean([emp(k).Col.Cost]);
    else
        emp(k).TotalCost=emp(k).Imp.Cost;
    end
end
end

function emp=InterEmpireCompetition(emp)
if numel(emp)==1
    return;
end
global ICASettings;
alpha=ICASettings.alpha;
TotalCost=[emp.TotalCost];
 [~, WeakestEmpIndex]=max(TotalCost);
WeakestEmp=emp(WeakestEmpIndex);
P=exp(-alpha*TotalCost/max(TotalCost));
P(WeakestEmpIndex)=0;
P=P/sum(P);
```

```
if any(isnan(P))
    P(isnan(P))=0;
    if all(P==0)
        P(:)=1;
    end
    P=P/sum(P);
end
if WeakestEmp.nCol>0
    [~, WeakestColIndex]=max([WeakestEmp.Col.Cost]);
    WeakestCol=WeakestEmp.Col(WeakestColIndex);

    WinnerEmpIndex=RouletteWheelSelection(P);
    WinnerEmp=emp(WinnerEmpIndex);

    WinnerEmp.Col(end+1)=WeakestCol;
    WinnerEmp.nCol=WinnerEmp.nCol+1;
    emp(WinnerEmpIndex)=WinnerEmp;

    WeakestEmp.Col(WeakestColIndex)=[];
    WeakestEmp.nCol=WeakestEmp.nCol-1;
    emp(WeakestEmpIndex)=WeakestEmp;
end
if WeakestEmp.nCol==0

    WinnerEmpIndex2=RouletteWheelSelection(P);
    WinnerEmp2=emp(WinnerEmpIndex2);

    WinnerEmp2.Col(end+1)=WeakestEmp.Imp;
    WinnerEmp2.nCol=WinnerEmp2.nCol+1;
    emp(WinnerEmpIndex2)=WinnerEmp2;

    emp(WeakestEmpIndex)=[];
end
end
```

Cost calculation function:

```
clear;
close all;
clc;

n=6; % number of generators;

cost_c(:,3)=[240;200;220;200;220;190]; % for unit 1, 2,
..., n
cost_c(:,2)=[7.0;10.0;8.5;11.0;10.5;12.0]; % for unit
1, 2, ..., n
cost_c(:,1)=[0.0070;0.0095;0.0090;0.0090;0.0080;0.0075]
; % for unit 1, 2, ..., n

% B-Coefficients
B_c=[0.0017,0.0012,0.0007,-0.0001,-0.0005,-0.0002;...
0.0012,0.0014,0.0009,0.0001,-0.0006,-0.0001;...
0.0007,0.0009,0.0031,0.0000,-0.0010,-0.0006;...
-0.0001,0.0001,0.0000,0.0024,-0.0006,-0.0008;...
-0.0005,-0.0006,-0.0010,-0.0006,0.0129,-0.0002;...
-0.0002,-0.0001,-0.0006,-0.0008,-0.0002,0.0150];
B0_c=0.001*[-0.3908,-0.1297,0.7047,0.0591,0.2161,-
0.6635];
B00_c=0.056;

P_base=100;
P_D=1263/P_base;

P_G_min=[100;50;80;50;50;50]/P_base; % for P1, P2, ...,
Pn (in MW)
P_G_max=[500;200;300;150;200;120]/P_base; % for P1, P2,
..., Pn (in MW)

lambda(1)=0;
epsi=0.001;
A_m=zeros(n,n);
C_m=zeros(n,1);
```

```
%%initializing lamda
a_total=(sum(cost_c(:,1).^-1))^-1;
b_total=a_total*sum((cost_c(:,1).^-1).*cost_c(:,2));

lambda(2)=a_total*P_D*P_base+b_total;

P_G(1:2,:)=zeros(2,n);

itt=2;
flag=1;
while (flag)
    itt=itt+1;
    for i=1:n
        for j=1:n
            if i==j
                A_m(i,j)=cost_c(i,1)*P_base/lambda(itt-
1)+2*B_c(i,i);
            else
                A_m(i,j)=-2*B_c(i,j);
            end
        end
    end

    for i=1:n
        C_m(i,1)=(1-B0_c(i))-cost_c(i,2)/lambda(itt-1);
    end

    P_G(itt-1,:)=A_m\C_m;
    for i=1:n
        if P_G(itt-1,i)>=P_G_max(i)
            P_G(itt-1,i)=P_G_max(i);
        elseif P_G(itt-1,i)<=P_G_min(i)
            P_G(itt-1,i)=P_G_min(i);
        end
    end
    %Claculating P_loss
    P_loss=P_G(itt-1,:)*B_c*P_G(itt-1,:)'+B0_c*P_G(itt-
1,:)'+B00_c;

    if(sum(P_G(itt-1,:))-P_loss-P_D<epsi &&
sum(P_G(itt-1,:))-P_loss-P_D>0)
        flag=0;
    else
        lambda(itt)=lambda(itt-1)+(lambda(itt-1)-
lambda(itt-2))*(P_loss+P_D-sum(P_G(itt-1,:)))...
```

```
                /(sum(P_G(itt-1,:))-sum(P_G(itt-2,:))+eps);
    end

end
% Transmission loss
P_loss=P_loss*P_base
% Power output of units
P_G_final=P_G(end,:)'*P_base
% Calculating fuel cost of units
fuel_cost=(P_G(end,:).^2)*cost_c(:,1)+P_G(end,:)*cost_c
(:,2)+sum(cost_c(:,3))
% Calculating error of power unblance
err_unblance=sum(P_G(end,:)'*P_base)-P_D*P_base-P_loss;
```

Main MATLAB script code for multi objective particle swarm optimization

```
clc;
clear;
close all;

%% Problem Definition
CostFunction=@(x) CostF(x);       % Cost Function
nVar=6;                   % Number of Decision Variables
VarSize=[1 nVar];         % Size of Decision Variables Matrix
VarMin=0;                 % Lower Bound of Variables
VarMax=1;                 % Upper Bound of Variables

%% MOPSO Parameters
MaxIt=300;                 % Maximum Number of Iterations
nPop=600;                  % Population Size
nRep=300;                  % Repository Size
w=0.5;                    % Inertia Weight
wdamp=0.99;               % Intertia Weight Damping Rate
c1=1;                     % Personal Learning Coefficient
c2=2;                     % Global Learning Coefficient
nGrid=7;                  % Number of Grids per Dimension
alpha=0.1;                % Inflation Rate
beta=2;                   % Leader Selection Pressure
gamma=2;                  % Deletion Selection Pressure
mu=0.1;                   % Mutation Rate

%% Initialization
empty_particle.Position=[];
empty_particle.Velocity=[];
empty_particle.Cost=[];
```

```
empty_particle.Best.Position=[];
empty_particle.Best.Cost=[];
empty_particle.IsDominated=[];
empty_particle.GridIndex=[];
empty_particle.GridSubIndex=[];
pop=repmat(empty_particle,nPop,1);
for i=1:nPop
    pop(i).Position=unifrnd(VarMin,VarMax,VarSize);
    pop(i).Velocity=zeros(VarSize);
    pop(i).Cost=CostFunction(pop(i).Position);
    % Update Personal Best
    pop(i).Best.Position=pop(i).Position;
    pop(i).Best.Cost=pop(i).Cost;

end

% Determine Domination
pop=DetermineDomination(pop);
rep=pop(~[pop.IsDominated]);
Grid=CreateGrid(rep,nGrid,alpha);
for i=1:numel(rep)
    rep(i)=FindGridIndex(rep(i),Grid);
end

%% MOPSO Main Loop
for it=1:MaxIt
    for i=1:nPop
        leader=SelectLeader(rep,beta);
        pop(i).Velocity = w*pop(i).Velocity ...
            +c1*rand(VarSize).*(pop(i).Best.Position-
pop(i).Position) ...
            +c2*rand(VarSize).*(leader.Position-
pop(i).Position);
        pop(i).Position = pop(i).Position +
pop(i).Velocity;
        pop(i).Position = max(pop(i).Position, VarMin);
        pop(i).Position = min(pop(i).Position, VarMax);
        pop(i).Cost = CostFunction(pop(i).Position);
        % Apply Mutation
        pm=(1-(it-1)/(MaxIt-1))^(1/mu);
        if rand<pm
            NewSol.Position= Mu-
tate(pop(i).Position,pm,VarMin,VarMax);
            NewSol.Cost=CostFunction(NewSol.Position);
            if Dominates(NewSol,pop(i))
                pop(i).Position=NewSol.Position;
```

```
                pop(i).Cost=NewSol.Cost;
            elseif Dominates(pop(i),NewSol)
                % Do Nothing
            else
                if rand<0.5
                    pop(i).Position=NewSol.Position;
                    pop(i).Cost=NewSol.Cost;
                end
            end
        end
        if Dominates(pop(i),pop(i).Best)
            pop(i).Best.Position=pop(i).Position;
            pop(i).Best.Cost=pop(i).Cost;
        elseif Dominates(pop(i).Best,pop(i))
            % Do Nothing
        else
            if rand<0.5
                pop(i).Best.Position=pop(i).Position;
                pop(i).Best.Cost=pop(i).Cost;
            end
        end

    end

    % Add Non-Dominated Particles to REPOSITORY
    rep=[rep
         pop(~[pop.IsDominated])]; %#ok

    % Determine Domination of New Resository Members
    rep=DetermineDomination(rep);

    % Keep only Non-Dminated Memebrs in the Repository
    rep=rep(~[rep.IsDominated]);

    % Update Grid
    Grid=CreateGrid(rep,nGrid,alpha);

    % Update Grid Indices
    for i=1:numel(rep)
        rep(i)=FindGridIndex(rep(i),Grid);
    end

    % Check if Repository is Full
    if numel(rep)>nRep
```

```
            Extra=numel(rep)-nRep;
            for e=1:Extra
                rep=DeleteOneRepMemebr(rep,gamma);
            end

        end

        % Plot Costs
        figure(1);
        PlotCosts(pop,rep);
        pause(0.01);

        % Show Iteration Information
        disp(['Iteration ' num2str(it) ': Number of Rep
    Members = ' num2str(numel(rep))]);

        % Damping Inertia Weight
        w=w*wdamp;

    end



    function fitness_values=CostF(x)
        P_base=100; % base power
        n=6; % number of units;
        %Fuel coost coefficients
        cost_c(:,1)= [0.0070;0.0095;0.0090;
    0.0090;0.0080;0.0075]; % for unit 1, 2, ..., n
        cost_c(:,2)=[7.0;10.0;8.5;11.0;10.5;12.0]; % for
    unit 1, 2, ..., n
        cost_c(:,3)=[240;200;220;200;220;190]; % for unit
    1, 2, ..., n
        % Valve-Point Loading Effects Coefficients
        cost_c(:,4)=[300;200;150;150;150;150]; % for unit
    1, 2, ..., n
        cost_c(:,5)=[0.031;0.042;0.063;0.063;0.063;0.063];
    % for unit 1, 2, ..., n

        % B-Coefficients
        B_c=[0.0017,0.0012,0.0007,-0.0001,-0.0005,-
    0.0002;...
        0.0012,0.0014,0.0009,0.0001,-0.0006,-0.0001;...
        0.0007,0.0009,0.0031,0.0000,-0.0010,-0.0006;...
        -0.0001,0.0001,0.0000,0.0024,-0.0006,-0.0008;...
        -0.0005,-0.0006,-0.0010,-0.0006,0.0129,-0.0002;...
        -0.0002,-0.0001,-0.0006,-0.0008,-0.0002,0.0150];
```

```
    B0_c=0.001*[-0.3908,-0.1297,0.7047,0.0591,0.2161,-
0.6635];
    B00_c=0.056;

    % Emission-Rates Coefficients
    Alpha=[4.091;2.543;4.258;5.426;4.258;6.131]; % for
unit 1, 2, ..., n
    Beta=-1*[5.554;6.047;5.094;3.550;5.094;5.555]; %
for unit 1, 2, ..., n
    Gamma=[6.490;5.638;4.586;3.380;4.586;5.151]; % for
unit 1, 2, ..., n
    Xi=[0.000200;0.000500;0.000001;0.002000;0.000001;
0.000010]; % for unit 1, 2, ..., n
    Delta=[2.857;3.333;8.000;2.000;8.000;6.667]; % for
unit 1, 2, ..., n

    % Ramp-Rate Limits
    Pnow=[440;170;200;150;190;110]/P_base; % for unit
1, 2, ..., n
    UR=[80;50;65;50;50;50]/P_base; % for unit 1, 2,
..., n
    DR=[120;90;100;90;90;90]/P_base; % for unit 1, 2,
..., n

    % Prohibited Zones
    PZL1=[210;90;150;80;90;75]/P_base; % 1st Lower Lim-
it of Prohibited Zone 1
    PZU1=[240;110;170;90;110;85]/P_base; % 1st Upper
Limit of Prohibited Zone 1
    PZL2=[350;140;210;110;140;100]/P_base; % 2nd Lower
Limit of Prohibited Zone 2
    PZU2=[380;160;240;120;150;105]/P_base; % 2nd Upper
Limit of Prohibited Zone 2

    P_D=1263/P_base;
    P_G_min=[100;50;80;50;50;50]/P_base; % for P1, P2,
..., Pn (in MW)
    P_G_max=[500;200;300;150;200;120]/P_base; % for P1,
P2, ..., Pn (in MW)
    for i=1:n
        P_G(i)=P_G_min(i)+(P_G_max(i)-P_G_min(i))*x(i);
    end
    P_loss=P_G*B_c*P_G'+B0_c*P_G'+B00_c;
    err_unblance=abs(sum(P_G)-P_loss-P_D);
    err_PZ1=(P_G'<PZL1 .* P_G'>P_G_min + P_G'>PZU1 .*
P_G'<PZL2 + P_G'>PZU2 .* P_G'<P_G_max);
```

```
    err_RL=~((P_G'>Pnow).*((P_G'-Pnow)<UR) +
(P_G'<=Pnow).*((Pnow-P_G')<DR));
    fuel_cost=(P_G.^2)*cost_c(:,1)+P_G*cost_c(:,2)+
sum(cost_c(:,3))+abs(cost_c(:,4)'*sin(cost_c(:,5).*
(P_G_min-P_G')));
    emmision=1e-2*(Alpha+P_G'.*Beta+P_G'.^2.*Gamma)+
sum(Xi.*exp(P_G'.*Delta));
    f1=fuel_cost+1e3*(sum(~err_PZ1)+1e-3*err_unblance+
sum(err_RL));
    f2=sum(emmision)+1e3*(sum(~err_PZ1)+1e-
3*err_unblance+ sum(err_RL));
    fitness_values=[f1
                        f2];
end



function pop=DetermineDomination(pop)
    nPop=numel(pop);
    for i=1:nPop
        pop(i).IsDominated=false;
    end
    for i=1:nPop-1
        for j=i+1:nPop
            if Dominates(pop(i),pop(j))
               pop(j).IsDominated=true;
            end
            if Dominates(pop(j),pop(i))
               pop(i).IsDominated=true;
            end


        end
    end
end
```

```
function Grid=CreateGrid(pop,nGrid,alpha)
    c=[pop.Cost];
    cmin=min(c,[],2);
    cmax=max(c,[],2);
    dc=cmax-cmin;
    cmin=cmin-alpha*dc;
    cmax=cmax+alpha*dc;
    nObj=size(c,1);
    empty_grid.LB=[];
    empty_grid.UB=[];
    Grid=repmat(empty_grid,nObj,1);
    for j=1:nObj
        cj=linspace(cmin(j),cmax(j),nGrid+1);
        Grid(j).LB=[-inf cj];
        Grid(j).UB=[cj +inf];
    end
end
```

```
function particle=FindGridIndex(particle,Grid)
    nObj=numel(particle.Cost);
    nGrid=numel(Grid(1).LB);
    particle.GridSubIndex=zeros(1,nObj);
    for j=1:nObj
        particle.GridSubIndex(j)=...
            find(particle.Cost(j)<
Grid(j).UB,1,'first');
    end
    particle.GridIndex=particle.GridSubIndex(1);
    for j=2:nObj
        particle.GridIndex=particle.GridIndex-1;
        particle.GridIndex=nGrid*particle.GridIndex;
        particle.GridIndex= particle.GridIndex+ parti-
cle.GridSubIndex(j);
    end
end
```

```
function leader=SelectLeader(rep,beta)
    % Grid Index of All Repository Members
    GI=[rep.GridIndex];
    % Occupied Cells
    OC=unique(GI);
    % Number of Particles in Occupied Cells
    N=zeros(size(OC));
    for k=1:numel(OC)
        N(k)=numel(find(GI==OC(k)));
    end
    % Selection Probabilities
    P=exp(-beta*N);
    P=P/sum(P);
    % Selected Cell Index
    sci=RouletteWheelSelection(P);
    % Selected Cell
    sc=OC(sci);
    % Selected Cell Members
    SCM=find(GI==sc);
    % Selected Member Index
    smi=randi([1 numel(SCM)]);
    % Selected Member
    sm=SCM(smi);
    % Leader
    leader=rep(sm);
end


function xnew=Mutate(x,pm,VarMin,VarMax)
    nVar=numel(x);
    j=randi([1 nVar]);
    dx=pm*(VarMax-VarMin);
    lb=x(j)-dx;
    if lb<VarMin
        lb=VarMin;
    end
    ub=x(j)+dx;
    if ub>VarMax
        ub=VarMax;
    end
    xnew=x;
    xnew(j)=unifrnd(lb,ub);
end
```

```
function b=Dominates(x,y)
    if isstruct(x)
        x=x.Cost;
    end
    if isstruct(y)
        y=y.Cost;
    end
    b=all(x<=y) && any(x<y);
end
```

```
function rep=DeleteOneRepMemebr(rep,gamma)
    % Grid Index of All Repository Members
    GI=[rep.GridIndex];
    % Occupied Cells
    OC=unique(GI);
    % Number of Particles in Occupied Cells
    N=zeros(size(OC));
    for k=1:numel(OC)
        N(k)=numel(find(GI==OC(k)));
    end
    % Selection Probabilities
    P=exp(gamma*N);
    P=P/sum(P);
    % Selected Cell Index
    sci=RouletteWheelSelection(P);
    % Selected Cell
    sc=OC(sci);
    % Selected Cell Members
    SCM=find(GI==sc);
    % Selected Member Index
    smi=randi([1 numel(SCM)]);
    % Selected Member
    sm=SCM(smi);
    % Delete Selected Member
    rep(sm)=[];
end
```

```
function PlotCosts(pop,rep)
%     pop_costs=[pop.Cost];
%     plot(pop_costs(1,:),pop_costs(2,:),'ko');
%     hold on;
    rep_costs=[rep.Cost];
    plot(rep_costs(1,:),rep_costs(2,:),'r*');
    xlabel('Fuel Cost ($/hr)');
    ylabel('Emission (ton/hr)');
    grid on;
    hold off;
end
```

MATLAB Script and function codes for Non-Sorted Genetic Algorithm II (NSAG II):

```
% Copyright (c) 2015, Yarpiz (www.yarpiz.com)
% All rights reserved. Please read the "license.txt"
for license terms.
%
% Project Code: YPEA120
% Project Title: Non-dominated Sorting Genetic Algo-
rithm II (NSGA-II)
% Publisher: Yarpiz (www.yarpiz.com)
%
% Developer: S. Mostapha Kalami Heris (Member of Yarpiz
Team)
%
% Contact Info: sm.kalami@gmail.com, info@yarpiz.com
%

clc;
clear;
close all;

%% Problem Definition
```

```
CostFunction=@(x) CostF(x);      % Cost Function
nVar=6;                  % Number of Decision Variables
VarSize=[1 nVar];        % Size of Decision Variables Matrix
VarMin=0;               % Lowe1r Bound of Variables
VarMax= 1;               % Upper Bound of Variables

% Number of Objective Functions
nObj=numel(CostFunction(unifrnd(VarMin,VarMax,VarSize))
);

%% NSGA-II Parameters
MaxIt=200;        % Maximum Number of Iterations
nPop=200;          % Population Size
pCrossover=0.7;                             % Crossover
Percentage
nCrossover=2*round(pCrossover*nPop/2);  % Number of
Parnets (Offsprings)
pMutation=0.4;                              % Mutation Per-
centage
nMutation=round(pMutation*nPop);            % Number of Mu-
tants
mu=0.02;                       % Mutation Rate
sigma=0.1*(VarMax-VarMin);  % Mutation Step Size

%% Initialization
empty_individual.Position=[];
empty_individual.Cost=[];
empty_individual.Rank=[];
empty_individual.DominationSet=[];
empty_individual.DominatedCount=[];
empty_individual.CrowdingDistance=[];
pop=repmat(empty_individual,nPop,1);
for i=1:nPop
    pop(i).Position=unifrnd(VarMin,VarMax,VarSize);
    pop(i).Cost=CostFunction(pop(i).Position);
end

% Non-Dominated Sorting
[pop, F]=NonDominatedSorting(pop);

% Calculate Crowding Distance
pop=CalcCrowdingDistance(pop,F);

% Sort Population
[pop, F]=SortPopulation(pop);
```

```
%% NSGA-II Main Loop

for it=1:MaxIt
    % Crossover
    popc=repmat(empty_individual,nCrossover/2,2);
    for k=1:nCrossover/2
        i1=randi([1 nPop]);
        p1=pop(i1);
        i2=randi([1 nPop]);
        p2=pop(i2);
        [popc(k,1).Position,
popc(k,2).Position]=Crossover(p1.Position,p2.Position);
        popc(k,1).Cost= CostFunc-
tion(popc(k,1).Position);
        popc(k,2).Cost= CostFunc-
tion(popc(k,2).Position);
    end
    popc=popc(:);

    % Mutation
    popm=repmat(empty_individual,nMutation,1);
    for k=1:nMutation
        i=randi([1 nPop]);
        p=pop(i);
        popm(k).Position=Mutate(p.Position,mu,sigma);
        popm(k).Cost=CostFunction(popm(k).Position);
    end

    % Merge
    pop=[pop
         popc
         popm]; %#ok

    % Non-Dominated Sorting
    [pop, F]=NonDominatedSorting(pop);

    % Calculate Crowding Distance
    pop=CalcCrowdingDistance(pop,F);

    % Sort Population
    pop=SortPopulation(pop);

    % Truncate
    pop=pop(1:nPop);
```

```
    % Non-Dominated Sorting
    [pop, F]=NonDominatedSorting(pop);

    % Calculate Crowding Distance
    pop=CalcCrowdingDistance(pop,F);

    % Sort Population
    [pop, F]=SortPopulation(pop);

    % Store F1
    F1=pop(F{1});

    % Show Iteration Information
    disp(['Iteration ' num2str(it) ': Number of F1 Mem-
bers = ' num2str(numel(F1))]);

    % Plot F1 Costs
    figure(1);
    PlotCosts(F1);
    pause(0.01);
end



function fitness_values=CostF(x)
    P_base=100; % base power
    n=6; % number of units;
    %Fuel coost coefficients
    cost_c(:,1)=[0.0070;0.0095;0.0090;0.0090;
0.0080;0.0075]; % for unit 1, 2, ..., n
    cost_c(:,2)=[7.0;10.0;8.5;11.0;10.5;12.0]; % for
unit 1, 2, ..., n
    cost_c(:,3)=[240;200;220;200;220;190]; % for unit
1, 2, ..., n

    % Valve-Point Loading Effects Coefficients
    cost_c(:,4)=[300;200;150;150;150;150]; % for unit
1, 2, ..., n
    cost_c(:,5)=[0.031;0.042;0.063;0.063;0.063;0.063];
% for unit 1, 2, ..., n

    % B-Coefficients
    B_c=[0.0017,0.0012,0.0007,-0.0001,-0.0005,-
0.0002;...
    0.0012,0.0014,0.0009,0.0001,-0.0006,-0.0001;...
    0.0007,0.0009,0.0031,0.0000,-0.0010,-0.0006;...
```

```
    -0.0001,0.0001,0.0000,0.0024,-0.0006,-0.0008;...
    -0.0005,-0.0006,-0.0010,-0.0006,0.0129,-0.0002;...
    -0.0002,-0.0001,-0.0006,-0.0008,-0.0002,0.0150];
    B0_c=0.001*[-0.3908,-0.1297,0.7047,0.0591,0.2161,-
0.6635];
    B00_c=0.056;

    % Emission-Rates Coefficients
    Alpha=[4.091;2.543;4.258;5.426;4.258;6.131]; % for
unit 1, 2, ..., n
    Beta=-1*[5.554;6.047;5.094;3.550;5.094;5.555]; %
for unit 1, 2, ..., n
    Gamma=[6.490;5.638;4.586;3.380;4.586;5.151]; % for
unit 1, 2, ..., n
    Xi=[0.000200;0.000500;0.000001;0.002000;
0.000001;0.000010]; % for unit 1, 2, ..., n
    Delta=[2.857;3.333;8.000;2.000;8.000;6.667]; % for
unit 1, 2, ..., n

    % Ramp-Rate Limits
    Pnow=[440;170;200;150;190;110]/P_base; % for unit
1, 2, ..., n
    UR=[80;50;65;50;50;50]/P_base; % for unit 1, 2,
..., n
    DR=[120;90;100;90;90;90]/P_base; % for unit 1, 2,
..., n

    % Prohibited Zones
    PZL1=[210;90;150;80;90;75]/P_base; % 1st Lower Lim-
it of Prohibited Zone 1
    PZU1=[240;110;170;90;110;85]/P_base; % 1st Upper
Limit of Prohibited Zone 1
    PZL2=[350;140;210;110;140;100]/P_base; % 2nd Lower
Limit of Prohibited Zone 2
    PZU2=[380;160;240;120;150;105]/P_base; % 2nd Upper
Limit of Prohibited Zone 2
    P_D=1263/P_base;
    P_G_min=[100;50;80;50;50;50]/P_base; % for P1, P2,
..., Pn (in MW)
    P_G_max=[500;200;300;150;200;120]/P_base; % for P1,
P2, ..., Pn (in MW)
    for i=1:n
        P_G(i)=P_G_min(i)+(P_G_max(i)-P_G_min(i))*x(i);
    end
    P_loss=P_G*B_c*P_G'+B0_c*P_G'+B00_c;
    err_unblance=abs(sum(P_G)-P_loss-P_D);
```

```
    err_PZ1=(P_G'<PZL1 .* P_G'>P_G_min + P_G'>PZU1 .*
P_G'<PZL2 +...
    P_G'>PZU2 .* P_G'<P_G_max);
    err_RL=~((P_G'>Pnow).*((P_G'-Pnow)<UR) +
(P_G'<=Pnow).*((Pnow-P_G')<DR));
    fuel_cost=(P_G.^2)*cost_c(:,1)+P_G*cost_c(:,2)+
sum(cost_c(:,3))+abs(cost_c(:,4)'*sin(cost_c(:,5).*(P_G
_min-P_G')));
    emission=1e-2*(Alpha+P_G'.*Beta+P_G'.^2.*Gamma)+
sum(Xi.*exp(P_G'.*Delta));

    %objective one
    f1=fuel_cost+1e3*(sum(~err_PZ1)+sum(err_RL))+
err_unblance;
    %objective two
    f2=sum(emission)+1e3*(sum(~err_PZ1)+sum(err_RL))+
err_unblance;
    %objective three
    f3=P_loss*100+1e3*(sum(~err_PZ1)+sum(err_RL))+
err_unblance;
    %fitness value
    fitness_values=[f1 f2 f3]';
end


function [pop, F]=NonDominatedSorting(pop)

% Copyright (c) 2015, Yarpiz (www.yarpiz.com)
% All rights reserved. Please read the "license.txt"
for license terms.
%
% Project Code: YPEA120
% Project Title: Non-dominated Sorting Genetic Algo-
rithm II (NSGA-II)
% Publisher: Yarpiz (www.yarpiz.com)
%
% Developer: S. Mostapha Kalami Heris (Member of Yarpiz
Team)
%
% Contact Info: sm.kalami@gmail.com, info@yarpiz.com
%

    nPop=numel(pop);
    for i=1:nPop
        pop(i).DominationSet=[];
        pop(i).DominatedCount=0;
    end
```

```
    F{1}=[];

    for i=1:nPop
        for j=i+1:nPop
            p=pop(i);
            q=pop(j);
            if Dominates(p,q)
                p.DominationSet=[p.DominationSet j];
                q.DominatedCount=q.DominatedCount+1;
            end
            if Dominates(q.Cost,p.Cost)
                q.DominationSet=[q.DominationSet i];
                p.DominatedCount=p.DominatedCount+1;
            end
            pop(i)=p;
            pop(j)=q;
        end
        if pop(i).DominatedCount==0
            F{1}=[F{1} i];
            pop(i).Rank=1;
        end
    end
    k=1;
    while true
        Q=[];
        for i=F{k}
            p=pop(i);
            for j=p.DominationSet
                q=pop(j);
                q.DominatedCount=q.DominatedCount-1;
                if q.DominatedCount==0
                    Q=[Q j]; %#ok
                    q.Rank=k+1;
                end
                pop(j)=q;
            end
        end
        if isempty(Q)
            break;
        end
        F{k+1}=Q; %#ok
        k=k+1;
    end
end
```

```
function b=Dominates(x,y)
% Copyright (c) 2015, Yarpiz (www.yarpiz.com)
% All rights reserved. Please read the "license.txt"
for license terms.
%
% Project Code: YPEA120
% Project Title: Non-dominated Sorting Genetic Algo-
rithm II (NSGA-II)
% Publisher: Yarpiz (www.yarpiz.com)
%
% Developer: S. Mostapha Kalami Heris (Member of Yarpiz
Team)
%
% Contact Info: sm.kalami@gmail.com, info@yarpiz.com
%
    if isstruct(x)
        x=x.Cost;
    end
    if isstruct(y)
        y=y.Cost;
    end
    b=all(x<=y) && any(x<y);
end
```

```
function pop=CalcCrowdingDistance(pop,F)
% Copyright (c) 2015, Yarpiz (www.yarpiz.com)
% All rights reserved. Please read the "license.txt"
for license terms.
%
% Project Code: YPEA120
% Project Title: Non-dominated Sorting Genetic Algo-
rithm II (NSGA-II)
% Publisher: Yarpiz (www.yarpiz.com)
%
% Developer: S. Mostapha Kalami Heris (Member of Yarpiz
Team)
%
% Contact Info: sm.kalami@gmail.com, info@yarpiz.com
%
    nF=numel(F);
    for k=1:nF
        Costs=[pop(F{k}).Cost];
        nObj=size(Costs,1);
        n=numel(F{k});
        d=zeros(n,nObj);
        for j=1:nObj
```

```
            [cj, so]=sort(Costs(j,:));
            d(so(1),j)=inf;
            for i=2:n-1
                d(so(i),j)=abs(cj(i+1)-cj(i-1))/
abs(cj(1)-cj(end));
            end
            d(so(end),j)=inf;
        end
        for i=1:n
            pop(F{k}(i)).CrowdingDistance=sum(d(i,:));
        end
    end
end
```

```
function [pop, F]=SortPopulation(pop)
% Copyright (c) 2015, Yarpiz (www.yarpiz.com)
% All rights reserved. Please read the "license.txt"
for license terms.
%
% Project Code: YPEA120
% Project Title: Non-dominated Sorting Genetic Algo-
rithm II (NSGA-II)
% Publisher: Yarpiz (www.yarpiz.com)
%
% Developer: S. Mostapha Kalami Heris (Member of Yarpiz
Team)
%
% Contact Info: sm.kalami@gmail.com, info@yarpiz.com
%
    % Sort Based on Crowding Distance
    [~, CDSO]=sort([pop.CrowdingDistance],'descend');
    pop=pop(CDSO);

    % Sort Based on Rank
    [~, RSO]=sort([pop.Rank]);
    pop=pop(RSO);

    % Update Fronts
    Ranks=[pop.Rank];
    MaxRank=max(Ranks);
    F=cell(MaxRank,1);
    for r=1:MaxRank
        F{r}=find(Ranks==r);
    end
end
```

```
function [y1, y2]=Crossover(x1,x2)
% Copyright (c) 2015, Yarpiz (www.yarpiz.com)
% All rights reserved. Please read the "license.txt"
for license terms.
%
% Project Code: YPEA120
% Project Title: Non-dominated Sorting Genetic Algo-
rithm II (NSGA-II)
% Publisher: Yarpiz (www.yarpiz.com)
%
% Developer: S. Mostapha Kalami Heris (Member of Yarpiz
Team)
%
% Contact Info: sm.kalami@gmail.com, info@yarpiz.com
%
    alpha=rand(size(x1));
    y1=alpha.*x1+(1-alpha).*x2;
    y2=alpha.*x2+(1-alpha).*x1;
end
```

```
function y=Mutate(x,mu,sigma)
% Copyright (c) 2015, Yarpiz (www.yarpiz.com)
% All rights reserved. Please read the "license.txt"
for license terms.
%
% Project Code: YPEA120
% Project Title: Non-dominated Sorting Genetic Algo-
rithm II (NSGA-II)
% Publisher: Yarpiz (www.yarpiz.com)
%
% Developer: S. Mostapha Kalami Heris (Member of Yarpiz
Team)
%
% Contact Info: sm.kalami@gmail.com, info@yarpiz.com
%

    nVar=numel(x);
    nMu=ceil(mu*nVar);
    j=randsample(nVar,nMu);
    if numel(sigma)>1
        sigma = sigma(j);
    end
    y=x;
    y(j)=x(j)+sigma.*randn(size(j));
end
```

```
function PlotCosts(pop)
% Copyright (c) 2015, Yarpiz (www.yarpiz.com)
% All rights reserved. Please read the "license.txt"
for license terms.
%
% Project Code: YPEA120
% Project Title: Non-dominated Sorting Genetic Algo-
rithm II (NSGA-II)
% Publisher: Yarpiz (www.yarpiz.com)
%
% Developer: S. Mostapha Kalami Heris (Member of Yarpiz
Team)
%
% Contact Info: sm.kalami@gmail.com, info@yarpiz.com
%

    Costs=[pop.Cost];
    plot3(Costs(1,:),Costs(2,:),Costs(3,:),'r*',
'MarkerSize',8);
    xlabel('Fuel Cost ($/hr)');
    ylabel('Emission (ton/hr)');
    zlabel('Transmission Loss (MW)');
    grid on;
end
```

References

1. Venkatesh, P., Gnanadass, R., Padhy, N.: Comparison and application of evolutionary programming techniques to combined economic emission dispatch with line flow constraints. IEEE Trans. Power Syst. **18**(2), 688–697 (2003)
2. Liang, Z.-X., Glover, J.: A zoom feature for a dynamic programming solution to economic dispatch including transmission losses. IEEE Trans. Power Syst. **7**(2), 544–550 (1992)
3. Walters, D., Sheble, G.: Genetic algorithm solution of economic dispatch with valve point loading. IEEE Trans. Power Syst. **8**(3), 1325–1332 (1993)
4. Park, J.-B., Jeong, Y.-W., Shin, J.-R., Lee, K.Y.: An improved particle swarm optimization for nonconvex economic dispatch problems. IEEE Trans. Power Syst. **25**(1), 156–166 (2010)
5. Dhanalakshmi, S., Kannan, S., Mahadevan, K., Baskar, S.: Application of modified NSGA-II algorithm to combined economic and emission dispatch problem. Int. J. Electr. Power Energy Syst. **33**(4), 992–1002 (2011)
6. Yuan, X., Ji, B., Chen, Z., Chen, Z.: A novel approach for economic dispatch of hydrothermal system via gravitational search algorithm. Appl. Math. Comput. **247**, 535–546 (2014)
7. Abido, M.: Multiobjective particle swarm optimization for environmental/economic dispatch problem. Electr. Power Syst. Res. **79**, 1105–1113 (2009)
8. Mondal, S., Bhattacharya, A., nee Dey, S.H.: Multi-objective economic emission load dispatch solution using gravitational search algorithm and considering wind power penetration. Electr. Power Energy Syst. **44**, 282–292 (2013)
9. Atashpaz-Gargari, E., Lucas, C.: Imperialist competitive algorithm: An algorithm for optimization inspired by imperialistic competition. In: IEEE Congress on Evolutionary Computation, Singapore (2007)
10. Azari, M.N., Mirsalim, M., Pahnehkolaei, S.M.A., Mohammadi, S.: Optimum design of a line-start permanent-magnet motor with slotted solid rotor using neural network and imperialist competitive algorithm. IET Electr. Power Appl. **11**(1), 1–8 (2017)
11. Moradi, M.H., Zeinalzadeh, A., Mohammadi, Y., Abedini, M.: An efficient hybrid method for solving the optimal sitting and sizing problem of DG and shunt capacitor banks simultaneously based on imperialist competitive algorithm and genetic algorithm. Int. J. Electr. Power Energy Syst. **54**, 101–111 (2014)
12. RezaeeJordehi, A.: Optimal allocation of FACTS devices for static security enhancement in power systems via imperialistic competitive algorithm (ICA). Appl. Soft Comput. **48**, 317–328 (2016)
13. Tang, Y., Zhao, L., Han, Z., Bi, X., Guan, X.: Optimal gray PID controller design for automatic voltage regulator system via imperialist competitive algorithm. Int. J. Mach. Learn. Cybernet. **7**(2), 229–240 (2016)
14. Enayatifar, R., Yousefi, M., Abdullah, A.H.: MOICA: a novel multi-objective approach based on imperialist competitive algorithm. Appl. Math. Comput. **219**(17), 8829–8841 (2013)
15. Abedinia, O., Amjady, N., Kiani, K., Shayanfar, H., Ghasemi, A.: Multiobjective environmental and economic dispatch using imperialist competitive algorithm. Int. J. Tech. Phys. Probl. Eng. (IJTPE) **11**, 63–70 (2012)
16. Chiang, C.-L.: Improved genetic algorithm for power economic dispatch of units with valve-point effects and multiple fuels. IEEE Trans. Power Syst. **20**(4), 1690–1699 (2005)

17. Wood, A., Wollenberg, B.: Power Generation, Operation, and Control. Wiley, Hoboken (2012)
18. Orero, S., Irving, M.: Economic dispatch of generators with prohibited operating zones: a genetic algorithm approach. IEE Proc. Gener. Transm. Distrib. **143**(6), 529–534 (1996)
19. Deb, K.: Multi-objective optimization Interactive and Evolutionary Approaches. Search Methodologies. Springer US, pp. 403–449 (2014)
20. Miettinen, K.: Nonlinear Multiobjective Optimization (1998)
21. Goldberg, D.R.J.: Genetic algorithms with sharing for multimodal function optimization. In: Proceedings of the First International Conference on Genetic Algorithms and Their Applications (1987)
22. Deb, K., Pratap, A., Agarwal, S., Meyarivan, T.: A fast and elitist multiobjective genetic algorithm: NSGA-II. IEEE Trans. Evol. Comput. **6**(2), 182–197 (2002)
23. Doerner, K., Gutjahr, W.J., Hartl, R.F., Strauss, C., Stummer, C.: Pareto ant colony optimization: a metaheuristic approach to multiobjective portfolio selection. Ann. Oper. Res. **131**(1), 79–99 (2004)
24. Yang, X.-S.: Bat algorithm for multi-objective optimisation. Int. J. Bio-Inspired Comput. **3**(5), 70–75 (2011)
25. Dhillon, J.S., Parti, S.C., Kothari, D.P.: Stochastic economic emission load dispatch. Electr. Power Syst. Res. **26**(3), 179–186 (1993)
26. Gaing, Z.-L.: Constrained dynamic economic dispatch solution using particle swarm optimization. In: IEEE, Denver, CO, USA (2004)
27. Abido, M.A.: Multiobjective particle swarm optimization for environmental/economic dispatch problem. Electr. Power Syst. Res. **79**(7), 1105–1113 (2009)

Voltage Control by Optimized Participation of Reactive Power Compensation Using Fixed Capacitor and STATCOM

Nitin Kumar Saxena

Abstract FACTS devices play a significant role in providing voltage control through adequate reactive power compensation under the conditions of load and input changes. In isolated wind diesel based hybrid electrical system, choosing adequate participation of reactive power compensation device becomes more important because of the following aspects; (i) unlike to grid connected system, additional sources are required for supplying reactive power, (ii) normally self excited induction generators are used for power generation through wind and these generators require reactive power for building up the voltage, (iii) wind generators power output is much affected by changes in input wind speed and these changes require additional reactive power to control the voltage, (iv) similar to input change, load changes also require additional reactive power to maintain the voltage level, (v) compensating device should respond fast for nullifying the voltage deviation in minimum time, (vi) the procedure adopted for reactive power compensation should be economically acceptable even for the last end user in the society. Therefore, the reactive power compensating devices for voltage control in isolated hybrid electric system should be participated optimally by considering these technical and economical aspects simultaneously. In this chapter, MATLAB (programming along with simulink model) based approach is demonstrated for voltage control through optimized participation of reactive power compensation using fixed capacitor as static and STATCOM as dynamic compensator.

Keywords Static Compensator · Dynamic Compensator · Reactive power compensators · Compensation cost · Ancillary services

---

N. K. Saxena (✉)
Electrical and Electronics Engineering, KIET Group of Institutions, Ghaziabad, India
e-mail: nitinsaxena.iitd@gmail.com

M. Pesaran Hajiabbas and B. Mohammadi-Ivatloo (eds.),
Optimization of Power System Problems, Studies in Systems, Decision and Control 262,
https://doi.org/10.1007/978-3-030-34050-6_13

1 Role of Reactive Power Compensation

Techno-economical studies in distributed power system have been presented by many researchers. These studies depict that the electrification through traditional centralized generating units is a real challenge for far located remote/rural areas because of geographical diversity, concentrated availability of natural resources and dispersed power demand. Electrification without long transmission lines can be a better option to such far located remote/rural consumers. Because of the remoteness of such non-electrified population, renewable energy can offer a cost effective and environmental friendly means of providing power. In recent years, production of clean energy (renewable ones) by private investors is encouraged. Government in almost all countries are also promoting public participation through several schemes for installing small units of power generation using hybrid systems [1]. Government of India is also promoting to private investors for installing distributed generating units because of technical and economical limitations of supplying grid connected power system at such far located rural areas [2]. Private investors' participation in installing renewable energy system (RES) can be better understood through Fig. 1. It has also been noticed that 88% renewable energy sources are installed by private investors in India [3].

Wind energy is the most promising form of renewable energy for generation of electric power but suffering from intermittent nature of the wind. To provide continuous and reliable power in such far located areas, renewable energy based generators along with conventional fuel based generators can be used without grid connection. Researchers have presented hybrid power generation models in which self excited induction generator (SEIG) and synchronous generator (SG) are used together for generating power through wind and diesel respectively [4–8]. In this

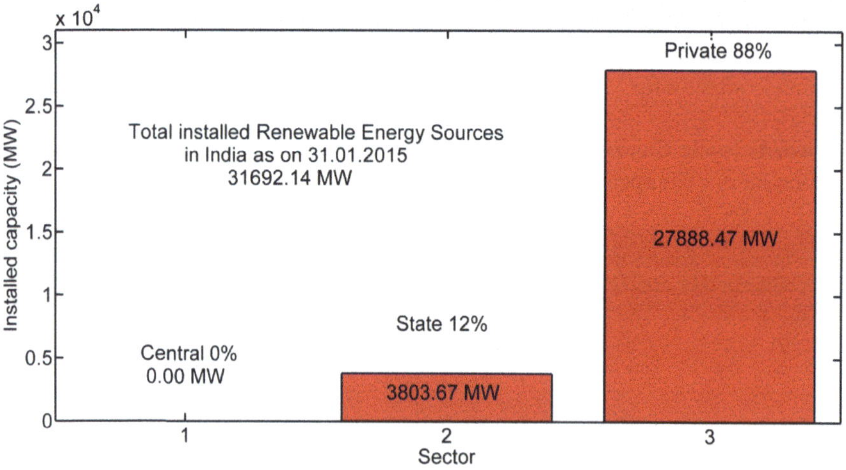

Fig. 1 Participation of different sectors for installing renewable energy system in India

chapter studies are focuses for such wind diesel based hybrid electrical system. Since these electrical systems are isolated from grid so, called isolated hybrid energy system (IHES).

This wind-diesel based IHES is one of the most promising systems to provide continuous, efficient, economical and reliable electrical energy and has a wide scope especially in developing countries. Due to grid isolation, hybrid configuration of generation units, random behaviour of consumers load and evolvement of public-private investors' participation, IHES has many technical and economical issues in their operations. Selection of adequate reactive power compensation for such IHES is one of them which is being focussed technically as well as economically in this chapter.

Problems in electrical systems can be broadly identified in two categories; (i) active power frequency ($P - f$) control, and (ii) reactive power voltage ($Q - V$) control. The $P - f$ and $Q - V$ controls are almost non-interactive in electrical systems because small changes in active power are mainly dependent on changes in generator speed and are almost independent of changes in terminal bus voltage, while small changes in terminal bus voltage are mainly dependent on machine excitation and are almost independent of changes in generator speed. Since, excitation control is a fast acting with less time constant encountered as that of generator field, while power frequency control is a slow acting with more time constant as contributed by turbine and generator moment of inertia. So, the time constant for $P - f$ control loop is much larger than that of the $Q - V$ control loop. Even in conventional grid connected power system, active power is exported on transmission line to load centre but reactive power required by load is produced closer to the requirement to avoid large transmission losses and voltage variations. It should also be noticed that the production of reactive power involves only capital cost but no fuel cost. Therefore, $P - f$ and $Q - V$ control loops are assumed to be decoupled in power system and the problem of reactive power voltage ($Q - V$) control can be focussed separately at load centres.

The IHES is designed with the help of diesel, a non renewable energy source to provide continuous and reliable supply and wind, a renewable energy source to provide environmental friendly energy supply. The self excited induction generator is used to extract power from wind while the synchronous generator is used to extract power from diesel in this isolated hybrid energy system. The major disadvantage of self excited induction generator is the requirement of reactive power for its operation. In grid-connected system, induction generator can be excited from either grid or capacitor banks, whereas in an isolated system, reactive power excitation along with load reactive power demand can only be achieved through reactive power compensators. Apart from steady state reactive power requirement, dynamic conditions also require reactive power for regulating the voltage response due to instant changes in load demand and input power in system. The automatic voltage regulator (AVR) of the synchronous generator, connected in parallel with induction generator in this IHES, may not be able to offset the reactive power mismatch as its prime function is to generate the real power for load keeping terminal voltage within limits with minimum over and under excitation.

Therefore, reactive power compensators are required for additional reactive power demand in system. Deficiency in this extra reactive power demand can cause severe problems of large voltage fluctuations at load terminal and therefore, affect the quality of supply. In absence of the proper voltage control, this may even damage the system stability. Voltage control problems are complex in nature especially for heavily loaded power system and unbalance in generation. Voltage control is one of the six ancillary services that is used to maintain the voltage profile through injecting or absorbing reactive power [9]. Therefore, proper reactive power compensation techniques are required to ease the voltage control problems in IHES. Hence, as a technical issue, if the system operator does not consider impact of reactive power on voltage control, it may move the system toward voltage instability point. Therefore, sufficient fast acting reactive power reserve is necessary to prevent unacceptable voltage deviation after any system disturbance or due to load uncertainty [10]. This is called $Q - V$ control loop problem and is mainly focussed for reactive power compensation and voltage control studies of this chapter for IHES.

Since, the optimal and adequate reactive power deployment in the competitive electricity markets is identified as one of the important ancillary services and is provided by the Independent System Operator (ISO). The procurement of reactive power as an ancillary service involves cost investment and thus needs to be remunerated. Effective regulatory policies are necessary to ensure an adequate supply of reactive power at reasonable cost whether in independent or integrated power system. The rules for procuring reactive power can affect whether adequate reactive power supply is available, as well as whether the supply is procured efficiently from the most reliable and lowest cost sources. Fast acting device for reactive power compensation gives better results of voltage regulation in system but at the same time they increase system compensation cost much. On the other side, static compensator has very low cost but alone cannot be suitable for reactive power compensation in system. Hence, economic analysis of reactive power compensation in IHPS is also an important aspect.

Therefore, a hybrid use of compensating devices; static as well as dynamic compensators, can be used for techno-economic solution of reactive power compensation in IHES for controlling the system voltage under specified limits. In this chapter, the technical benefits from the hybrid participations of static and dynamic reactive power compensators in voltage control studies are mainly focussed. A MATLAB program is developed for choosing the best possible participations of fixed capacitor and STATCOM for voltage control studies in system during steady state and dynamic conditions.

2 Introduction to Reactive Power Compensators

In isolated hybrid electrical system, reactive power compensation plays a key role in controlling the system voltage. The reactive power support, essential to maintain the voltage profile and stability of the system, is one of the six ancillary services specified in the FERC order no. 888 [11]. Reference [12] explains two types requirement of reactive power for system operation; (i) under steady state and (ii) under dynamic conditions. Reference [13, 14] assumes that the generators are obligated to provide a certain amount of reactive power (up to generator's mandatory limit) without any payment. Synchronous generator is primarily used to generate real power to system therefore only mandatory limit reactive power is supposed to be released form it without considering opportunity cost through it. In Reference [15] voltage response is explored with external rotor resistance along with excitation capacitor for autonomous SEIG. Wang et al. [16] proposed an analysis to predict both minimum and maximum values of capacitance required for self-excitation of a three phase induction generator.

In hybrid electrical system reactive power compensation becomes complex due to the parallel operation of different generators along with load influence. Standalone operation of a squirrel-cage induction generator based WECS with regulated output voltage and frequency requires either an asynchronous link (ac–dc–ac) power electronic converter or a matrix converter. The excitation capacitor bank of large rating has to be implemented with thyristors rectifier because thyristors rectifier can only absorb active and reactive powers. This makes the system efficiency low. Therefore, shunt connected VSI with a capacitor and a switched resistor in the dc bus is proposed alternatively in [17]. References [18, 19] proposed a hybrid exciter in which one set of a parallel connected three-phase fixed frequency pulse width modulation (PWM) inverter fed from a battery and fixed capacitor bank is used. In order to avoid the problem of mismatch of reactive power generation and absorption in system switch capacitor may be used in place of fixed capacitor. But switched capacitor can only give discrete solutions for avoiding reactive power mismatch in system. A variable reactive power source is required match the generation and absorption of reactive power. Three SVC models are explained for reactive power compensation in a hybrid system [20].

The STATCOM device is the static counterpart of the rotating synchronous condenser but it generates/absorbs reactive power at a faster rate. The STATCOM employs a voltage source inverter (VSC), which internally induces inductive or capacitive reactive power as required. In principle, it performs the same voltage regulation function as the classical SVC but in a more robust manner and is also advantageous than that of SVC [6]. It goes on well advanced energy storage facilities, which open the door for a number of new applications, such as energy markets and network security [21]. Reference [22] proposes STATCOM transient stability and power flow models as improved versions of models previously proposed in the literature. Reference [23] proposes a new method, called the flatness-based adaptive control (FBAC), for STATCOM voltage regulation.

2.1 Introduction to Reactive Power Market

Private investors in deregulated and isolated mode of electricity markets in many countries worldwide bring new perspectives for small businesses specializing in energy generation. Wind power generation has better options for investment and therefore, attracts the private sector [24]. Size optimization is among the most important studies in order to achieve efficient and economical utilization in the hybrid system [25]. Reference [26] discusses the problem in dealing two objectives simultaneously (costs and unmet load) which are usually in conflict, since a reduction in design costs implies a rise in unmet load and vice versa. Reference [27] suggests that voltage and reactive power support are linked to each other and reactive support is distinguished as an ancillary service, which can facilitate active power transportation in system. Modal analysis technique is proposed for the management of reactive power generation to improve the voltage stability margin in [28]. Reference [29] addresses the problem of how to pay the voltage support providers; and how to allocate the incurred costs to the users. Reference [30] focuses on two aspects; voltage profile management and reactive dispatch and voltage regulation in isolated system.

In the new open access environment, in pursuit of profit, the power producer has incentive to sell active power as much as possible. A generator can sell its active power if only there is enough reactive power to support it. Otherwise, the generator is no longer able to sell active power due to system security constraints. So, it is essential to establish a mechanism for financial compensation of the reactive power ancillary service [31]. Different methods are used in different electricity markets for reactive power procurement. As main philosophy of the electricity markets, the system operator tries to provide reactive power with the lowest possible cost. Also, because of important role of reactive power in network operation and security, many researchers have considered technical issues as well as economic issues. Reference [32] provides a techno-economic analysis to decide configuration of autonomous system on the basis of power quality, system overall cost, payback time and emission of green house gases. The effects of load variation on system configuration and cost are also examined.

Although reactive power costs constitute only about 1% of total power industry costs [33], it's still important to make it clearly analyzed when the reactive power market is concerned. According to economics aspects, total cost of any commodities can be divided in two components; fixed cost component and a variable cost component. The capital investment of equipments are categorised into fixed cost category while costs connected to the output quantity are categorised into variable cost category. Without any fuel cost to generate reactive power, the variable cost of generators include maintenance and operation cost and opportunity cost. Equipments like synchronous condensers, shunt capacitors, STATCOMs, and SVCs don't produce real power, so they don't have opportunity cost.

Reactive power cost curve of a synchronous condenser can be formulated including operating cost and investment cost. Reference [34] deals with evaluation of capacitive reactive power cost under the deregulation environment. For the cost assessment of reactive power, the duration curve of reactive power demand is introduced to take into account the investment costs. Capacitor reactive power cost function is given in [35]. Cost functions of UPFC, TCSC and SVC are given in polynomial form in [36]. Furthermore, cost functions are incorporated for bids of suppliers and consumers and investment costs of FACTS devices. Cost function is defined as the sum of capital cost and installation cost.

Model presented in many papers optimize the certain objective function (e.g., reactive power production cost minimization or social welfare maximization) using optimal power flow (OPF) models and use of fixed capacitor and FACTS device is proposed for future work. The resources for reactive power such as synchronous generators, synchronous condensers, capacitor banks, reactors, Flexible AC Transmission System (FACTS) devices are owned by the independent generators or local suppliers. Reference [37] defines costs for the service performed by these devices, then it proposes an optimal co-ordination method which allows distributors to select, for every operating condition, the more profitable combination of reactive sources in order to maintain the network voltage levels within a desired range and to minimise the regulation action global cost.

2.2 Selection of Dynamic and Static Compensator

In most of the present researches available on wind diesel based IHES, the main thrust is on technical benefits using fast acting compensating devices for reactive power compensation and voltage control while economic issues of reactive power compensation are not focussed by researchers yet. The system dynamic responses can be suppressed in least time within the permissible range by use of FACTS devices namely; Static VAR Compensator (SVC) or Static Compensator (STATCOM). FACTS device produces better responses in terms of system voltage control compare to the conventional compensation devices viz. fixed capacitor (FC), switched capacitor and synchronous compensator. Although SVC/STATCOM give better results as discussed in various research papers [4, 20], their cost is very high compared to fixed capacitor. It is well known that a fair pricing of such a service can lead to market liquidity which in turn results in approaching the optimal condition. Therefore, for a competitive market environment, the economic viability should also be considered with engineering requirements. Getting the benefits through Government promotional schemes, private investors can develop isolated units for continuous power supply to far located remote area. At such rural/remote areas, the end users are not too developed economically to pay more tariff rates for generating companies (Gencos). Even most of the time Government has to subsidize the power to such consumers and therefore, power supply continuity is a prime concern however its quality degradation may be allowed up to some extent to keep the cost low. And therefore, method to reduce the compensation cost for the consumer is proposed in this chapter.

Reactive power demand in system can be classified by two categories; (i) fixed demand, and (ii) variable demand. Fixed demand includes steady state load demand and induction generator excitation. Variable demand includes load and induction generator reactive power demand due to sudden changes in system conditions. The characteristic comparison of voltage control equipment, i.e., fixed capacitor (FC), SVC, and STATCOM (ST), is mentioned in Table 1 [38]. For voltage control, reactive power is required through compensators in IHES. This compensation may be

Table 1 Characteristics comparison of voltage support equipment

Equipment	Equipment type	Response speed	Voltage support	Operating cost
FC	Static compensator	Slow	Poor, drops with V^2	Very low
SVC	Dynamic compensator	Fast	Poor, drops with V^2	Moderate
STATCOM	Dynamic compensator	Fast	Fair, drops with V	High

achieved using static, dynamic or combination of both types of compensators. As in Table 1, dynamic reactive power compensators (SVC and STATCOM) give best results for voltage control [2, 4, 6], but their cost is very high compared to that of static reactive power compensator (FC). On the other side, the cost of static compensator is very low, but they alone are not capable of providing the adequate solution of voltage regulation.

Before deciding the compensation techniques, following observations must be noticed;

1. Reactive power is required for steady state and dynamic conditions. Steady state requirement can be supplied by static and/or dynamic compensating device but dynamic conditions can only be supplied by dynamic compensating device [12].
2. Static compensators are cheaper but cannot regulate to system voltage for fast acting changes in system [4].
3. Dynamic compensators have good characteristic for regulating the system voltage during sudden changes but their use make system's compensation cost very high [38].
4. Participation of static compensator with dynamic compensator can be allowed up to the extent where system voltage response remain within its pre defined acceptable range [30].

Hence, a combination of static as well as dynamic compensating devices can be used to mitigate system fixed demand while fast respond dynamic compensators are necessarily required in suppressing the dynamic demand of the system in minimal time. The methodology may be adopted for deciding the participation of static and dynamic compensators including these two behaviours of demand. The concept for reducing STATCOM size with a fixed capacitor for self-excited induction generator is compared in Ref. [39] for full rating of STATCOM alone and half rating shared with fixed capacitor along with STATCOM. Since the IHESs have both fixed demand and variable demand so static compensator alone cannot be installed to mitigate the effect of sudden changes in load and wind input power. These changes may result in a serious problem of large voltage fluctuation without proper reactive power compensation. Dynamic compensator alone can mitigate the voltage control issues but the compensation cost becomes high through it. To overcome this problem, STATCOM and fixed capacitors, both are installed in the system together to control reactive power and to minimize voltage fluctuations. The proper selection of both static as well as dynamic compensators simultaneously may provide the optimum solution between the system voltage control and cost of compensation.

Therefore, the selection of STATCOM and fixed capacitors ratings depend on the two aspects; overall cost of compensation and system voltage response [40]. In this chapter, a cost analysis is done for reactive power participation in isolated hybrid electrical system through fixed capacitor as static and STATCOM as dynamic compensator.

3 Reactive Power Compensation Cost Analysis

This chapter presents the technical benefits from the hybrid participations of static and dynamic reactive power compensators in voltage control studies. A method of reactive power compensation pricing is proposed by including static and dynamic compensators in system. Concepts about reactive power compensation as ancillary service in power system, method of cost formulation and cost formulae for different compensating devices are discussed in this section.

3.1 Reactive Power as an Ancillary Service

Any end user of power system is implicitly a consumer of ancillary services who is demanding continuous and quality of power supply. For power producers, ancillary services are mainly defined by the basic contributions they make to fulfil the system functions. Besides the supply of active power, they supply or absorb reactive power and control the voltage as well as contribute to maintaining the system frequency. According to North American Electric Reliability Council (NERC), ancillary services can be categorized into three categories;

Category-1: Services required for routine operation
Category-2: Services needed to avoid blackout
Category-3: Services needed to restore after blackout.

In category 1, voltage control has prime importance in the system along with other ancillary services like system control, regulation, load following, energy imbalance. In isolated hybrid power system, voltage is controlled by the compensation of reactive power with the help of synchronous generators, static and dynamic compensator. These reactive power devices have several characteristics for consideration such as their dynamics, response speed, voltage changing ability, capital costs, operating costs, and opportunity costs.

It has been suggested that the independent generators or customers install their own reactive support resources and the ISO should enter into contracts with those independent generators or customers for such provision. These reactive support resources may be synchronous generators, synchronous condensers, capacitor banks, reactors, static VAR compensators and FACTS devices. Perceived demand conditions, mix of the load and availability of reactive power resources should be considered for procurement of reactive power services [1].

Reference [41] describes that reactive power through generator and synchronous condenser is recognized as "ancillary services." Changes at the policy level are necessary to include other reactive power sources such as capacitors, reactors, SVCs and FACTS devices etc., as ancillary services. This would enlarge the market and increase the competition, and inevitably increase the market efficiency and fairness.

In this chapter, it is being assumed that IHESs are owned by private investors who are committed to provide electricity on cheaper rates for far located remote area based end users. It is also assumed that power quality degradation within permissible range can be acceptable for reducing the cost in such remote areas. Available and possible pricing options for reactive power compensation are discussed in this section first. Since reactive power can be supplied by synchronous generator, FC and STATCOM in system, the cost issues by them are also being discussed here.

3.2 Pricing Options in Reactive Power Compensation

Reactive power pricing was started with the Commission's Order No. 888, its Open Access Rule, issued in April 1996. In that order, the Commission concluded that "reactive supply and voltage control from generation sources" is one of six ancillary services that transmission providers must include in an open access transmission tariff. The main aim for reactive power procurement is to ensure the adequate supplies of reactive power (including reactive reserves) in the system at least cost for steady state and dynamic conditions. Optimization of reactive power cost is required because; the static compensators having lowest cost cannot always be reliable and adequate producers of reactive power as dynamic compensators. Dynamic compensation might be expensive reactive power procurement but they must sometimes be purchased even if cheap reactive power sources are available. Two general ways have been suggested for providing reactive power though generators compensation in literature.

3.2.1 Capacity Payment Option

In capacity payment option generator is paid in advance for the capability of producing or consuming reactive power. The payment could be made through a bilateral contract or through a generally applicable tariff provision. Once the generator is paid, it could be obligated to produce or consume reactive power up to the limits of its commitment without further compensation when instructed by the system operator. To ensure that the generator follows instructions in real time, the generator could face penalties for failing to produce or consume when instructed. Currently, this is the most common method for compensating reactive power providers. Four methods are suggested for capacity payment option as;

- A cost-based payment

- Capacity market payment
- Prices determined through auction
- Pay nothing.

3.2.2 Real-Time Price Option

In real-time price option, generator is paid in real-time for the reactive power that it actually produces or consumes. Under this option, the generator is paid only for what it produces or consumes, but it pays no penalty for failing to produce when instructed. Four methods are also suggested for real time capacity payment option as;

- Pay nothing
- Unit-specific opportunity costs
- Market clearing prices determined through auction
- Prices (or a pricing formula) announced in advance.

Reactive power spot pricing can be adopted by including the features of both capacity payment option and real time capacity payment option. A method of reactive power compensation cost analysis is proposed by including static and dynamic compensators in system keeping compensation through synchronous generator constant and equal to its mandatory limit. This proposed method includes two important aspects for reactive power compensation; first, to encourage efficient and reliable investment for steady state reactive power demand and second, to encourage production and consumption of reactive power from exciting infrastructure for dynamic state demand.

3.3 Synchronous Generator as Reactive Power Service Provider

Synchronous generators are basically used for active power generation; however, they are also able to provide reactive power for security purposes. The synchronous generator's capacity is limited by the armature current, field current and under-excitation limits. The stable operating point of a generator is always restricted to its capability curve boundaries, which are defined according to armature and field winding heating limits. Synchronous generator may generate the reactive power in three regions namely; mandatory cost, cost of loss and opportunity cost, as shown in Fig. 2.

When synchronous generator releases reactive power in mandatory cost region, it does not receive any payment for reactive power production. In cost of loss region, generator is entitled to receive two components of payments availability component and cost of loss component. In opportunity cost region, generator is entitled

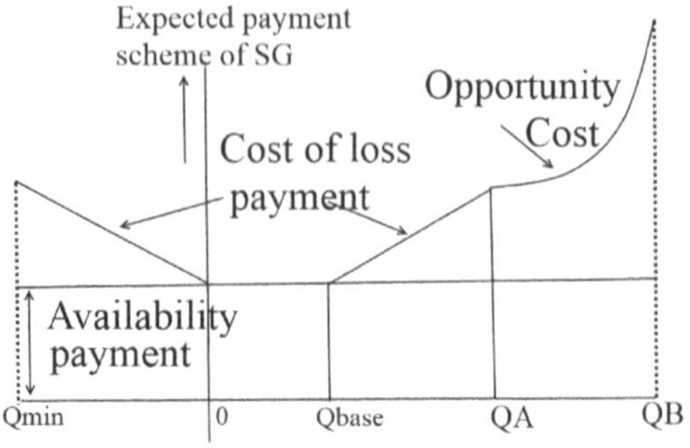

Fig. 2 Reactive power scheme in synchronous generator

to receive payment with its opportunity cost of reduced real power production [38]. Mathematically,

Expectation of Payment Function (EPF) of synchronous generator for reactive power,

$$EPF = Mandatory\,cost + cost\,of\,loss + opportunity\,cost \tag{1}$$

A generator's cost of producing reactive power can sometimes include opportunity costs associated with forgone real power production. Opportunity costs arise because there can be a trade-off between the amount of reactive power and real power that a generator can produce. When a generator is operating at certain limits, a generator can increase its production or consumption of reactive power only by reducing its production of real power as the winding of the synchronous generator is designed for a particular rating of current. Further, this method is somewhat complex, and is only cost effective when a large amount of compensation is needed [9, 20]. For calculating the different points shown in Fig. 2 for reactive power scheme of synchronous generator, i.e. mandatory reactive power (Q_{base}), cost of loss reactive power (Q_A), reserve reactive power ($Q_A - Q_{base}$), and opportunity cost reactive power ($beyond\ Q_A$), following mathematical expressions can be used [42].

For synchronous generator, mandatory reactive power can be calculated by the Eq. (2) in which P_{SG} is the rated real power of generator at $\cos\theta_{SG}$ lagging power factor.

$$Q_{mandatory} = P_{SG,pu} \tan\theta_{SG} \tag{2}$$

From field current limit equations of synchronous generator,

$$\mathrm{P_{SG}} = \frac{3\mathrm{VE_q}}{\mathrm{X_s}}\mathrm{Sin}\delta \tag{3}$$

$$Q_{SG} = \frac{3VE_q}{X_s}\mathrm{Cos}\delta - \frac{3V^2}{X_s} \tag{4}$$

Squaring and adding the Eqs. (3) and (5),

$$P^2_{SG,pu} + \left(Q_{SG,pu} + \frac{3V^2}{X_s}\right)^2 = \left(\frac{3VE_q}{X_s}\right)^2 \quad (5)$$

For estimating the cost of loss reactive power by synchronous generator, Eq. (5) can be solved for getting the value of $Q_{SG,pu}$. So,

$$Q_{cost\ of\ loss} = Q_{SG,pu} \quad (6)$$

$$Q_{reserve} = Q_{cost\ of\ loss} - Q_{mandatory} \quad (7)$$

The lost opportunity cost can be determined above the rated reactive power requirements and below the maximum limit reactive power generation from the generators.

Since, the synchronous generator should not be entitled to receive any payment for reactive power production in mandatory cost region. It is assumed that synchronous generator provides reactive power equal to mandatory limit only in the study.

3.4 Fixed Capacitor as Reactive Power Service Provider

The function of cost for capacitor is assumed to be proportional to the amount of the reactive power output purchased and equal to the product of depreciation rate and amount of the reactive power output purchased [11]. Fixed capacitor function of cost (C_{FC}) can be expressed as,

$$C_{FC} = rQ_{FC}\ in\ \$/\mathrm{H} \quad (8)$$

where, symbol *r*, defines the cost or depreciation rate of fixed capacitor Q_{FC}, is the amount of reactive power supplied by fixed capacitor to the system. The rating of the Q_{FC} is in MVAr. The depreciation rate is calculated by the ratio of investment cost and the operation hours of FC. The fixed cost for life span of 15 years is considered as per general practice [35],

$$C_{FC} = 0.132 * Q_{FC}\ in\ \$/\mathrm{H} \quad (9)$$

Example 1 A delta connected capacitor bank having per phase capacitance of 200 μF is connected with a electrical system. The generated voltage with this electrical system is 400 V at 50 Hz. Find the MVAr generated by this capacitor bank. Also find the compensation cost through FC if life span of this capacitor bank is assumed to be 15 years.

Solution For delta connected capacitor bank,

$$V_P = V_L = 400\,\text{V}$$

Reactance for the capacitor is,

$$X_C = \frac{1}{2\pi f C}$$

Reactive power developed by the capacitor bank,

$$Q_{FC} = \frac{3V_P^2}{X_C} \times 10^{-6}\,\text{MVAR}$$

The compensation cost through FC for life span of 15 years is,

$$C_{FC} = 0.132 * Q_{FC}\,\$/\text{H}$$

Using the mathematical expressions given above, MATLAB codes are written for getting the solutions of this example as below.

```
%%% MATLAB codes for compensation cost through Fixed
Capacitor (FC)
>>clear all;
>>clc;
>>vl=400;
>>f=50;
>>c=200e-6;
>>vp=vl;
>>xc=1/(2*pi*f*c);
>>qfc=(3*vp^2/xc)*10^-6          % MVAr generated
>>cost_qfc=0.132*qfc          %   compensation   cost

through FC
```

The results for the program given above are;
For MVAr generated = 0.0302 MVAR
Compensation cost through FC = 0.0040 $ per H.

3.5 STATCOM as Reactive Power Service Provider

An empirical method is available to obtain function of cost for STATCOM. The curve is plotted between the investment costs and the ratings for the different installation of STATCOM. An expression is developed for this as a function of STATCOM rating based on the quadratic polynomial curve fitting method. ST average operating life is also taken same as that of FC i.e. 15 years. The rating of the Q_{ST} is in MVAr in Eq. (10). The expression for STATCOM function of cost (C_{ST}) is given as [43];

$$C_{ST} = \frac{1000 * Q_{ST}}{8760 * 15}\left(0.0002466 Q_{ST}^2 - 0.2243 Q_{ST} + 150.527\right) in\ \$/\mathrm{H} \tag{10}$$

Example 2 In previous Example 1, if same MVAR are being supplied by STATCOM. Find the compensation cost through STATCOM.

Solution The expression for compensation cost through STATCOM,

$$C_{ST} = \frac{1000 * Q_{ST}}{8760 * 15}(0.0002466 Q_{ST}^2 - 0.2243 Q_{ST} + 150.527)\$/\mathrm{H}$$
$$Q_{ST} = 0.0302\,\mathrm{MVAR}$$

Using the mathematical expressions given above, MATLAB codes are written for getting the solutions of this example as below.

```
%%% MATLAB codes for compensation cost through STATCOM
(ST)
>>clear all;
>>clc;
>>qst=0.0302;              % MVAr generated
% compensation cost through ST
>>cost_qst=((1000*qst)/(8760*15))*((0.0002466*qst*qst)-
(0.2243*qst)+150.527)
```

The results for the program given above are;
For MVAr generated = 0.0302 MVAR
Compensation cost through FC = 0.0346 \$ per H.

4 Reactive Power Compensation Scheme in Ihes

A block diagram for wind-diesel based IHES is given in Fig. 3. Self excited induction generator coupled with wind turbine, synchronous generator coupled with diesel genset, fixed capacitor and STATCOM as reactive power compensators and a load are connected in parallel to a common bus line to define an isolated hybrid electric system. Mathematically, under steady state,

$$\Delta P_L = \Delta P_{IG} + \Delta P_{SG} \tag{11}$$

Fig. 3 Basic configuration of isolated hybrid power system with compensation schemes

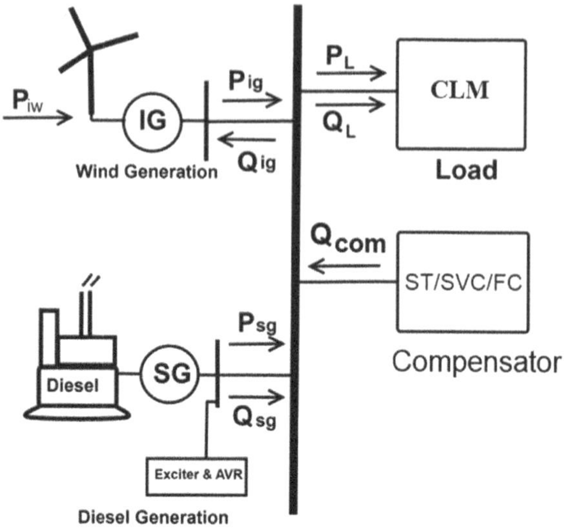

$$\Delta Q_L + \Delta Q_{IG} = \Delta Q_{SG} + \Delta Q_{Com} \tag{12}$$

From Eqs. (11) and (12), it can be depicted that induction generator and synchronous generator both will manage for any change in the real power requirement and change in reactive power may occur due to demand of either induction generator or load or both together. As depicted in Eq. (12), this reactive power requirement may be supplied to IHES by either synchronous generator or compensator or both together. In present study, author is interested to investigate the cost of compensation through static and dynamic compensators only, so it is assumed that synchronous generator is generating only mandatory reactive power. The reactive power demand of system is fulfilled by compensators in response to change in system voltage when subjected to small disturbances.

In available studies, STATCOM alone was carried out for reactive power compensation in IHES due to the technical advantage of fast response of it. Although STATCOM has better compensation performance but it gives compensation at a very high cost. So, STATCOM as dynamic compensator alone does not give economic solution for voltage control in IHES. The compensation cost of fixed capacitor as static compensator is very low, but they alone are not capable of providing the adequate solution of voltage regulation. The compensation cost can be reduced by introducing static compensation with dynamic compensation on compromising with voltage response within permissible range. Hence, optimization technique is introduced in this chapter that provides economic solution of reactive power compensation for an isolated hybrid electric system.

Mathematically, total reactive power compensated by reactive power compensators is given as in Eq. (13). Since static and dynamic compensators both are being participated for reactive power compensation, this reactive power Q_{com} must be equal to the sum of reactive power generated by static compensator (FC) and dynamic compensator (ST) as represented in Eq. (14);

$$Q_{com} = Q_{IG} + Q_{Load} - Q_{SG} \tag{13}$$

$$Q_{com} = Q_{FC} + Q_{ST} \tag{14}$$

Since, a fast acting device is necessarily required for compensation so that system may reach its steady state with less settling time under dynamic conditions. Variable demand is satisfied by dynamic compensation i.e. STATCOM only while steady state fixed demand can be satisfied either by STATCOM alone or combination of fixed capacitor with STATCOM. Therefore, system total compensation at any instant of time is given by Eq. (15).

Therefore, mathematically,

$$Q_{com} = \left\{Q_{FC}^{ss} + Q_{ST}^{ss}\right\} + Q_{ST}^{ts} \tag{15}$$

In a restructured environment, in spite of the fact that the cost of reactive power may be dominantly linked with the price of active energy as well as other services, it is considered as an ancillary service which is priced separately. It is further assumed that isolated hybrid electrical system is designed by an independent supplier who used to decide participation of reactive power compensators on the basis of their cost, rating, and system voltage response. For cost analysis, compensation cost function is defined for fixed capacitor and STATCOM in succeeding sections. Equation (15) gives the participation of static and dynamic compensators during steady state and only dynamic compensator during dynamic condition. Total reactive power compensation must satisfy Eq. (13) always in system. Therefore, total compensation cost is formulated in Eq. (16) and it can be evaluated using the cost function of fixed capacitor and STATCOM. It is assumed that the cost of reactive power in system includes only the reactive power production cost of STATCOM and fixed capacitors as explained in preceding section.

$$C(Q_{com}) = \left\{C_{FC}\left(Q_{FC}^{ss}\right) + C_{ST}\left(Q_{ST}^{ss}\right)\right\} + C_{ST}\left(Q_{ST}^{ts}\right) \tag{16}$$

5 Simulink Model Representation for IHES

A basic block diagram of wind diesel based IHES is presented in Fig. 3. In this section, the transfer functions of each component, which are used to develop the simulink model for electrical system shown in Fig. 4, are presented in their corresponding subsections. The s-domain quantities/expressions are represented with the s symbol in parenthesis with quantity. This simulink model will support in developing the

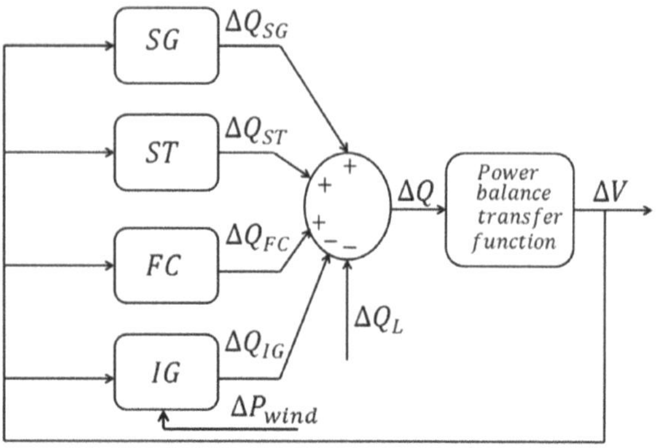

Fig. 4 Representation of voltage-reactive power balance equation in IHES

voltage and reactive power responses of the system component and these responses will be used in getting the optimal values reactive power with the help of available reactive power compensators. Since the study is being focussed for reactive power compensation and voltage control of the electrical system in this chapter. A reactive power balance equation given in Eq. (12) is taken for the study only and hence, simulink block diagram is developed for the transfer functions of change in reactive power with voltage for each component and for complete model.

5.1 Modelling for Reactive Power Balance in IHES

According to the energy policies and recommendations of international standard IEC 60038, the voltage permissible range at load end is $\pm 10\%$; thus other devices connected in system should respond fast to achieve desirable voltage. This demand is maintained by releasing extra reactive power from synchronous generator, STATCOM and fixed capacitor. But, this disturbance will cause a voltage change due to which reactive power required by induction generator and load will also vary. The net reactive-power surplus,

$$\Delta Q = \Delta Q_{SG} + \Delta Q_{FC} + \Delta Q_{ST} - \Delta Q_{IG} - \Delta Q_L \tag{17}$$

The governing voltage reactive power balance equation for IHES is well established in Ref. [44] and presented in Eq. (18) below.

$$\Delta Q = \left(s\frac{V}{\omega X_m} + D_v\right)\Delta V \tag{18}$$

D_v is defined as the transfer function of change in reactive power with voltage change for load. The procedure for estimating this is explained in detail in Ref. [40]. Equations (17) and (18) can be clubbed to form a complete linear model of

IHES as shown in Fig. 4. For block diagram model represented in Fig. 3, system attains disturbances through load reactive power and wind input real power change. Figure 4 represents synchronous generator, STATCOM, fixed capacitor and induction generator as subsystem in IHES. As in Fig. 4, linear model of each subsystem is required for developing simulink model and therefore, linear model for all system components are discussed in succeeding subsections here.

5.2 Synchronous Generator Model Equations

Synchronous generator is the most popular diesel operated genset in small scale power generation. The synchronous generator is equipped with governor and exciter [5]. The exciter is a device which feeds dc supply to the main generator field. IEEE has proposed following standard models for representation of excitation systems for system studies.

1. Type 1 Excitation System—Continuously Acting Exciter And Regulator
2. Type 2 Excitation System–Rotating Rectifier System
3. Type 3 Excitation System—Static with Terminal Current and Potential Sources
4. Type 4 Excitation System—Non Continuous acting.

It is elaborated that IEEE excitation system of type-1 is the most popularly used excitation system with diesel genset and the same is used in this study too [45]. A linear model of synchronous generator with ΔV as input and ΔQ_{SG} as output is developed in Ref. [46] and is presented in Fig. 5.

The values of constants K_1, K_2, K_3 and K_4, shown in Fig. 5, are given in Eq. (19)–(22) from Ref. [40].

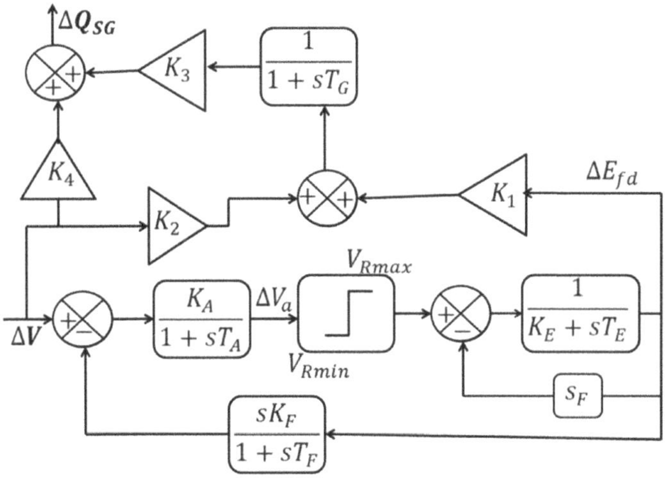

Fig. 5 Representation of linear model *of synchronous generator*

$$K_1 = \frac{X'_d}{X_d} \tag{19}$$

$$K_2 = \frac{\left(X_d - X'_d\right)}{X_d} \cos\delta \tag{20}$$

$$K_3 = \frac{V\cos\delta}{X'_d} \tag{21}$$

$$K_4 = \frac{E'_q\cos\delta - 2V}{X'_d} \tag{22}$$

The standard values of these parameters are chosen in model and are defined as in Ref. [47];

Voltage regulator gain constant; $K_A = 40$
Voltage regulator time constant; $T_A = 0.05$ s
Exciter gain constant; $K_E = 1.0$
Exciter time constant; $T_E = 0.5$ s
Stabilizing circuit gain constant; $K_F = 0.5$
Stabilizing circuit time constant; $T_F = 0.75$ s
Saturation function; $S_F = 0$.

5.3 Induction Generator Model Equations

A linear model of induction generator is developed for approximate equivalent circuit of induction generator as shown in Fig. 6.

$$R_{eq} = R_s + R_r \tag{23}$$

$$X_{eq} = X_s + X_r \tag{24}$$

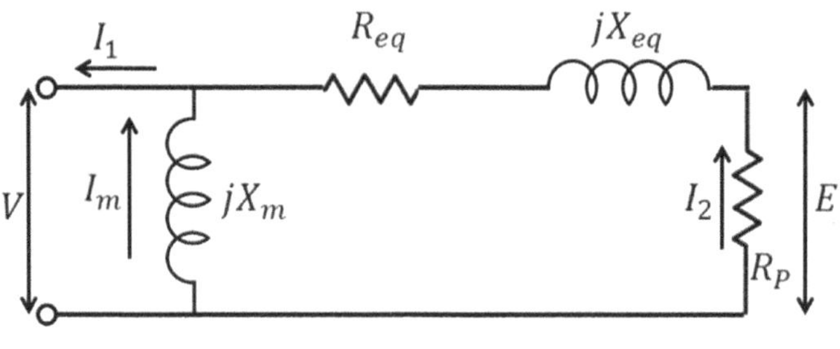

Fig. 6 Reduced approximate equivalent circuit for induction generator

$$R_y = R_P - R_{eq} \tag{25}$$

In Fig. 6, applying nodal analysis at R_P terminal

$$\frac{E - V}{R_{eq} + jX_{eq}} = \frac{E}{R_P} \tag{26}$$

On solving,

$$E = \left[\frac{R_P R_y}{R_y^2 + X_{eq}^2} + j \frac{R_P X_{eq}}{R_y^2 + X_{eq}^2} \right] V \tag{27}$$

Induction generator apparent power S_{ig} would be;

$$S_{ig} = V I_1^* \tag{28}$$

$$S_{ig} = V(I_2 + I_m)^* \tag{29}$$

$$S_{ig} = V I_2^* - V I_m^* \tag{30}$$

$$S_{ig} = V\left(\frac{E - V}{R_{eq} + jX_{eq}}\right)^* - V\left(\frac{V}{jX_m}\right)^* \tag{31}$$

Substituting values and solving for separating real and imaginary terms,

$$S_{ig} = \left[\frac{R_y}{R_y^2 + X_{eq}^2} V^2 \right] - j\left[\left\{ \frac{X_{eq}}{R_y^2 + X_{eq}^2} + \frac{1}{X_m} \right\} V^2 \right] \tag{32}$$

S_{ig} in Eq. (32) gives expression of total electric power generated by induction generator. Real part is the active power developed by induction generator. Since imaginary term of expression is negative in magnitude. This shows that the reactive power is absorbed by the induction generator. Also reactive power expression has two terms; first term denotes the power absorbed by induction generator and second term denotes the reactive power required for magnetization in induction generator.

The wind input power P_{wind} is given by,

$$P_{wind} = Re\{E I_2^*\} = Re\left\{ E \left(\frac{E - V}{R_{eq} + jX_{eq}} \right)^* \right\} \tag{33}$$

On solving,

$$P_{wind} = \frac{R_y}{R_y^2 + X_{eq}^2} V^2 \tag{34}$$

From Eq. (32), the reactive power which is absorbed by the induction generator

$$Q_{IG} = \frac{X_{eq}}{R_y^2 + X_{eq}^2} V^2 \tag{35}$$

Self Excited Induction generator may work under two basic conditions: operation of the wind energy conversion system (WECS) at constant speed or variable speed in terms of change in wind input real power.

In the case of constant speed/slip operation Eq. (32) can be rewritten in s plane as,

$$\Delta Q_{IG} = \frac{2V X_{eq}}{R_y^2 + X_{eq}^2} \Delta V \tag{36}$$

$$K_5 = \frac{2V X_{eq}}{R_y^2 + X_{eq}^2} \tag{37}$$

$$\Delta Q_{IG} = K_5 \Delta V \tag{38}$$

If the induction generator is operating for variable speed/slip then the term R_y is not constant and its value will depend on the slip. Therefore, the expression for the reactive power will not depend only on the voltage but also on the input power available at blade of the induction generator. Solving for small perturbation in the case of variable speed/slip operation, the equation can be written in s plane as [48],

$$\Delta Q_{IG}(s) = K_6 \Delta P_{wind}(s) + K_7 \Delta V(s) \tag{39}$$

$$K_6 = \frac{X_{eq}}{R_P - \{(R_y^2 + X_{eq}^2)/2R_y\}} \tag{40}$$

$$K_7 = \frac{2V}{R_y^2 + X_{eq}^2}\left[X_{eq} - \frac{R_P X_{eq}}{R_P - \{(R_y^2 + X_{eq}^2)/2R_y^2\}}\right] \tag{41}$$

Equations (38) and (39) represent expressions for linear model of induction generator. These expressions have been developed for constant and variable speed respectively as shown in Fig. 7a, b. For IHES in this chapter, variable speed model of induction generator is used.

5.4 *Fixed Capacitor Model Equations*

Reactive power and voltage relation for fixed capacitor is a well established one. Equation (42) provides design information about capacitance per phase in the system. Change in reactive power of fixed capacitor with voltage variation is given in Eq. (43)

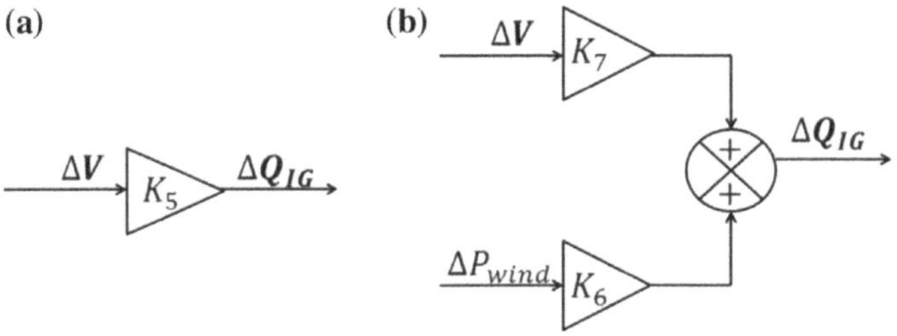

Fig. 7 Linear model for **a** constant slip and **b** variable slip model of induction generator

Fig. 8 Linear model for fixed capacitor

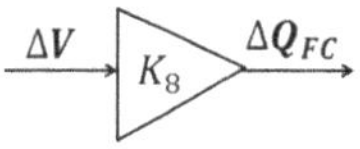

for small perturbation. Linear model for fixed capacitor is represented in Fig. 8 [46].

$$Q_{FC} = \frac{V^2}{X_C} \quad (42)$$

$$\Delta Q_{FC}(s) = K_8 \Delta V(s) \quad (43)$$

$$K_8 = \frac{2V}{X_C} \quad (44)$$

5.5 *STATCOM Model Equations*

STATCOM controls the reactive current flow by adjusting firing angles of thyristor for suitable control of the inverter voltage with respect to the bus voltage and finally, STATCOM controls reactive power generation or absorption in system. For the power flow modelling of the STATCOM, the reactive power expression is given in Eq. (45) [21],

$$Q_{ST} = (kV_{dc})^2 B_{ST} - kV_{dc} V B_{ST} \cos\alpha \quad (45)$$

STATCOM reactive power depends upon two main variables V and α. Based on the Eq. (45), the linear STATCOM equation for small disturbance is given below [44],

$$\Delta Q_{ST}(s) = K_{11} \Delta\alpha(s) + K_{12} \Delta V(s) \quad (46)$$

where,

$$K_{11} = kV_{dc}VB_{ST}\text{Sin}\alpha \quad (47)$$

$$K_{12} = -kV_{dc}B_{ST}\text{Cos}\alpha \quad (48)$$

The small signal models of STATCOM used in dynamic analysis can be designed using three blocks namely regulator, thyristor firing delay and phase sequence delay [49]. A proportional integral controller based linear model STATCOM are given in Fig. 9.

Example 3 Find the cost of loss, mandatory and reserve reactive power for 111 kVA, 400 V, 50 Hz synchronous generator having 0.9 lagging power factor following parameters are specified,

$$\begin{array}{ll} R_a = 0.002 & \mathrm{K_A} = 40 \\ X'_d = 0.15 & \mathrm{T_A} = 0.05\,\mathrm{s} \\ X_d = 1 & \mathrm{K_E} = 1.0 \\ T'_{d0} = 5 & \mathrm{T_E} = 0.5\,\mathrm{s} \\ X_q = 1 & \mathrm{K_F} = 0.5 \\ X_s = 1 & T_F = 0.75\,\mathrm{s} \\ X'_s = 0.15 & \end{array}$$

Solution From the given parameters,

$$S_{SG} = 111\,\text{kVA}$$

$$\text{Cos}\,\theta_{SG} = 0.9$$

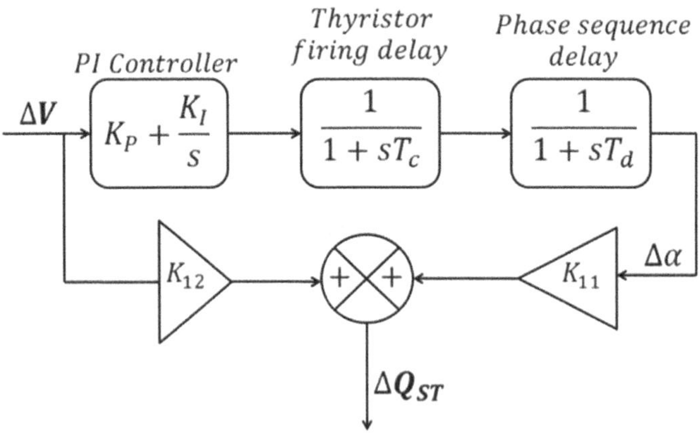

Fig. 9 Linear model of STATCOM

$$P_{SG} = S_{SG} \text{Cos}\,\theta_{SG} = 100\,\text{kW}$$

Let base power and base voltage as given below,

$$P_{SG,Base} = 100$$

$$V_{SG,Base} = 400$$

So,

$$P_{SG,pu} = 1$$

$$V_{SG,pu} = 1$$

Mandatory reactive power can be calculated as,

$$Q_{mandatory} = P_{SG,pu} \tan\theta_{SG}$$

For estimative cost of loss reactive power solve the expression for getting the value of $Q_{SG,pu}$,

$$P^2_{SG,pu} + \left(Q_{SG,pu} + \frac{3V^2}{X_s}\right)^2 = \left(\frac{3VE_q}{X_s}\right)^2$$

$$Q_{cost\ of\ loss} = Q_{SG,pu}$$

$$Q_{reserve} = Q_{cost\ of\ loss} - Q_{mandatory}$$

Using the mathematical expressions given above, MATLAB codes are written for getting the solutions of this example as below.

```
%%% MATLAB codes for cost of loss, mandatory and re-
serve reactive power for SG
clear all
clc
vb=400;
xs=1;
xq=1;
ra=0.002;
psgbase=100;
sgpf=0.9; % lagging power factor (LPF)
v=complex(400,0); % volatge in polar form
vpu=abs(v)/vb; % per unit line voltage
psg=100;            % kW of SG
psgpu= psg/psgbase; % per unit kW power
phi=acosd(sgpf); % phase angle
ia=psgpu/(sqrt(3)*vpu*cosd(phi));
jayee=atand(((vpu*sind(phi))+(ia*xq))/((vpu*cosd(phi))+
(ia*ra)));
delta=jayee-phi;
iasg=complex(ia*cosd(phi),ia*sind(-1*phi));
eq_complex=(vpu+1i*0)+(iasg*complex(0,xs));
eq=abs(eq_complex);
qsgpu=psgpu*tand(phi); % per unit reactive power of SG
q_mandatory=qsgpu;
f1=psgpu^2;
f2=3*vpu*vpu/xs;
f3=(3*vpu*eq/xs)^2;
syms x;
y=((f1)+(x+(f2)).^2-f3); % formula from field current
limit of SG
z=double(solve(y));
if z(1,1)>=0
 z=z(1,1);
end
 if z(2,1)>=0
 z=z(2,1);
 end
q_cost_of_loss=z;
q_reserve=q_cost_of_loss-q_mandatory;
% per phase actual value
q_cost_of_loss=(q_cost_of_loss/3)*psgbase
q_mandatory=(q_mandatory/3)*psgbase
q_reserve=(q_reserve/3)*psgbase
```

The results for the above given MATLAB program are,

$Q_{mandatory} = 16.1441$ kVAR per phase

$Q_{cost\ of\ loss} = 36.3693$ kVAR per phase

$Q_{reserve} = 20.2252$ kVAR per phase.

Example 4 For developing the linear model of synchronous generator as shown in Fig. 5, evaluate the constants K_1, K_2, K_3 and K_4 with the help of synchronous generator parameters as given in previous Example 3.

Solution All the parameters given in Example 3 are used to find the value of constants K_1, K_2, K_3 and K_4. All these constant are being calculated in per unit quantities and the base power and base voltage for these calculations is being assumed equal to SG rating. Therefore, MATLAB codes are written for getting the solutions of this example as below.

```
%%% MATLAB codes for evaluating the value of constants
K1, K2, K3 and K4
clear all;
clc;
%input data for system design (all powers are kW, kVAR)
v-complex(400,0); vpu=1;
psg=100;
psg_base=100;
sgpf=0.9;
phi=acosd(sgpf);
qsg=psg*tand(phi);
psgpu=psg/psg_base;
ra=0.002;
xd_dash=0.15; xd=1;
tdo_dash=5;
xq=1;
xs=1;
xs_dash=0.15;
ke=1;
te=0.5;
ka=40;
ta=0.05;
kf=0.5;
tf=0.75;

%%%% SG model constants calculation for simulink model
zsg=1;                   % per unit impedance of SG
ia=(psg/psg_base)/(sqrt(3)*vpu*cosd(phi));
za-
yee=atand(((vpu*sind(phi))+(ia*xq*zsg))/((vpu*cosd(phi)
)+(ia*ra*zsg)));
delta=zayee-phi;
iasg=complex(ia*cosd(phi),ia*sind(-1*phi));
eq_complex=(vpu)+(iasg*(1i*xs*zsg));
eq=abs(eq_complex);
```

```
id=(eq-(vpu*cosd(delta)))/(xd*zsg);
eq_dash=eq-((xd-xd_dash)*zsg*id);
tg=(tdo_dash*(xd_dash*zsg)/(xd*zsg));
k1=(xd_dash*zsg)/(xd*zsg)
k2=(xd-xd_dash)*zsg*cosd(delta)/(xd*zsg)
k3=vpu*cosd(delta)/(xd_dash*zsg)
k4=((eq_dash*cosd(delta))-(2*vpu))/(xd_dash*zsg)
```

The results for the above given MATLAB program are,

$$K_1 = 0.15$$
$$K_2 = 0.775$$
$$K_3 = 6.0787$$
$$K_4 = -7.3421$$

Example 5 Calculate the full load and no load reactive power requirement for the induction generator with following specifications; $V = 400$ V, $P_{IG} = 150$ kW, $\cos\theta_{IG} = 0.9$, $P = 2$, $s = 0.04$, $f = 50$, $\eta = 90\%$. Consider line voltage as base voltage and real power of generator as base power.

Solution The calculations are being done for the equivalent circuit diagram of induction generator as given in Fig. 6. Most of the mathematical expressions used for the calculation purposes in writing MATLAB codes are imported from the Ref. [50]. For better understanding of the codes readers are suggested to read this paper.

```
%%% MATLAB codes for evaluating the reactive power re-
quirement for the induction generator
clear all;
clc;
%%%% base values for pu calculation
vb=400;        % Base voltage
sigb=150;              % Base power
zig_base=(vb*vb)/(sigb*1000); % Base impedance
%input data for system design
v=complex(400,0); vpu=1;
pig=150;             igpf=0.9;            theta=acosd(igpf);
qig=pig*tand(theta);
f=50; pole=2; s=0.04; eff=0.9; p_mech=(pig*eff);
%%%%% IG equivalent circuit parameters
i1=(pig*10^3)/(sqrt(3)*abs(v)*igpf); % for current re-
fer equivalent circuit of IG
I1=complex(i1*cosd(theta),i1*sind(-theta));
```

```
zeq=(abs(v)/sqrt(3))/I1;
zeq_complex=zeq*cosd(theta)+1i*zeq*sind(theta);
im_zeq=imag(zeq_complex);
sigma=(1-igpf)/(1+igpf);
xm=(abs(v)/sqrt(3))/(abs(I1)*sqrt(sigma));
Xm=complex(0,xm);
Im=((v)/sqrt(3))/Xm;
I2=I1-Im;                    % -1 multiplied with I1 as it
is generator
i2=abs(I2);
Rp=(p_mech*10^3)/(3*i2*i2);
r=Rp*s/(1-s);
req=2*r;
rsum=req+Rp;
ry=Rp-req;
syms xeq
Nr=(rsum*rsum*xm)+(xeq*xm*(xeq+xm));
Dr=(rsum^2)+(xeq+xm)^2;
form_x=im_zeq-(Nr/Dr);
z=double(solve(form_x));
a=z(1,1);
b=z(2,1);
if a>=0
 z=a;
end
if b>=0
 z=b;
end
xeq=z;
x=xeq/2;
% converting parameters in pu values
xeq=xeq/zig_base;
x=x/zig_base;
req=req/zig_base;
r=r/zig_base;
xm=xm/zig_base;
Rp=Rp/zig_base;
ry=ry/zig_base;
%%%% no load and full load current of IG
io=(vpu/sqrt(3))/(1i*xm);
i2=(vpu/sqrt(3))/(req+Rp+1i*xeq);
io=abs(io);
i2=abs(i2);
% per unit values of full load and no load reactive
power of IG
```

Therefore, the results for the above given MATLAB program are,

Full load reactive power requirement for the induction generator = 38.4334 kVAR per phase.

No load reactive power requirement for the induction generator 14.9368 kVAR per phase.

Example 6 For developing the linear models of induction generator as shown in Fig. 7, evaluate the constants K_5, K_6 and K_7 with the help of induction generator as specified in previous Example 5. Also calculate the wind input power by neglecting the constant losses.

Solution MATLAB codes given in Example 5 are followed up to the estimation of equivalent circuit parameters i.e. (from line "clear all" to "ry = ry/zig_base;"). The continuation command to this for evaluating the constants K_5, K_6 and K_7 are summarized below in the solution. All these constant are being calculated in per unit quantities and the base power and base voltage for these calculations is being assumed equal to IG rating. All the mathematical expression used in writing MATLAB codes are imported from the Sect. 5.3. Therefore, MATLAB codes are written for getting the solutions of this example as below.

```
qig_fl=(io^2*xm)+(i2^2*xeq);
qig_nl=(io^2*(xm+xeq));
% actual values of full load and no load reactive power
of IG
qig_fl=qig_fl*sigb
qig_nl=qig_nl*sigb
```

```
%%% MATLAB codes for evaluating the value of constants
K5, K6 and K7 and input wind power
% copy the program of example 13.5 just before writing
the codes given below
% pu parameters at IG base
xeq=xeq/zig_base;
x=x/zig_base;
req=req/zig_base;
r=r/zig_base;
xm=xm/zig_base;
Rp=Rp/zig_base;
ry=ry/zig_base;
Pwind=Rp*vpu^2/(ry^2+xeq^2);
Pconstantloss=0;
Piw=(Pwind+Pconstantloss)                          % input
wind power estimation
k5=2*vpu*xeq/(ry^2+xeq^2)
k6=xeq/(Rp-((ry^2+xeq^2)/(2*ry)))
k7=(2*vpu/(ry^2+xeq^2))*(xeq-(Rp*xeq/(Rp-
((ry^2+xeq^2)/(2*ry)))))
```

Therefore, the results for the above given MATLAB program can be summarized as,

$$\text{Wind input power to induction generator} = 0.7996\,\text{pu}$$

$$K_5 = 1.2617$$

$$K_6 = 3.9024$$

$$K_7 = -4.9792$$

Example 7 A single unit wind and single unit diesel generator are coupled together to develop an isolated hybrid electrical system. If induction generator and synchronous generator are used for wind driven power generator and diesel driven power generator respectively. A 200 kW at 0.9 lagging power factor load is being supplied by this IHES in which IG and SG are used for supplying power 150 kW at 0.9 lagging power factor and 250 kW at 0.9 lagging power factor respectively. Other specifications for SG and IG are same as in Example 3 and 5 respectively. Calculate the reactive power requirement through reactive power compensator assuming that SG is generating only mandatory reactive power.

Solution This example is designed to understand the reactive power balance for IHES during steady state conditions. For the operation of this IHES, reactive power is required for load and IG which can be supplied by synchronous generator. But it has been cleared in the problem that SG will generate only the amount of reactive power equal to mandatory reactive power. So, rest reactive power requirement can only be fulfilled by any compensator. Mathematically,

$$Q_{com} = Q_L + Q_{IG} - Q_{SG}$$

Q_{SG} is the mandatory reactive power which has been obtained in Example 3. Q_{IG} is the full load reactive power requirement of induction generator that has already been evaluated in Example 5.

$$Q_{SG} = 16.1441\ \text{kVAR per phase}$$

$$Q_{IG} = 38.4334\ \text{kVAR per phase}$$

$$Q_L = \frac{P_L \tan\left(\cos^{-1}(load\ power\ factor)\right)}{3} = 32.2881\ \text{kVAR per phase}$$

Therefore,

$$Q_{com} = 62.6495\ \text{kVAR per phase}$$

Example 8 Repeat the Example 7 and find the per unit value of reactive power from compensator for load 250 kW at 0.9 lagging power factor.

Solution To keep all the quantities at the same base, load power is taken as base power.

```
%%% MATLAB codes for evaluating the per unit value of
reactive power from compensator
clear all
clc
sbase=250;
qsg=16.1441/sbase; % As obtained in example 13.3
qig=38.4334/sbase; % As obtained in example 13.5
pl=250/(3*sbase);
lpf=0.9;
ql=pl*tand(acosd(lpf))
qcom=ql+qig-qsg
```

Therefore, the results for the above given MATLAB program can be summarized as,

$$\text{Reactive power required from compensator} = 0.2506\,\text{pu kVAR per phase}$$

Example 9 Consider a 250 kW, 0.9 lagging power factor exponential type static load is connected with the IHES as explained in Example 7. If the exponential factor for static load is 3, find the transfer function D_v for this load as in Eq. (18).

Solution The detail explanation for obtaining the transfer function of exponential type static load function is beyond the scope of this chapter. In Ref. [50], the detail documentation for obtaining it is well presented and the same is being used here.

```
%%% MATLAB codes for evaluating transfer function of
exponential type static load function
clear all
clc
lpf=0.9;
vpu=1;              % base voltage
Pl=250/3;            % per phase load power
s_base=250;                 % base power
q=3;          % exponential factor
Ql=Pl*tand(acosd(lpf));
Ql_pu=Ql/s_base;
num=q*Ql_pu;
den=vpu;
TFsl=tf(num,den)  % estimation of Dv as in ref. [51]
```

Therefore, the results for the above given MATLAB program can be summarized as,

$$\text{Load transfer function } D_v = 0.4843$$

Example 10 If reactive power required in Example 8 through reactive power compensator is given by fixed capacitor only, evaluate the constant K_8 for FC as shown in Fig. 8.

Solution Reactive power required from compensator is 0.2506 pu kVAR per phase. If the same reactive power is supplied through FC only at voltage 1.0 pu, the constant K_8 can be evaluated as $K_8 = \frac{2V}{X_C}$, where capacitive reactance per phase can be defined as $X_C = \frac{V^2}{Q_{FC}}$. It is assumed that capacitor bank is delta connected so phase voltage is same as line voltage.

```
%%% MATLAB codes for evaluating constant k8 for FC
clear all
clc
v=1;
qfc=0.2506;
xc=v^2/qfc
k8=2*v/xc
```

Therefore, the results for the above given MATLAB program can be summarized as,

Capacitive reactance per phase, $X_C = 3.9904$ pu ohm

$$K_8 = 0.5012$$

Example 11 Repeat the Example 10, if reactive power required through reactive power compensator is given by STATCOM only. Evaluate the constants K_{11} and K_{12} for STATCOM as shown in Fig. 9.

Solution For the power flow modelling of the STATCOM, the reactive power expression is given in Eq. (45).

$$Q_{ST} = (kV_{dc})^2 B_{ST} - kV_{dc} V B_{ST} \cos\alpha$$

STATCOM constants K_{11} and K_{12} can be expressed as in Eqs. (47) and (48). The required variable for STATCOM has been evaluated by the author in Ref. [44]. The same methodology is being followed for evaluating these parameters in this example and the corresponding expressions are imported in MATLAB codes as given below.

```
%%% MATLAB codes for evaluating constant k11 and k12
for ST
clear all
clc
p=12; % Number of pulses
vpu=1;
a_st=1.2; % modulation index for ST
fs=10e3; % Switching frequency for ST
f=50;
qst=0.2506; % Reactive power requirement
m=p*sqrt(6)/(6*pi);
k=1/m;
vac=vpu;
vdc=vac/k;
is=(qst)/(sqrt(3)*vac);
icr=(5/100)*(2*sqrt(2)*is);
lac=(sqrt(2)*vac/(6*a_st*fs*icr));
B=1/(2*pi*f*lac);
alpha=acosd(((k*vdc)^2*B-(qst))/(k*vdc*vac*B));
k11=k*vdc*vac*B*sind(alpha)
k12=-(k*vdc*B*cosd(alpha))
```

Therefore, the results for the above given MATLAB program can be summarized as,

$$K_{11} = 1.2646$$

$$K_{12} = -3.0653$$

6 Importance of Dynamic Compensator for Voltage Control

For IHES showing in Fig. 3, steady state reactive power can be generated for IHES either by static or by dynamic compensator as depicted by two Examples 10 and 11. Equation (15) explains that dynamic condition reactive power requirement cannot be generated by static compensator. To verify this statement and to check any feasibility of using single static compensators for dynamic changes, only fixed capacitor is connected as reactive power compensator. A simulink model is developed in MATLAB simulink toolbox window for the IHES components as shown in Fig. 4 except the STATCOM block. All the constants and parameters are well estimated in the preceding examples of this chapter. For analyzing the FC behaviour for dynamic conditions, a 10% step disturbances are given at t = 1 s in input power and load.

Figure 10 clearly demonstrates how the voltage collapse in presence of FC as

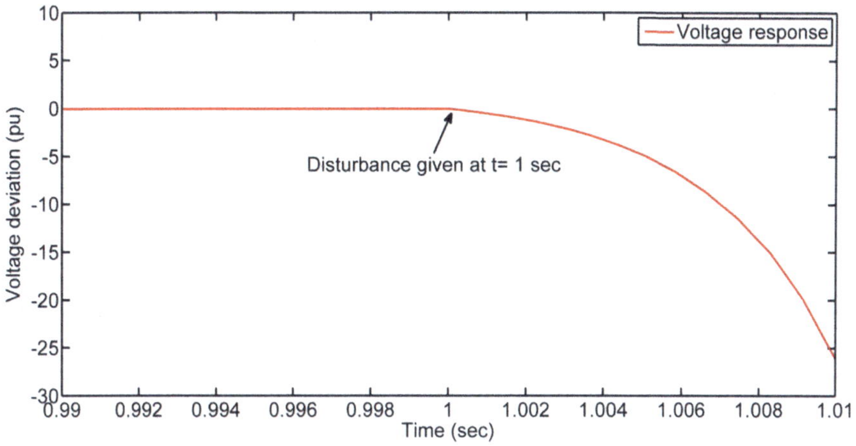

Fig. 10 Voltage collapse for load pattern 1 in presence of static compensator only

only compensator for IHES at t = 1 s. Till t = 1 s IHES demands only steady state reactive power that can easily be supplied by FC. But, as soon as the disturbances occur in IHES at t = 1 s, FC alone is not capable to support the system for this dynamic compensation requirement. On the other side, if cost of compensation is not a constraint for adopting reactive power compensation method, STATCOM can alone be used for providing reactive power compensator. A complete simulink diagram for IHES with ST only as reactive power compensator is presented in Fig. 11.

7 Optimization of Reactive Power Participation

It can be calculated from preceding section that optimized participation of both static compensator (FC) and dynamic compensator (ST) can be used to get technically and economically accepted solution of voltage control for any IHES. A method for getting optimum participations of reactive power compensation is proposed in this section with the help of Fig. 12 in which FC and ST are being connected for supplying reactive power for IG and Load in presence of SG.

It has already been discussed that cost of fixed capacitor is very low compare to STATCOM but fixed capacitors do not respond for system under dynamics conditions. STATCOM alone can provide an adequate solution of reactive power compensation for system voltage control but it makes system very costly. For dynamic conditions, reactive power can only be generated through fast acting dynamic compensating device but steady state reactive power requirement can be planned through participation of static as well as dynamic compensators so that overall compensation cost may be reduced. The role of static compensation deforms the voltage response

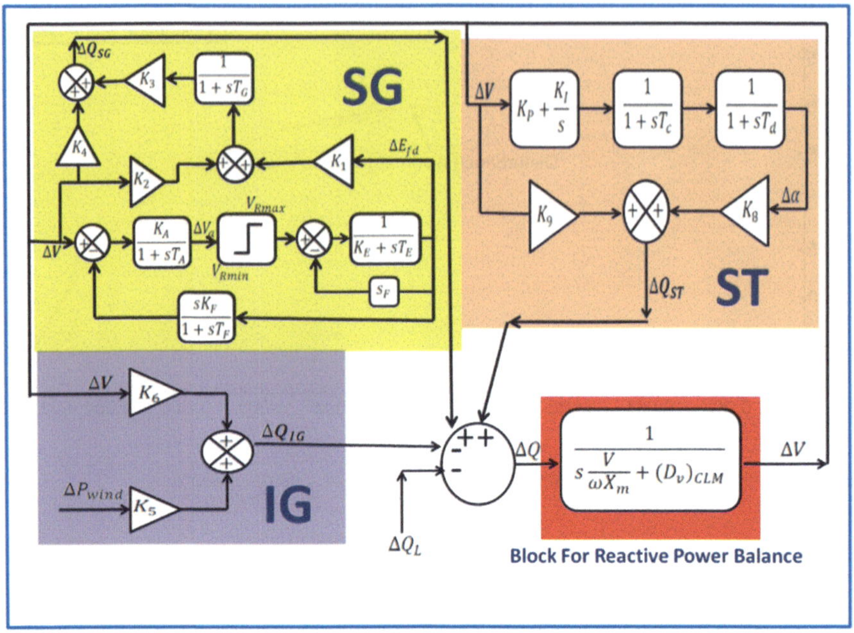

Fig. 11 Simulink model for IHPS with SG, IG, ST and CLM

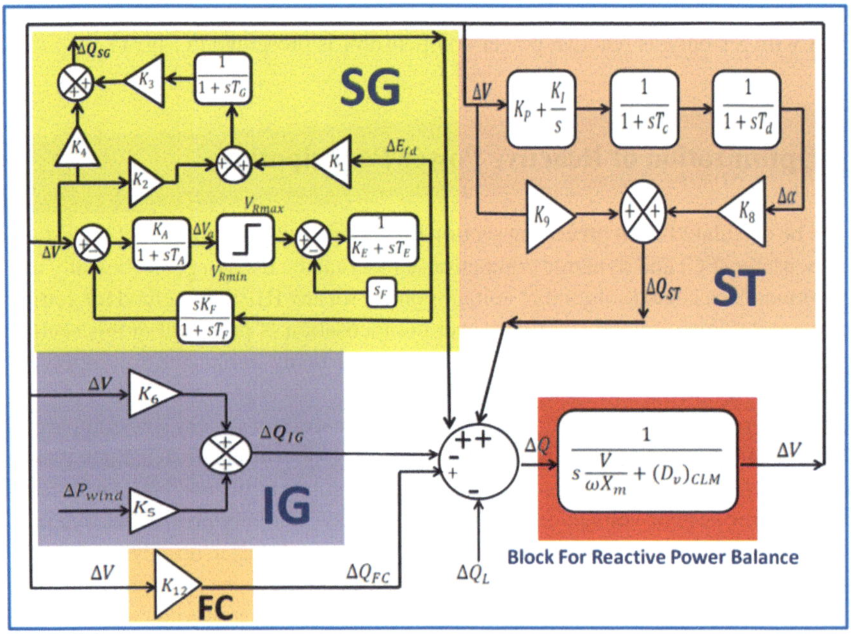

Fig. 12 Simulink block diagram with static and dynamic reactive power compensator

of the system and hence participation of fixed capacitor with STATCOM should be optimized up to the extent of voltage variations within the permissible range.

Therefore, an optimization problem is formulated for reactive power participation using static and dynamic compensators in system but considering two important aspects;

1. Minimizing the cost of compensation under steady state through participation of fixed capacitor as static compensator along with STATCOM as dynamic compensator, and
2. Participation of static compensator with dynamic compensation up to the extent where system voltage responses remain in its pre defined acceptable range.

The proposed approach allows minimizing cost function given in Eq. (16). As discussed earlier, dynamic state reactive power is supplied by STATCOM only therefore term $C_{ST}\left(Q_{ST}^{ts}\right)$ does not require to add in optimization problem. Therefore, an objective function J which represents cost function of reactive power compensation as in Eq. (49) is optimized to find best solution of reactive power compensation in the system. Functions of cost for STATCOM and fixed capacitor depend upon the reactive power released by them;

Objective function

$$J = C_{FC}\left(Q_{FC}^{ss}\right) + C_{ST}\left(Q_{ST}^{ss}\right) \tag{49}$$

Dynamic equations of fixed capacitor and STATCOM are represented by Eqs. (9) and (10). Their corresponding linear models as given in Figs. 8 and 9 are used in system's simulink model as in Fig. 4. This simulink model is used to find voltage response of system and reactive power responses for all system components in presence of both static and dynamic reactive power compensator together. Voltage responses for different participations between fixed capacitor and STATCOM can be tracked and used to decide the final acceptable system response.

Mathematically, reactive power must remain balance in the system and corresponding expressions for equality constraints can be written as;

Equality constraints

$$Q_{demand} = Q_{release} \tag{50}$$

$$Q_{demand} = Q_{IG} + Q_{L} - Q_{ST} \tag{51}$$

$$Q_{release} = Q_{FC}^{ss} + Q_{ST}^{ss} \tag{52}$$

Equation (52) explains that steady state reactive power is fulfilled by fixed capacitor and STATCOM. Optimized value of reactive power through STATCOM and fixed capacitor will be chosen with pre-acceptable range of voltage response. Reactive power released by STATCOM and fixed capacitors must be within the range as in Eqs. (53) and (54). Equations (55) and (56) define pre-acceptable range of voltage response for the system.

Inequality constraints

$$0 \leq Q_{ST}^{ss} \leq Q_{demand} \tag{53}$$

$$0 \leq Q_{FC}^{ss} \leq Q_{demand} \tag{54}$$

$$V_{min} \leq \Delta V \leq V_{max} \tag{55}$$

$$settling\,time \leq settling\,time_{acceptable} \tag{56}$$

Stability constraints
System voltage deviates due to load and input disturbances from its steady state value. The voltage deviation should reach to zero as earliest by the additional reactive power generation by the static and dynamic compensators. In other words, system should remain stable.

Acceptable range of voltage response should be decided first as described in Eqs. (55) and (56). Previously available papers suggest use of STATCOM only for reactive power compensation in isolated hybrid power system. In this work, optimum values of static and dynamic compensators are obtained. Therefore, two cases can be defined for comparative study of compensation cost analysis.

- Case I: STATCOM alone is used for reactive power compensation in the system.
- Case II: Participation of both fixed capacitor and STATCOM is used for reactive power compensation in the system.

To find the participation of compensators for getting system voltage response within the predefined acceptable range, a reference voltage is required. Though, this decision must be based on mutual acceptance of power quality between end user and electricity producers (private investors) in terms of system voltage quality. In this optimizing problem, characteristic parameters from voltage response are obtained in case I and these are used as a reference for deciding acceptable range of parameters for case II.

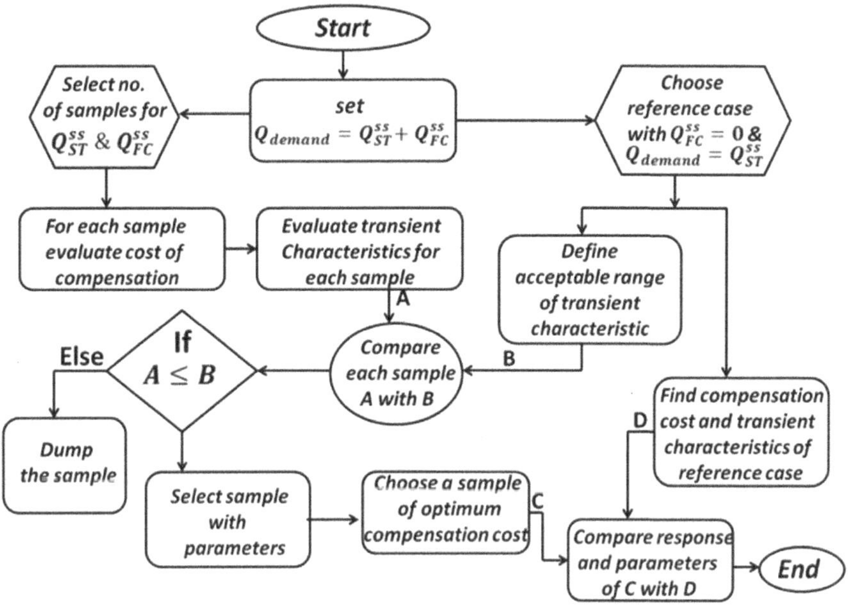

Fig. 13 Flow chart for reactive power compensation cost determination

For deciding the optimize participation of static and dynamic compensators, number of samples for reactive power generation from fixed capacitor and STATCOM are developed satisfying Eq. (52) by gradually increasing reactive power generation through fixed capacitor and decreasing reactive power generation through STATCOM.

For each sample, system voltage response is tracked and compared with predefined values and finally, reactive power participation is selected for optimum value of compensation cost. A flowchart showing algorithm for proposed optimization of compensation cost is presented in Fig. 13.

Example 12 Develop a simulink block diagram for an IHES with 150 kW IG and 100 kW SG. Plot the voltage response for the system if 10% step disturbance occurs in connected load of rating 250 kW and wind input power. Other ratings and specifications are same as in preceding examples.

Solution Figure 4 represents the complete simulink diagram for IHES. This simulink diagram is developed on MATLAB simulink model window with the help of required constants that are developed as in preceding examples. It should be noted that all the constants and parameters must be estimated on a common base values and therefore load values of voltage and power is taken as base power now. The constant are;

$k_1 = 0.1500$; $k_2 = 0.7750$; $k_3 = 2.4315$; $k_4 = -2.9368$; $k_5 = 0.7570$; $k_6 = 3.9024$; $k_7 = -2.9875$; $k_{11} = 1.2646$; $k_{12} = -3.0653$

A 10% step disturbances in wind input power and load reactive power is produced in simulink model with the help of source block parameter from the simulink library browser. STATCOM block diagram represented in Fig. 9 demonstrates regulator block. In Ref. [44], complete detail of STATCOM is given and same is used here. PI controller is used as regulator and the value of constants can be estimated using ISE criterion [51]. Algorithm to develop voltage response can be summarized as follows;

1. Develop reactive power required form STATCOM after evaluating the reactive power for SG, IG, and Load as in Examples 3, 5 and 7.
2. Evaluate gain constants for SG as estimated in Example 4.
3. Evaluate gain constants for IG as estimated in Example 6.
4. Evaluate gain constants for ST as estimated in Example 11.
5. Develop the constants for reactive power balance equation as given in Eq. (18).
6. Club all the IHES components together as presented in Fig. 4 at MATLAB simulink model window.
7. Define input wind power and load disturbance in simulink model disturbances.
8. Estimate the PI controller gain constants for STATCOM linear model.
9. Run the MATLAB codes having all the above discussed information. Simulink model is called in this MATLAB program and all the required information can be imported to simulink model through MATLAB codes.
10. Develop the required responses for the IHES like voltage response etc.

Figure 14 explains the voltage response for IHES when only STATCOM is used for supplying reactive power compensation. Due to disturbances at time 1 s, voltage starts to deviate and demand additional reactive power from the system to stabilize

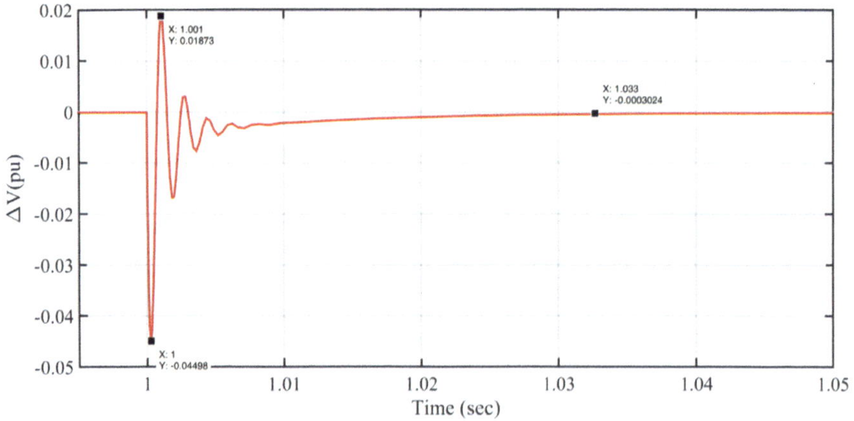

Fig. 14 Voltage response for ST as reactive power compensator

the system voltage. To control the voltage STATCOM starts acting and generate additional required reactive power to control the voltage. The maximum voltage deviation is 0.01873 pu and minimum deviation is −0.04498 pu while system voltage steles down at time 1.033 s.

Example 13 Find the reactive power compensation cost for IHES in Example 12.

Solution The expression for compensation cost through STATCOM,

$$C_{ST} = \frac{1000 * Q_{ST}}{8760 * 15}\left(0.0002466 Q_{ST}^2 - 0.2243 Q_{ST} + 150.527\right)\$/\mathrm{H}$$

$$Q_{ST} = 0.2506 \text{ pu MVAR}$$

Using the mathematical expressions given above, MATLAB codes are written for getting the solutions of this example as below.

```
%%% MATLAB codes for compensation cost through STATCOM
(ST)
>>clear all;
>>clc;
>>Q_st=0.2506;              % pu MVAr generated
>> pl_base =250;
% compensation cost through ST by converting pu in MVAR
>>cost_Qst_ref=(1000.*(Q_st.*pl_base./1000)./(8760.*15)
).*((0.0002466.*Q_st.*Q_st)-(0.2243.*Q_st)+150.527)
```

The results for the program given above are;

$$\text{Compensation cost through ST} = 0.0717\ \$\ per\ \mathrm{H}$$

Example 14 For the voltage response developed in Example 12, find the transient parameters such as voltage dip, voltage rise and rise time.

Solution For the voltage response developed in Example 12, transient parameters can be evaluated by using the MATLAB codes as written below;

```
% evalaution of the transient parameters for the known
voltage response
sim('NAME_OF_MDL_FILE_STORED') % Syntax for running the
simulink model form MATLAB code
% From volatge response stored in workspace of simulink
model, time and volatge array can be seperated as
plot_t=sim_v.time;
plot_v=sim_v.signals.values;
% Reference voltage parametrs can be evaluated by these
two commands
V_par1=lsiminfo(plot_v,plot_t)
V_par2=stepinfo(plot_v,plot_t)
% syntax to get all parametrs saprately
V_data11=struct2cell(V_par1)
V_data1=cell2mat(V_data11)
V_data22=struct2cell(V_par2)
V_data2=cell2mat(V_data22)
Settlingtime=V_data1(1)
Voltagedip=V_data1(2)
Voltagerise=V_data1(4)
Risetime=V_data2(1)
```

Therefore, the results for the above given MATLAB program can be summarized as,

$$\text{Settling time} = 1.0199\,\text{s}$$
$$\text{Voltages dip} = -0.0450\,\text{pu}$$
$$\text{Voltage rise} = 0.0187\,\text{pu}$$

Example 15 Consider the settling time, voltage dip and voltage rise obtained in Example 14 as a reference parameters. Define the predefined acceptable range of voltage response (i.e. inequality constraints) for optimization procedure.

Solution Energy policies and recommendations of international standard IEC 60038, the voltage permissible range at load end is $\pm 10\%$. Voltage response produced for IHES through STATCOM only as a reactive power compensator is assumed as reference response for achieving optimization participation of FC and ST. Therefore,

Reference value for voltage dip = −0.0450 pu
Reference value for voltage rise = 0.0187 pu
Reference value for settling time = 1.0199 s

The predefined acceptable range of voltage response (i.e. inequlity constraints) for optimization procedure is assumed to be,

Acceptable voltage rise ⇐ voltage rise + 0.05
Acceptable voltage dip ⇐ |voltage dip| + 0.05
Acceptable voltage settling time ⇐ settling time + 0.01

Example 16 Find the optimized participation for reactive power compensation using FC along with ST for this IHES keeping voltage control in its predefined acceptable range.

Solution In Example 13, compensation cost is given when ST is only used for reactive power compensation. Example 14 provides the voltage response transient parameters for the Fig. 14 given as the solution of Example 12. As explained earlier that dynamic condition requirement can only be given by ST only while to reduce the compensation cost FC can be introduced with ST for fulfilling the steady state reactive power demand. To make a technical acceptable solution for participations of static and dynamic compensator together, an acceptable solution should be decided first. The acceptable range of voltage response (i.e. inequality constraints) for optimization procedure is estimated in Example 15. Equality constraint for optimization problem is that total reactive power requirement should be the sum of reactive power from FC and reactive power from ST. Procedure for getting the optimized solution of compensation cost is defined through flow chart given in Fig. 13. A MATLAB code is developed in which numbers of samples are generated by increasing the reactive power from FC gradually and decreasing the reactive power from ST gradually keeping the sum of required compensation constant always. For each sample, voltage response is achieved and compared with the transient responses reference value as in Example 14. All the samples having their transient responses with in predefined acceptable range are sorted from the total number of samples. Out of these sorted samples, a sample having least compensation cost is selected as optimized participation of reactive power compensation. Therefore, MATLAB codes can be written for getting the solutions of this example as below;

```
% MATLAB code is continued after clubbing all preceding
examples codes together
% variable "loop_ref" shows an array for all reference
parameters
loop_ref=[cost_Qst_ref,cost_Qfc_ref,settlingtime_ref,ma
xvoltage_ref, minvoltage_ref];
 % For producing number of samples satisfying equality
constraints
ql; qig; qsg;
qcom=ql+qig-qsg;
Qst_upper=qcom; Qst_lower=0;
Qfc_upper=qcom; Qfc_lower=0;
Qdemand=qig+ql; % total reactive power demand in the
system
count1=0; sample=1000; % 1000 samples are considered
Qsg=qsg;
Qfc=linspace(Qfc_lower,Qfc_upper,sample); % initialize
the reactive power from FC
 % for producing the samples for FC and ST participa-
tion satisfying equality constraints
for n1=1:sample
 x111=Qfc(n1);
 x333=-x111-Qsg+Qdemand;
 if x333<=Qst_upper && x333>=Qst_lower
 count1=count1+1;
 Q_fc(count1)=x111;
 Q_st(count1)=x333;
 end
 end
count1; % it gives number of samples
possible_participation=[Q_st' Q_fc']; % the values of
each participation between ST and FC
%%%% cost calculation for each sample
cost_Qfc=0.132.*(Q_fc.*(pl_base)./1000);
cost_Qst=(1000.*(Q_st.*pl_base./1000)./(8760.*15)).*((0
.0002466.*Q_st.*Q_st)-(0.2243.*Q_st)+150.527);
cost=cost_Qfc+cost_Qst;
cost_Q=[Q_st' Q_fc' cost_Qfc' cost_Qst' cost'] % matrix
for cost comparison for each sample data
% transient study parameters for all sample data
for n=1:(count1-1)
costofQ(n)=cost(n);
% Fixed Capacitor constant for each sample
xc=vac*vac/(Q_fc(n));
k12=2*vpu/xc;
```

```
% STATCOM constants for each sample
m=p*sqrt(6)/(6*pi);
k=1/m;
vdc=vac/k;
is=(Q_st(n))/(sqrt(3)*vac); % in Ampere
icr=(5/100)*(2*sqrt(2)*is);
lac=(sqrt(2)*vac/(6*a_st*fs*icr)); % in henry
B=1/(w*lac);
alpha=acosd(((k*vdc)^2*B-(Q_st(n)))/(k*vdc*vac*B));
k11=k*vdc*vac*B*sind(alpha);
k12=-(k*vdc*B*cosd(alpha));
% variables for running simulink model of IHES
qsg;
qst=Q_st(n);
qfc=Q_fc(n);
 sim('simforpaperkW');
 n
plot_t=sim_v.time;
plot_v=sim_v.signals.values;
 % To check the voltage response stable condition
z=size(plot_v);
z=z(1,1);
counter=0;
vforlocalmaxima=plot_v;
for zz=2:z-1
 if vforlocalmaxima(zz)>0
 if vforlocalmaxima(zz)>vforlocalmaxima(zz+1) && vfor-
localmaxima(zz)>vforlocalmaxima(zz-1)
 counter=counter+1;
 local_maxima(counter)=vforlocalmaxima(zz);
 end
 end
end
if local_maxima(1)>local_maxima(2)
loop_1=lsiminfo(plot_v,plot_t);
loop_2=stepinfo(plot_v,plot_t);
loop_data11=struct2cell(loop_1);
loop_data1=cell2mat(loop_data11);
loop_data22=struct2cell(loop_2);
loop_data2=cell2mat(loop_data22);
loop_setlingtime_opt(n)=loop_data1(1);
loop_minvoltage_opt(n)=loop_data1(2);
loop_maxvoltage_opt(n)=loop_data1(4);
loop_risetime_opt(n)=loop_data2(1);
loop_overshoot_opt(n)=loop_data2(5);
```

```
loop_undershoot_opt(n)=loop_data2(6);
else
loop_setlingtime_opt(n)=100;
loop_minvoltage_opt(n)=100;
loop_maxvoltage_opt(n)=100;
loop_risetime_opt(n)=100;
loop_overshoot_opt(n)=100;
loop_undershoot_opt(n)=100;
end
end
pi_sample_array=[costofQ'        loop_settlingtime_opt'
loop_maxvoltage_opt' loop_minvoltage_opt']
%%%%% for pi based sample sorting
count3_loop=0;
for count2_loop=1:count1-1
 if
loop_maxvoltage_opt(count2_loop)<=maxvoltage_ref+0.05
&&
abs(loop_minvoltage_opt(count2_loop))<=abs(minvoltage_r
ef)+0.05                                             &&
loop_setlingtime_opt(count2_loop)<=settlingtime_ref+.01
 count3_loop=count3_loop+1;
 loop_Q_st(count3_loop)=Q_st(count2_loop);
 loop_Q_fc(count3_loop)=Q_fc(count2_loop);
 loop_cost_Qfc(count3_loop)=cost_Qfc(count2_loop);
 loop_cost_Qst(count3_loop)=cost_Qst(count2_loop);
 loop_cost(count3_loop)=cost(count2_loop);

loop_MMaxvalue_Vr(count3_loop)=loop_maxvoltage_opt(coun
t2_loop);

loop_MMinvalue_Vr(count3_loop)=loop_minvoltage_opt(coun
t2_loop);

loop_rrisetimevalue(count3_loop)=loop_risetime_opt(coun
t2_loop);

loop_ssettlingtimevalue(count3_loop)=loop_setlingtime_o
pt(count2_loop);
 end
 end
count3_loop;
loop_sample_array1=[loop_cost' loop_ssettlingtimevalue'
loop_MMaxvalue_Vr' loop_MMinvalue_Vr']
 %loop_oovershoot_opt' loop_uundershoot_opt'
```

```
mincost = min(loop_cost);
for z=1:count3_loop
 nit=loop_cost(z);
 if nit==mincost
 break
 end
end
z;
% Hence solution for optimize participations of FC and
ST are
qst=loop_Q_st(z)
qfc=loop_Q_fc(z)
qst_cost=loop_cost_Qst(z)
qfc_cost=loop_cost_Qfc(z)
total_cost=loop_cost(z)
```

Therefore, the results for the above given MATLAB program can be summarized as,

Reactive power from ST = 0.0936 pu MVAR
Reactive power from FC = 0.1570 pu MVAR
Compensation cost from ST = 0.0268 \$ per h
Compensation cost from FC = 0.0052 \$ per h
Total Compensation cost for optimized participation = 0.0320 \$ per h

Example 17 Compare the voltage response for IHES for reference case having ST only and optimized participation of reactive power compensation using FC and ST together. Also compare the settling time, voltage dip and voltage rise, reactive powers from FC and ST and compensation costs in both the cases.

Solution The voltage responses are obtained by expanding the MATLAB codes given in Example 16. The voltage response comparison is given in Fig. 15 and the settling time, voltages dip and voltage rise in both the cases are tabulated in Table 2.

8 Conclusion and Future Scope

This chapter explains the economical benefits of hybrid participations of static and dynamic reactive power compensators in voltage control studies for IHES. It is elaborated how the rating of a STATCOM can be reduced with the use of an FC so that the overall compensation cost may be reduced up to the extent of voltage variation within the permissible range. The use of dynamic compensator (STATCOM) alone gives a technically viable solution, but the introduction of static compensator (FC) can provide a technically and economically viable solution. Results explain how the total compensation cost can be reduced by introducing static compensator along

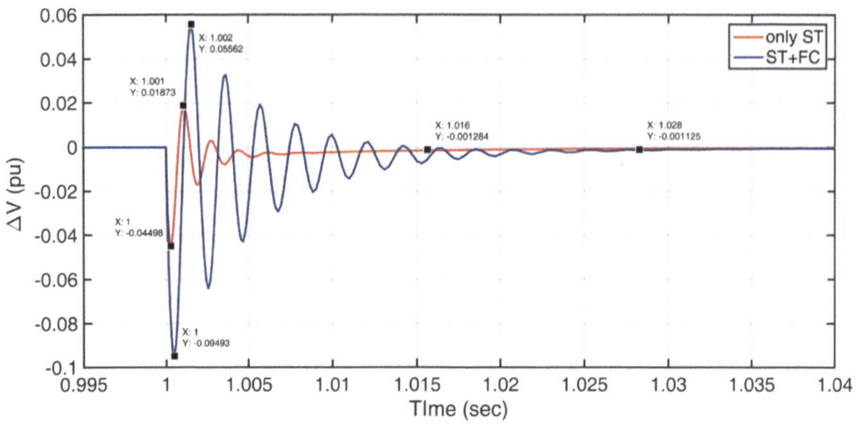

Fig. 15 Voltage response comparison for ST alone and ST & FC together as reactive power compensator

Table 2 Comparative study of transient parameters for voltage response

Response parameters	ST only	ST + FC
Settling time (s)	1.016	1.028
Voltage dip (pu)	−0.04498	−0.09493
Voltage rise (pu)	0.01873	0.05562
Reactive power from FC (pu, MVAR)	0	0.1570
Reactive power from ST (pu, MVAR)	0.2506	0.0936
Compensation cost from FC ($ per hour)	0	0.0052
Compensation cost from ST ($ per hour)	0.0717	0.0268
Total compensation cost ($ per hour)	0.0717	0.0320

with dynamic compensator for generating reactive power for steady state conditions. MATLAB codes are developed for choosing the optimized participations of FC and STATCOM for voltage control studies in IHES. In this chapter, main objective was towards the achievement of optimize participation of reactive power compensation using FC and STATCOM for getting an economical viability of reactive power compensation as ancillary service for the remote area situated consumers. Still, this study has enormous future research scopes in the area of reactive power compensation and voltage control in terms of advance methods to estimate exact parameters of induction generator, modelling of other FACTS device for getting compensation techniques technically and economically, study with online load scheduling, estimation of gain constants of regulator used in FACTS devices by advanced algorithm, introducing to IHES in grid connected system etc.

Key Terms and Their Definitions

Static Compensator: Compensating devices for those reactive power generations can not be changed depending on the time.
Dynamic Compensator: Compensating devices for those reactive power generations can be changed easily as per requirement with time.
Reactive power compensators: To control the system voltage, an additional reactive power is supplied to the system. Such devices are called reactive power compensator.
Compensation cost: Cost asked by the power seller for providing the compensation in system. The actual compensation cost depends on the type of the device used because each device has its own cost function.
Ancillary services: Ancillary services are defined as the additional services provided by the power seller to upgraded the power quality for utility.

MATLAB Code
MATLAB Codes are given within the chapter with examples.

References

1. A Technical Report by Ministry of Power, Government of India.: Goods and services for implementation of rural electrification project for decentralized distributed generation under Rajiv Gandhi Grameen Vidyutikaran Yojana (RGGVY), New Delhi (2009)
2. A Technical Report by Central Electricity Authority, Ministry of Power, Government of India. Executive Summary Power Sector, New Delhi (2015)
3. Saxena, N.K., Kumar, A.: Cost based reactive power participation for voltage control in multi units based isolated hybrid power system. J. Electr. Syst. Inf. Technol. **3**(3), 442–453 (2016)
4. Sharma, P., Saxena, N.K., Ramakrishna, K.S.S., Bhatti, T.S.: Reactive power compensation of isolated wind-diesel hybrid power systems with STATCOM and SVC. Int. J. Electr. Eng. Inf. **2**(3) (2010)
5. Ahmed, M.K., Ali, M.Y.: Robust control of an isolated hybrid wind–diesel power system using linear quadratic Gaussian approach. Electr. Power Energy Syst. **33**, 1092–1100 (2011)
6. Sharma, P., Bhatti, T.S.: Performance investigation of isolated wind–diesel hybrid power systems with WECS having PMIG. IEEE Trans. Ind. Electron. **60**(4), 1630–1637 (2013)
7. Vachirasricirikul, S., Ngamroo, I., Kaitwanidvilai, S.: Coordinated SVC and AVR for robust voltage control in a hybrid wind-diesel system. Energy Convers. Manag. **51**, 2383–2393 (2010)
8. Bansal, R.C., Bhatti, T.S., Kothari, D.P.: A novel mathematical modelling of induction generator for reactive power control of isolated hybrid power systems. Int. J. Model. Simul. **24**(1), 1–7 (2004)
9. Zhao, D., Ni, Y., Zhong, J., Chen, S.: Reactive power and voltage control in deregulated environment. In: Proceedings of IEEE/PES Transmission and Distribution Conference & Exhibition: Asia and Pacific Dalian, China (2005)
10. Kargariana, A., Raoofatb, M., Mohammadi, M.: Probabilistic reactive power procurement in hybrid electricity markets with uncertain loads. Electr. Power Syst. Res. **82**, 68–80 (2012)

11. Chung, C.Y., Chung, T.S., Yu, C.W., Lin, X.J.: Cost-based reactive power pricing with voltage security consideration in restructured power systems. Electr. Power Syst. Res. **70**, 85–91 (2004)
12. Shivakumar, P., Thirukkovai, S., Yogeshraj, K., Abdullah, A.: Reactive power regulation of wind diesel hybrid system using modified AVR. Procedia Eng. **38**, 3152–3165 (2012)
13. El Araby, E.E., Yorino, N.: A hybrid PSO technique for procuring VAR ancillary service in the deregulated electricity markets. Electr. Power Energy Syst. **32**, 664–670 (2010)
14. Calderaroa, V., Coniob, G., Galdia, V., Piccoloa, A.: Reactive power control for improving voltage profiles: a comparison between two decentralized approaches. Electr. Power Syst. Res. **83**, 247–254 (2012)
15. Devabhaktuni, S., kumar, S.V.J.: Performance analysis of wind turbine driven self-excited induction generator with external Rotor capacitance. Int. J. Adv. Eng. Sci. Technol. **10**(1), 1–6 (2011)
16. Wang, L., Ching-Huei, L.: A novel analysis on the performance of an isolated self-excited induction generator. IEEE Trans. Energy Convers. **12**(2), 109–117 (1997)
17. Kasal, G.K., Singh, B.: Voltage and frequency controllers for an asynchronous generator-based isolated wind energy conversion system. IEEE Trans. Energy Convers. **21**(2) (2006)
18. Karthikeyan, A., Nagamani, C., Ilango, G.S., Sreenivasulu, A.: Hybrid, open-loop excitation system for a wind turbine-driven stand-alone induction generator. IET Renew. Power Gener. **3**(2), 144–151 (2009)
19. Sundar, M.V., Karthik, P.S.A., Nagamani, C., Karthikeyan, A.: Optimal sizing of reactive power support in a standalone hybrid excited induction generator system. In: Proceedings of IEEE Fifth Power India Conference, Murthal, Haryana (2012)
20. Bansal, R.C.: Automatic reactive-power control of isolated wind-diesel hybrid power systems. IEEE Trans. Industr. Electron. **53**(4), 2006 (2006)
21. Kouadri, B., Tahir, Y.: Power flow and transient stability modeling of a 12-pulse STATCOM. J. Cyber Inform. **7**, 9–25 (2008)
22. Canizares, C.A.: STATCOM modeling for voltage and angle stability studies. Electr. Power Energy Syst. **25**, 1–20 (2003)
23. Xu, Y., Li, F., Jin, Z., Huang, C.: Flatness-based adaptive control (FBAC) for STATCOM. Electr. Power Syst. Research, **122** (76–85) (2015)
24. Sandhu, K.S., Jain, S.P.: Steady state operation of self-excited induction generator with varying wind speeds. Int. J. Circ. Syst. Sig. Process. **2**(1) (2008)
25. Farhad, S., Ritwik, M., Ghosh, A., Ledwich, G., Firuz, Z.: Operation and control of a hybrid microgrid containing unbalanced and nonlinear loads. Electr. Power Syst. Res. **80**, 954–965 (2010)
26. Jose, L., Agustın, B., Lopez, R.L.: Multiobjective design and control of hybrid systems minimizing costs and unmet load. Electr. Power Syst. Res. **79**, 170–180 (2009)
27. Mozafari, B., Ranjbar, A.M., Amraee, T., Shirani, A.R.: A competitive market structure for reactive power procurement. Iran. J. Sci. Technol. **30**(B2), 259–276 (2006)
28. Rabiee, A.: MVAR management using generator participation factors for improving voltage stability margin. J. Appl. Sci. **9**(11), 2123–2129 (2009)
29. Silva, E.L., Hedgecock, J.J., Mello, J.C.O., Luz, J.C.F.: Practical cost-based approach for the voltage ancillary service. IEEE Trans. Power Syst. **16**(4) (2001)
30. Gil, J.B., Roman, T.G.S., Rios, J.J.A., Martin, P.S.: Reactive power pricing: a conceptual framework for remuneration and charging procedures. IEEE Trans. Power Syst. **15**(2) (2000)
31. Halbhavi, S.B., Karki, S., Kulkarni, S.G.: Reactive power pricing: problems & a proposal for a competitive market. Int. J. Innov. Eng. Technol. **1**(2), 22–27 (2012)
32. Tao, M., Yang, H., Lu, L.: Study on stand-alone power supply options for an isolated community. Electr. Power Energy Syst. **65**, 1–11 (2015)
33. Kirby, B., Hirst, E.: Ancillary Service Details: Voltage Contro. Oak Ridge National Laboratory, Oak Ridge (1997)
34. Moon, Y.H., Park, J.D., Jung, C.S., Kook, H.J.: (2001). Cost evaluation for capacitive reactive power under the deregulation environment. In: Proceedings of IEEE Power Engineering Society Winter Meeting, Columbus, USA

35. Murty, V.V.S.N., Kumar, A.: Comparison of optimal capacitor placement methods in radial distribution system with load growth and ZIP load model. Front. Energy **7**, 197–213 (2013)
36. Cai, L.J., Erlich, I., Stamtsis, G.: Optimal choice and allocation of FACTS devices in deregulated electricity market using genetic algorithms. In: Proceedings of IEEE Power Systems Conference and Exposition, New York, USA (2004)
37. Caldon, R., Rossetto, F., Scala, A.: Reactive power control in distribution networks with dispersed generators: a cost based method. Electr. Power Syst. Res. **64**, 209–217 (2003)
38. Zhong, J.: On Some Aspects of Design of Electric Power Ancillary Service Market. Doctoral dissertation, Chalmers University of Technology, Goteborg, Sweden (2003)
39. Singh, B., Murthy, S.S., Gupta, S.: Analysis and design of STATCOM-based voltage regulator for self-excited induction generators. IEEE Trans. Energy Convers. **19**(4), 783–790 (2004)
40. Saxena, N.K., Kumar, A.: Analytical comparison of static and dynamic reactive power compensation in isolated wind diesel system using dynamic load interaction model. Electr. Power Compon. Syst. **53**(5), 508–519 (2015)
41. Zhong, J., Bhattacharya, K.: Reactive power management in deregulated power systems- a review. In: Proceedings of IEEE Power Engineering Society Winter Meeting, 2, pp. 1287–1292 (2002)
42. Kothari, D.P., Nagrath, I.J.: Electric Machine. Tata-McGraw-Hill, India (2006)
43. Wenjuan, Z.: Optimal Sizing and Location of Static and Dynamic Reactive Power Compensation. Doctoral dissertation, The University of Tennessee, Knoxville (2007)
44. Saxena, N.K., Kumar, A.: Reactive power control in decentralized hybrid power system with STATCOM using GA, ANN and ANFIS methods. Int. J. Electr. Power Energy Syst. **83**, 175–187 (2016)
45. Xhu, N., Vadari, S., Hwang, D.: Analysis of a static VAR compensator using the dispatcher training simulator. IEEE Trans. Power Syst. **10**(3), 1234–1242 (1995)
46. Saxena, N.K., Kumar, A.: Dynamic Reactive Power Compensation and Cost Analysis for Isolated Hybrid Power System. Electric Power Components and Systems (in press)
47. Kundur, P.: Power System Stability and Control. Tata-Mcgraw-Hill, India (2006)
48. Bansal, R.C.: Automatic Reactive Power Control of Autonomous Hybrid Power System. Doctoral dissertation, Indian Institute of Technology, Delhi, India (2002)
49. Hingorani, N.G., Gyugyi, L.: Understanding FACTs: Concepts and Technology of Flexible AC Transmission Systems. IEEE Power Engineering Society, New York (2000)
50. Saxena, N.K., Kumar, A.: Estimation of composite load model with aggregate induction motor dynamic load for an isolated hybrid power system. Front. Energy **9**(4), 472–485 (2015)
51. Saxena, N.K., Kumar, A.: Reactive power compensation of isolated hybrid power system with load interaction using ANFIS tuned STATCOM. Front. Energy **8**(2), 261–268 (2014)

Backward-Forward Sweep Based Power Flow Algorithm in Distribution Systems

Farkhondeh Jabari, Farnaz Sohrabi, Pouya Pourghasem and Behnam Mohammadi-Ivatloo

Abstract As we know, Newton-Raphson method cannot find optimum operating points of radial and meshed distribution systems due to high R/X ratio of feeders. To solve this problem, backward-forward sweep (BFS) load flow algorithm is presented by scholars. This chapter aims to present MATLAB codes of BFS power flow method in a benchmark distribution grid. Feeder capacity and voltage magnitude limit are considered in finding a good operating point for test grid. Input data such as bus and line information matrices are presented in MATLAB codes. Simulations are conducted on IEEE-33 bus radial distribution system. Feeder current, bus voltage magnitude, active and reactive power flowing in or out of buses, total real power losses system are found as outputs of BFS load flow approach.

Keywords Backward-forward sweep · Load flow analysis · Feeder current capacity · Bus voltage limit · Active power losses

Nomenclature

$\dot{J}_i^{k+1}$	Injected current to node i in $(k+1)$th iteration
$\dot{V}^{(K)}$	Voltage of node i in kth iteration
$\dot{S}$	Power injected to node i
$\dot{Y}$	Parallel admittance of node i

F. Jabari (✉) · F. Sohrabi · P. Pourghasem · B. Mohammadi-Ivatloo
Faculty of Electrical and Computer Engineering, University of Tabriz, Tabriz, Iran
e-mail: f.jabari@tabrizu.ac.ir

F. Sohrabi
e-mail: sohrabi.farnaz@gmail.com

P. Pourghasem
e-mail: pouya.pourghasem@gmail.com

B. Mohammadi-Ivatloo
e-mail: bmohammadi@tabrizu.ac.ir

M. Pesaran Hajiabbas and B. Mohammadi-Ivatloo (eds.),
Optimization of Power System Problems, Studies in Systems, Decision and Control 262,
https://doi.org/10.1007/978-3-030-34050-6_14

$\dot{I}_i^{k+1}$	Current of branch i in $(k + 1)$th iteration
$\dot{I}_j^{k+1}$	Current of branch j in $(k + 1)$th iteration
Z_j	Impedance of branch j
$\dot{V}_j^{k+1}$	Voltage of bus i in $(k + 1)$th iteration
$\dot{V}_j^{k+1}$	Voltage of bus j in $(k + 1)$th iteration
F_{loss}	Total real power losses as objective function
$g_{i,j}$	Conductance of branch i to j
V_i	Voltage magnitude of bus i
V_j	Voltage magnitude of bus j
θ_i	Voltage angle of bus i
θ_j	Voltage angle of bus j
n_l	Total number of branches
I_b	Current of branch b

1 Introduction

As known, load flow analysis of distributed power systems is used for finding feeder current value, bus voltage magnitude and angle, active and reactive power losses, steady-state voltage stability assessment, etc. [1, 2]. Conventional Newton-Raphson and fast decoupled power flow methods fail in finding operating point of distribution systems because of [3]:

- High R/X ratio of feeders
- Weak meshed distribution systems
- Unbalanced system.

Hence, backward-forward sweep (BFS) algorithm is proposed by scholars to solve this problem [4]. In this approach, there is no need to Jacobian matrix of Newton-Raphson method [5]. Moreover, computational burden and calculation time of BFS algorithm is lower than those of conventional Newton-Raphson and fast decoupled power flow methods [6]. In [7], BFS load flow method uses a load-impedance matrix for fast and computationally efficient calculation of bus voltage magnitude in radial and weakly meshed distribution grids. Latin hypercube sampling based Cholesky decomposition (LHS-CD) technique is developed by Kabir et al. [8] for probabilistic load flow analysis of distribution systems under uncertain generations of photovoltaic (PV) cells. It is found that LHS-CD is faster than Monte Carlo simulations [9], point estimation method [10, 11], Gram–Charlier expansion [12], and Cornish Fisher expansion [13, 14], fuzzy [15, 16], interval arithmetic [17] algorithms. In [18], BFS based three phase power flow approach is implemented on radial distributed systems using Kirchhoff's current and voltage laws to compute the voltage magnitude of the upstream buses based on active and reactive power injections of end buses. Then, the voltage magnitude of the downstream nodes are updated in forward sweep stage. In [19], ZIP load model is incorporated into linear three phase power flow calculation of balanced and unbalanced distribution systems. In [20, 21], various connections of transformers are considered in BFS power flow analysis. In [22], an interval arithmetic based backward-forward sweep algorithm is used to model two lower and upper bounds of uncertain loads for balanced radial distribution system power flow analysis. However, upper and lower ranges estimated by interval arithmetic approach tend to be large especially in long iterative computations while ignoring the correlation of different variables [23].

As reviewed, different deterministic and probabilistic power flow algorithms have recently been introduced by researchers to find the bus voltage profile and the branch current vector of buses and feeders in radial and meshed distribution systems. But, MATLAB codes of BFS load flow algorithm has not been presented yet. Therefore, this chapter presents a comprehensive MATLAB code for power flow calculation of radial and meshed distribution grids. Other sections of this chapter are organized as follows: A comprehensive problem formulation is presented in Sect. 2. Simulation results and discussions are provided in Sect. 3. Finally, concluding remarks appear in Sect. 4.

2 Forward-Backward Sweep Power Flow

In this section, backward forward approach is introduced. By considering a sample distribution system as shown in Fig. 1, the injected current to the *i*th node can be calculated as Eq. (1).

$$\dot{J}_i^{k+1} = \left(\frac{\dot{S}_i}{\dot{V}_i^{(k)}}\right)^* - \dot{Y}_i \dot{V}_i^{(k)} \tag{1}$$

where,

$\dot{J}_i^{k+1}$: Injected current to node i in $(k+1)$th iteration
$\dot{V}_i^{(k)}$: Voltage of node i in kth iteration
$\dot{S}_i$: Power injection of node i
$\dot{Y}_i$: Parallel admittance of node i.

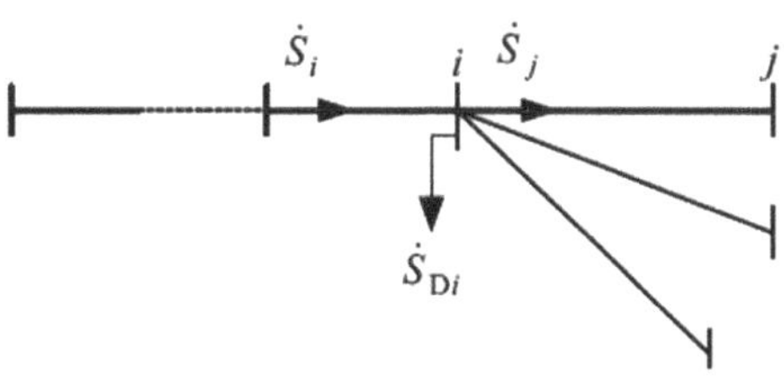

Fig. 1 A simple radial distribution network

Backward sweep
In this step, sum of all currents injecting to branch i is obtained from Eq. (2).

$$\dot{I}_i^{k+1} = -\dot{J}_i^{k+1} + \sum_{j \in C_i} \dot{I}_j^{k+1} \tag{2}$$

where,

$\dot{I}_i^{k+1}$: Current of branch i in $(k + 1)$th iteration
$\dot{I}_j^{k+1}$: Current of branch j in $(k + 1)$th iteration

Forward sweep
In this process, according to obtained current in backward process, new value of bus voltage is formulated by Eq. (3).

$$\dot{V}_j^{k+1} = \dot{V}_i^{k+1} - \dot{I}_j^{k+1} \times Z_j \tag{3}$$

In which,

Z_j: Impedance of branch j
$\dot{V}_i^{k+1}$: Voltage of bus i in $(k + 1)$th iteration
$\dot{V}_j^{k+1}$: Voltage of bus j in $(k + 1)$th iteration

Finally, in order to finalize power flow calculations, limitation (4) should be satisfied; otherwise, Eqs. (2) and (3) will be repeated. Total active power loss is calculated as Eq. (4).

$$F_{loss} = \sum_{\substack{i,j=1 \\ i \neq j}}^{n_l} g_{i,j}\left[V_i^2 + V_j^2 - 2V_i V_j \cos(\theta_i - \theta_j)\right] \tag{4}$$

where,

F_{loss}: Total real power loss as objective function
$g_{i,j}$: Conductance of branch i to j
V_i and V_j: Voltage magnitude of buses i and j, respectively
θ_i and θ_j: Voltage angle of buses i and j, respectively
n_l: Total number of branches.

3 Case Study and Discussions

In this section, a backward-forward sweep load flow algorithm is coded using MATLAB software. According to MATLAB codes are presented in MATLAB Code section, a 33-bus radial distribution system [24] is considered to for implementation of BFS sweep method. The single line diagram of IEEE 33-bus radial distribution system is illustrated in Fig. 2. The bus data matrix is defined as “bdata.not.per.unit”. The fist column of this matrix refer to number of nodes. Active and reactive power consumptions in each bus is presented at second and third columns of bus data matrix in kW and kVAr, respectively. Similarly, “ldata.not.per.unit” is line data matrix of 33-bus radial benchmark network. In each row of branch information matrix, number of starting and ending point of each line is determined using bus numbers. The third and fourth columns of the line data matrix represent the resistance and reactance of each branch in Ohm, respectively. Figure 3 shows the voltage profile of the test system after implementation of BFS power flow algorithm. Moreover, the bus voltage angle and the current flowing in or out of buses are reported in Tables 1 and 2, respectively. Active and reactive power flows through each feeder are presented in Table 3. As obvious from this table, total active and reactive power losses are equal to 129.6 kW and 94.9 kVAr, respectively.

4 Conclusion

Because of high R/X ratio of feeders and weak or unbalanced structure of radial distribution systems, Newton-Raphson and fast decoupled power flow calculations fail in finding the feasible operating point of the system. Meanwhile, backward-forward sweep is a useful tool to solve this problem. In this chapter, the mathematical formulation and the MATLAB codes of the backward-forward sweep load flow analysis was presented. The bus information matrix consist of active and reactive power consumption/injection of the buses were considered as the input load-generation data. Moreover, the line data matrix composed of the resistance and the reactance of the branches were presented as the impedance parameter of the feeders. The bus voltage magnitude and angle, the feeder current magnitude and phase, the active and reactive power flows, total real and reactive power losses were obtained using BFS algorithm. MATLAB codes not only carries out the power flow calculation, but also is able to draw the single line diagram of the test system based on the line and bus data matrices. In addition, the codes can find the best operating point of the meshed distribution grids by entering the bus and the line information matrices and the base MVA and voltage level. As the future trend, it is recommended that the MATLAB codes will be implemented on a meshed distribution system to demonstrate the robustness and applicability of the BFS algorithm in finding the good solutions for different distribution systems.

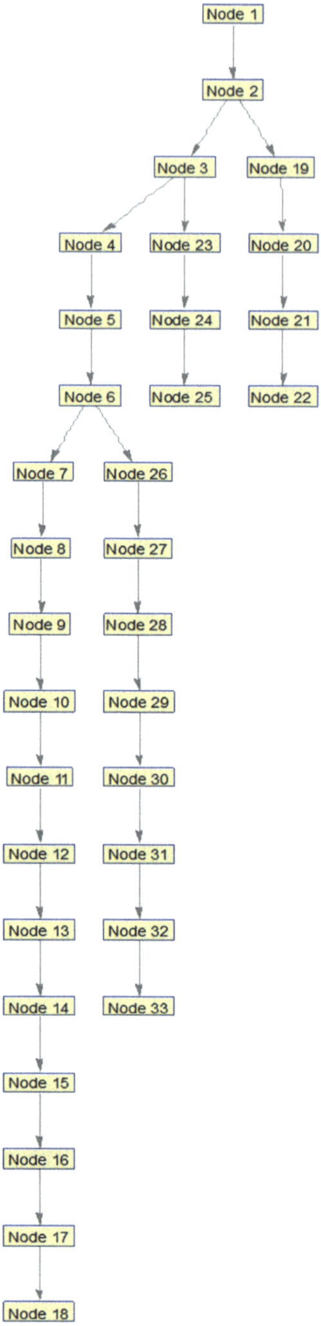

Fig. 2 Single line diagram of IEEE 33-bus radial distribution system

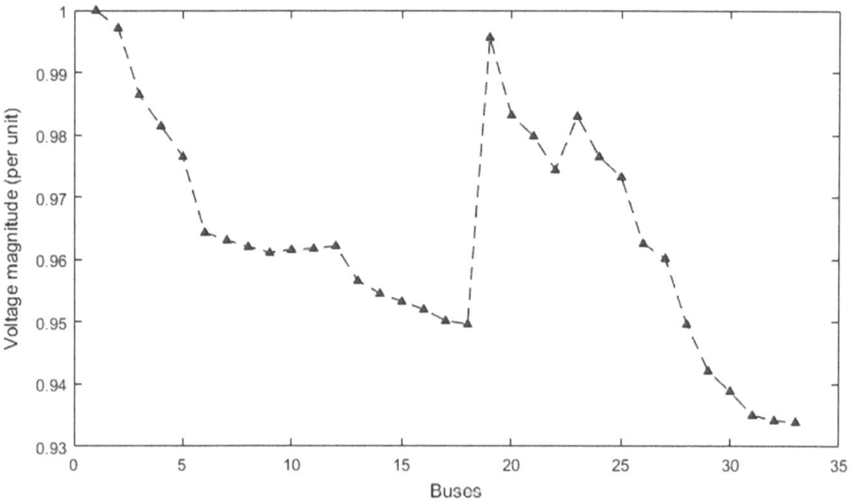

Fig. 3 The voltage profile of the test system after application of BFS sweep load flow method

Table 1 The bus voltage angle obtained from BFS algorithm

Bus number	Voltage angle (°)
1	0
2	0.0393
3	0.2956
4	0.5004
5	0.7074
6	0.6961
7	0.4520
8	0.4807
9	−2.5124
10	−2.4837
11	−2.4834
12	−2.4890
13	−2.7604
14	−2.9852
15	−3.0938
16	−3.1625
17	−3.3823
18	−3.4107
19	−0.0564
20	−0.7814
21	−1.0699
22	−1.6163
23	0.2010

(continued)

Table 1 (continued)

Bus number	Voltage angle (°)
24	−0.0696
25	−0.2028
26	0.8114
27	0.9753
28	1.2299
29	1.4654
30	1.7705
31	1.5377
32	1.4739
33	1.4524

Table 2 The feeder current obtained from BFS algorithm

From bus	To bus	Branch current (A) (*100)
1	2	0.0000 + 0.0000i
2	3	1.0030 − 0.6015i
3	4	0.9130 − 0.4040i
4	5	1.2250 − 0.8117i
5	6	0.6156 − 0.3048i
6	7	0.6230 − 0.2050i
7	8	2.0792 − 1.0331i
10	9	2.0816 − 1.0339i
11	10	0.6213 − 0.2168i
12	11	0.6210 − 0.2166i
12	13	0.4636 − 0.3184i
13	14	0.6185 − 0.3723i
14	15	0.6215 − 0.3755i
15	16	1.2431 − 0.8588i
16	17	0.6275 − 0.1157i
17	18	0.6265 − 0.2211i
2	19	0.6274 − 0.2223i
19	20	0.9396 − 0.4391i
20	21	0.9038 − 0.4020i
21	22	0.9136 − 0.4108i
3	23	0.9161 − 0.4137i
23	24	0.9199 − 0.4188i
24	25	0.9161 − 0.5076i
6	26	4.3002 − 2.0498i
26	27	4.3130 − 2.0598i
27	28	0.6245 − 0.2569i
28	29	0.6263 − 0.2570i
29	30	0.6332 − 0.2063i

(continued)

Table 2 (continued)

From bus	To bus	Branch current (A) (*100)
30	31	1.2797 – 0.7326i
31	32	2.1930 – 6.3693i
32	33	1.6105 – 0.7349i
22	12	2.2566 – 1.0520i

Table 3 Active and reactive power flows in or out of buses

From bus	To bus	Active power (kW)	Reactive power (kVAr)
1	2	10.9824	5.6818
2	3	30.1114	15.3367
3	4	2.9507	1.5027
4	5	9.6823	4.9314
5	6	5.4544	4.7085
6	7	4.1038	13.5655
7	8	14.5289	4.8014
10	9	1.3028	0.9234
11	10	2.1453	0.7093
12	11	0.1168	0.0386
12	13	0.0366	0.0288
13	14	0.0541	0.0712
14	15	0.1185	0.1054
15	16	1.1386	0.8315
16	17	1.4557	1.9435
17	18	0.3699	0.2901
2	19	0.0517	0.0493
19	20	0.2450	0.2207
20	21	0.0248	0.0290
21	22	4.7677	6.3038
3	23	2.4861	1.6987
23	24	3.9653	3.1312
24	25	3.0843	2.4134
6	26	1.0963	0.5584
26	27	0.3837	0.1954
27	28	10.5402	9.2931
28	29	7.2755	6.3383
29	30	4.1860	2.1322
30	31	6.3257	6.2517
31	32	0.4272	0.4979
32	33	0.1968	0.3060
22	12	0.0649	0.0649
Whole system	129.6	94.9	

MATLAB Code

```
clear all
close all
clc
bdata.not.per.unit=[%Bus P(Kw) Q(Kvar)
    1 0 0
    2 120 72
    3 108 48
    4 144 96
    5 72 36
    6 72 24
    7 240 120
    8 240 120
    9 72 24
    10 72 24
    11 54 36
    12 72 42
    13 72 42
    14 144 96
    15 72 12
    16 72 24
    17 72 24
    18 108 48
    19 108 48
    20 108 48
    21 108 48
    22 108 48
    23 108 60
    24 504 240
    25 504 240
    26 72 30
    27 72 30
    28 72 24
    29 144 84
    30 240 720
    31 180 84
    32 252 120
    33 72 48 ];
ldata.not.per.unit=[
    % Inbus Outbus Resistance(ohm) Reactance(ohm)
    1 2 0.0922 0.0470
    2 3 0.4930 0.2511
```

```
    3 4 0.3660 0.1864
    4 5 0.3811 0.1941
    5 6 0.8191 0.7070
    6 7 0.1872 0.6188
    7 8 0.7114 0.2351
    8 9 1.0300 0.7400
    9 10 1.0440 0.7400
    10 11 0.1966 0.0650
    11 12 0.3744 0.1238
    12 13 1.4680 1.1550
    13 14 0.5416 0.7129
    14 15 0.5910 0.5260
    15 16 0.7463 0.5450
    16 17 1.2890 1.7210
    17 18 0.7320 0.5740
    2 19 0.1640 0.1565
    19 20 1.5042 1.3554
    20 21 0.4095 0.4784
    21 22 0.7089 0.9373
    3 23 0.4512 0.3083
    23 24 0.8980 0.7091
    24 25 0.8960 0.7011
    6 26 0.2030 0.1034
    26 27 0.2842 0.1447
    27 28 1.0590 0.9377
    28 29 0.8042 0.7006
    29 30 0.5075 0.2585
    30 31 0.9744 0.9630
    31 32 0.3105 0.3619
    32 33 0.3410 0.5302];
sizbdata=size(bdata.not.per.unit);
busnum=sizbdata(1,1);
sizldata=size(ldata.not.per.unit);
branchnum=sizldata(1,1);
%per unit calculation:
Sbase=10^3;
Vbase=12.66*10^3;
Zbase=Vbase^2/Sbase;
bdata=bdata.not.per.unit;
ldata=ldata.not.per.unit;
for n=1:busnum
    bdata(n,2)=(bdata(n,2)*1000)/Sbase;
    bdata(n,3)=(bdata(n,3)*1000)/Sbase;
end
```

```
for n=1:branchnum
    ldata(n,3)=ldata(n,3)/Zbase;
    ldata(n,4)=ldata(n,4)/Zbase;
end
%per unit calculation finished
terminatebus=zeros(busnum,1);
intermediatebus=zeros(busnum,1);
junctionbus=zeros(busnum,1);
junctionnum=zeros(busnum,1);
refbus=0;
busI=zeros(busnum,1);
v=ones(1,busnum);
I=zeros(busnum,busnum);
for k=1:busnum
    co=0;
    l=0;
    for n=1:branchnum
        if ldata(n,1)==k
            co=co+1;
        end
    end
    if co==0
        terminatebus(k,1)=k;
    elseif co>=2
        junctionbus(k,1)=k;
        junctionnum(k,1)=co;
    elseif co==1
        for m=1:branchnum
            l=l+1;
            if ldata(m,2)==k
                intermediatebus(k,1)=k;
                break
            elseif l==branchnum
                refbus=k;
            end
        end
    end
end
junctionbus;
intermediatebus;
terminatebus;
refbus;
junctionnum;
tempterminatebus=terminatebus;
controljunctionnum=zeros(busnum,1);
k=0;
c=0;
```

```
itecount=0;
for s=1:15 %iteration
    itecount=itecount+1;
    %backward sweep
    while c==0
        k=k+1;
        juncnum=0;
        n=0;
        stop=0;
        previousI=0;
        if tempterminatebus(k,1)==k
            while(n<branchnum)&&(stop==0)
                n=n+1;
                if ldata(n,2)==k
                    a=ldata(n,1);
                    if a==refbus
                        c=1;
                    end
                    I(a,k)=busI(k)+(bdata(k,2)-
1i*bdata(k,3))/conj(v(k))+previousI;
                    previousI=I(a,k);
                    tempterminatebus(k,1)=0;
                    if junctionbus(a,1)==a
                        busI(a)=busI(a)+I(a,k);
controljunctionnum(a,1)=controljunctionnum(a,1)+1;
                        if
controljunction-num(a,1)==junctionnum(a,1)
                            tempterminatebus(a,1)=a;
                        end
                        break
                    end
                    k=a;
                    n=0;
                end
            end
            k=0;
        end
    end
    %end of backward sweep
    %forward sweep
    count=0;
    beforev=v;
    newldata=ldata;
    stop1=0;
    forwardbus=zeros(busnum,1);
    c=0;
    stopif=0;
```

```
while stop1==0
    for k=1:branchnum
        if(newldata(k,1)==refbus)&&(stopif==0)
            c=refbus;
            if junctionbus(refbus,1)==refbus
                forwardbus(refbus,1)=1;
            end
            stop2=0;
            while 1
                a=newldata(k,2);
                z=newldata(k,3)+1i*newldata(k,4);
                v(a)=v(c)-z*I(c,a);
                newldata(k,:)=0;
                if junctionbus(a,1)==a
                    forwardbus(a,1)=1;
                end
                if terminatebus(a,1)==a
                    stopif=1 ;
                    stop1=1;
                    break
                end
                for n=1:branchnum
                    if ldata(n,1)==a
                        c=a;
                        k=n;
                        break
                    end
                end
            end
        end
    end
end
stop3=0;
while stop3==0
    stopif1=0;
    for k=1:busnum
        if (forwardbus(k,1)==1)&& (stopif1==0)
            stop4=0;
            while stop4==0
                stopif2=0;
                hhh=0;
                counter=0;
                for n=1:branchnum
                    counter=counter+1;
                    if (newldata(n,1)==k)&&
```

```
(stopif2==0)
                                            hhh=1;
                                            c=k;
                                            a=newldata(n,2);
z=newldata(n,3)+1i*newldata(n,4);
                                            v(a)=v(c)-z*I(c,a);
                                            newldata(n,:)=0;
                                            if junctionbus(a,1)==a
                                                forwardbus(a,1)=1;
                                            end
                                            if terminatebus(a,1)==a
                                                if newldata==0
                                                    stopif1=1;
                                                    stop3=1;
                                                    stopif2=1;
                                                    stop4=1;
                                                    break
                                                end
                                                stopif2=1;
                                                stop4=1;
                                            end
                                            k=a;
                                            stopif2=1;
                                        end
                                        if (counter==branchnum) &&
(hhh==0)
                                            stop4=1;
                                        end
                                    end
                                end
                            end
                        end
                    end
    %end of forward sweep
    for p=2:busnum
        if abs(v(p)-beforev(p))<=0.000001
            count=count+1;
        end
    end
    if count==(busnum-1)
        break
    end
end
%data display
```

```
voldisp=zeros(busnum,3);
curdisp=zeros(branchnum,3);
co=0;
for n=1:busnum
    for k=1:busnum
        if I(n,k)~=0
            co=co+1;
            curdisp(co,1)=n;
            curdisp(co,2)=k;
            curdisp(co,3)=I(n,k);
        end
    end
end
disp('curdisp=');
disp(' from bus to bus current(pu)');
disp(curdisp);
fprintf('\n');
for n=1:busnum
    voldisp(n,1)=n;
    voldisp(n,2)=abs(v(n));
    voldisp(n,3)=angle(v(n))*180;
end
disp('bus number voltage(pu) angle(degree)');
disp(voldisp);
%data display end
%network display
co=0;
W=zeros(1,branchnum);
a=zeros(1,branchnum);
b=zeros(1,branchnum);
for n=1:busnum
    for k=1:busnum
        if I(n,k)~=0
            co=co+1;
            a(co)=n;
            b(co)=k;
        end
    end
end
DG = sparse(a,b,true,busnum,busnum);
view(biograph(DG))
%network display end
%loss calculation
Ploss=0; Qloss=0;
for k=1:branchnum
```

```
    z=curdisp(k,1);
    y=curdisp(k,2);
    for g=1:branchnum
        if ldata(g,1)==z && ldata(g,2)==y
            Ploss=Ploss+abs(curdisp(k,3))^2*ldata(g,3);
%ploss=r*|I|^2
            Qloss=Qloss+abs(curdisp(k,3))^2*ldata(g,4);
%qloss=x*|I|^2
            break
        end
    end
end
%end of loss calculation
Inew=zeros(busnum,1);
for l=1:busnum
    Inew(l,1)=(bdata(l,2)-1i*bdata(l,3))/conj(v(l));
end
Active_loss(i,j)=abs(Ploss);
Reactive_loss(i,j)=abs(Qloss);
end
end
```

References

1. de Araujo, L.R., Penido, D.R.R., Júnior, S.C., Pereira, J.L.R., Garcia, P.A.N.: Comparisons between the three-phase current injection method and the forward/backward sweep method. Int. J. Electr. Power Energy Syst. **32**(7), 825–833 (2010)
2. Díaz, G., Gómez-Aleixandre, J., Coto, J.: Direct backward/forward sweep algorithm for solving load power flows in AC droop-regulated microgrids. IEEE Trans. Smart Grid **7**(5), 2208–2217 (2016)
3. Rupa, J.M., Ganesh, S.: Power flow analysis for radial distribution system using backward/forward sweep method. Int. J. Electr. Comput. Electron. Commun. Eng. **8**(10), 1540–1544 (2014)
4. Augugliaro, A., Dusonchet, L., Favuzza, S., Ippolito, M., Sanseverino, E.R.: A new backward/forward method for solving radial distribution networks with PV nodes. Electr. Power Syst. Res. **78**(3), 330–336 (2008)
5. Bompard, E., Carpaneto, E., Chicco, G., Napoli, R.: Convergence of the backward/forward sweep method for the load-flow analysis of radial distribution systems. Int. J. Electr. Power Energy Syst. **22**(7), 521–530 (2000)
6. Luo, G.-X., Semlyen, A.: Efficient load flow for large weakly meshed networks. IEEE Trans. Power Syst. **5**(4), 1309–1316 (1990)
7. Ghatak, U., Mukherjee, V.: A fast and efficient load flow technique for unbalanced distribution system. Int. J. Electr. Power Energy Syst. **84**, 99–110 (2017)
8. Kabir, M., Mishra, Y., Bansal, R.: Probabilistic load flow for distribution systems with uncertain PV generation. Appl. Energy **163**, 343–351 (2016)

9. Carpinelli, G., Caramia, P., Varilone, P.: Multi-linear Monte Carlo simulation method for probabilistic load flow of distribution systems with wind and photovoltaic generation systems. Renew. Energy **76**, 283–295 (2015)
10. Delgado, C., Domínguez-Navarro, J.A.: Point estimate method for probabilistic load flow of an unbalanced power distribution system with correlated wind and solar sources. Int. J. Electr. Power Energy Syst. **61**, 267–278 (2014)
11. Morales, J.M., Perez-Ruiz, J.: Point estimate schemes to solve the probabilistic power flow. IEEE Trans. Power Syst. **22**(4), 1594–1601 (2007)
12. Zhang, P., Lee, S.T.: Probabilistic load flow computation using the method of combined cumulants and Gram-Charlier expansion. IEEE Trans. Power Syst. **19**(1), 676–682 (2004)
13. Usaola, J.: Probabilistic load flow with wind production uncertainty using cumulants and Cornish-Fisher expansion. Int. J. Electr. Power Energy Syst. **31**(9), 474–481 (2009)
14. Ruiz-Rodriguez, F., Hernández, J., Jurado, F.: Probabilistic load flow for photovoltaic distributed generation using the Cornish-Fisher expansion. Electr. Power Syst. Res. **89**, 129–138 (2012)
15. Pourahmadi-Nakhli, M., Seifi, A.R., Taghavi, R.: A nonlinear-hybrid fuzzy/probabilistic load flow for radial distribution systems. Int. J. Electr. Power Energy Syst. **47**, 69–77 (2013)
16. Kalesar, B.M., Seifi, A.R.: Fuzzy load flow in balanced and unbalanced radial distribution systems incorporating composite load model. Int. J. Electr. Power Energy Syst. **32**(1), 17–23 (2010)
17. Das, B.: Uncertainty modelling of wind turbine generating system in power flow analysis of radial distribution network. Electr. Power Syst. Res. **111**, 141–147 (2014)
18. Chang, G.W., Chu, S.Y., Wang, H.L.: An improved backward/forward sweep load flow algorithm for radial distribution systems. IEEE Trans. Power Syst. **22**(2), 882–884 (2007)
19. Garces, A.: A linear three-phase load flow for power distribution systems. IEEE Trans. Power Syst. **31**(1), 827–828 (2016)
20. Zhuding, W., Fen, C., Jingui, L.: Implementing transformer nodal admittance matrices into backward/forward sweep-based power flow analysis for unbalanced radial distribution systems. IEEE Trans. Power Syst. **19**(4), 1831–1836 (2004)
21. Kocar, I., Lacroix, J.: Implementation of a modified augmented nodal analysis based transformer model into the backward forward sweep solver. IEEE Trans. Power Syst. **27**(2), 663–670 (2012)
22. Das, B.: Radial distribution system power flow using interval arithmetic. Int. J. Electr. Power Energy Syst. **24**(10), 827–836 (2002)
23. Vaccaro, A., Cañizares, C.A., Bhattacharya, K.: A range arithmetic-based optimization model for power flow analysis under interval uncertainty. IEEE Trans. Power Syst. **28**(2), 1179–1186 (2013)
24. Venkatesh, B., Ranjan, R., Gooi, H.: Optimal reconfiguration of radial distribution systems to maximize loadability. IEEE Trans. Power Syst. **19**(1), 260–266 (2004)

The manufacturer's authorised representative in the EU is Springer Nature Customer Service Centre GmbH, Europaplatz 3, 69115 Heidelberg, Germany. If you have any concerns regarding our products, please contact ProductSafety@springernature.com

Printed and bound by CPI Group (UK) Ltd, Croydon, CR0 4YY
30/09/2026
02225449-0014